CALIFORNIA NATURAL HISTORY GUIDES

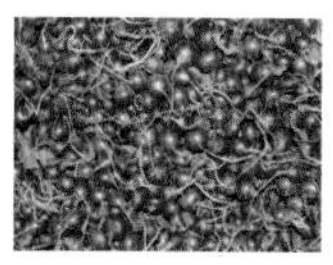

FIELD GUIDE TO CALIFORNIA AGRICULTURE

California Natural History Guides

Phyllis M. Faber and Bruce M. Pavlik, General Editors

Field Guide to
CALIFORNIA AGRICULTURE

Paul F. Starrs
Peter Goin

UNIVERSITY OF CALIFORNIA PRESS
Berkeley Los Angeles London

The opening dedication of this book is to Berkeley geographer James J. Parsons (1915–1997). With his inspirational work and affection for the byways and society of California agriculture, Jim was a reliable presence in our journeys, and we have tried to do right by his memory.

An even greater vote of thanks is owed to our families. They put up with absences, tolerated every manner of obscure allusions, helped us enjoy the produce we brought home, laughed at our jokes—whether photographic or literary—and permitted us to keep going back for more.

California Natural History Guides No. 98

University of California Press, one of the most distinguished university presses in the United States, enriches lives around the world by advancing scholarship in the humanities, social sciences, and natural sciences. Its activities are supported by the UC Press Foundation and by philanthropic contributions from individuals and institutions. For more information, visit www.ucpress.edu.

University of California Press
Berkeley and Los Angeles, California

University of California Press, Ltd.
London, England

Library of Congress Cataloging-in-Publication Data

Starrs, Paul F., 1957–.
Field guide to California agriculture / Paul Starrs, Peter Goin.
p. cm. — (California Natural History Guide Series ; no. 98)
Includes bibliographical references and index.
ISBN 978-0-520-24764-2 (cloth : alk. paper) — ISBN 978-0-520-26543-1 (pbk. : alk. paper)
1. Agriculture—California. I. Goin, Peter, 1951- II. Title.

S451.C2S73 2010
630.9794—dc22 2009039092

Manufactured in China
19 18 17 16 15 14 13 12 11 10
10 9 8 7 6 5 4 3 2 1

The paper used in this publication meets the minimum requirements of ANSI/NISO Z39.48-1992 (R 1997) (*Permanence of Paper*). ∞

Cover photograph: Kale grown in California. Photo by Peter Goin.
Note: All unattributed photographs captioned on pp. 428–429.

The publisher gratefully acknowledges the generous contributions to this book provided by

the Gordon and Betty Moore Fund
in Environmental Sciences

and

the General Endowment Fund of the
University of California Press Foundation

CONTENTS

Colors of California Agriculture

For the colors on the two pages that precede this one, read top left to bottom right, as you would a normal text document.

Left side, line 1 Alfalfa Flower, Almond Tree Grove, Amaranthus, Arkansas Black Apple, Artichoke, Asian Pear, Asparagus, Bean Field

Left side, line 2 Bing Cherry, Black Bean, Black Bull, Black-Eyed Pea, Black Fig, Blackberry, Blood Orange, Blueberry

Left side, line 3 Bok Choy, Broccoli Crown, Broccoli Flower, Brown Turkey Fig, Brunette Beauty Date, Brussels Sprouts, Burnt Prairie, Butter Lettuce

Left side, line 4 Cabbage Field, Cadenera Sweet Orange, Cara Cara Orange, Carrot, Celery, Chandler Grapefruit, Chili Pepper, Chinese Eggplant

Left side, line 5 Cilantro, Clementine Fina, Cling Peach, Collard Green, Compost Pile, Corn Stalk, Cotton Bale, Cotton Blossom

Left side, line 6 Cucumber, Dairy Cow, Dairy Feed, Eggshell, Farm Worker Denim, Farm Worker Hat, Faux Plastic Grape, Fennel Field

Left side, line 7 Fig Leaf, Fingered Citron, Flame Grapefruit, Free-Range Fryer, French Plum, Fukumoto Navel Orange, Goat, Golf Course

Left side, line 8 Gophinator, Grafted Grapevine, Grain Pile, Grazed-land, Green Apple, Green Bell Pepper, Green Olive, Green Tomato

Left side, line 9 Harvested Sudan Grass Field, Hass Avocado, Hay, Honeydew Melon 01, Honeydew Melon 02, Horseradish Leaf, Horseradish Root, Irrigation Water 01

Left side, line 10 Irrigation Water 02, Irrigation Water 03, John Deere Tractor, Kale, Kiwi, Kohlrabi, Kumquat, Latex Glove

Right side, line 1 Leeks, Lemon, Loquat, Mandarinquat, Marijuana, Mary Ellen Sweet Lime, Medical Marijuana, Mellow Gold Grapefruit

Right side, line 2 Michal Mandarin Orange, Millsweet Lemon, Monterey Mushroom, Moro Blood Orange, New Holland Tractor, Nopal, Nopal Fruit, Old Growth Grapevine

Right side, line 3 Onion, Orange, Organic Dried Haichya Persimmon, Organic Garlic, Organic Purple Cauliflower, Oroblanco GFT Hybrid White Grapefruit, Oyster, Packing Box

Right side, line 4 Palm Tree, Paved Farm Road, Peach, Persimmon, Pesticide Latex Glove, Peterbuilt Dump Truck, Pinot Noir, Pistachio

Right side, line 5 Pomegranate, Portable Toilet, Poultry Shed, Pummelo X Ruby Blood #1 Grapefruit Hybrid, Pumpkin, Radish, Rainier Cherry, Raisin Grape

Right side, line 6 Raspberry, Red Apple, Red Bell Pepper, Red Chard, Red Lettuce, Red Onion, Red Truck, Red Valencia Orange

Right side, line 7 Rice field, Ripening Pear, Roma Tomato, Rubidoux Pummelo, Safflower 01, Safflower 02, Sciabica's Unfiltered Extra Virgin Olive Oil, Sheep Wool

Right side, line 8 Silage, Spinach Field, Strawberry, Sudan Grass, Sugarbeet Field, Sulfur, Sunflower, Sweet Alyssum

Right side, line 9 Sweet Potato, Tahitian Pummelo, Tango Mandarin, Thompson Seedless Grape, Thornless Mexican Lime, Tin Farm Building, Tray Olive, Trellis

Right side, line 10 Valencia Orange, Vaniglia Sanguigno Orange, Walnut Tree Grove, White Potato, Worker Housing, Yellow Bell Pepper, Yellow Chard, Zucchini Blossom

ACKNOWLEDGMENTS

In six years, with thousands of miles of fieldwork that provided perhaps 250 fieldside conversations, we never met a single person who failed to remark on how intriguing and valuable a book of this sort might be. Out there are believers in the importance of agriculture who admire raising food, drink, and fiber from the land. Their respect is not for the money made, the work wielded, or the technology marshaled, but comes because they recognize that agriculture makes the world a more interesting—and in many senses a more delectable—place.

We recognize the importance of a multitude of key informants, some of whom went far out of their way to make our lives as field workers pleasant, comfortable, and interesting. Firm long-term friends now include Louise Fisher of Strathmore and Earl and Miriam Rutz of Pauma Valley. Tipton and Sandy Holloway of Live Oak reminded us at a crucial early point in our work just how much two people could know about agriculture—and local lifeways—in the Sacramento Valley. Friends at McCormack Sheep and Grain of Rio Vista shared in triumphs and the occasional debacle as we explored myriad practices in California agriculture.

Sue Conley at Cowgirl Creamery and John Finger at Hog Island Oyster Company were each valued early contacts, as was Warren Weber at Star Route Farms. Mike and Sally Gale at Chileno Beef (and U-pick apples) were eminently gracious. Terrance Welch at Phil Foster Ranches made time to show us around, as did a variety of other growers, managers, and workers who welcomed us onto the sites under their stewardship: Dennis Albrecht (and, through him, Challenge Dairy),

John Verwey at Johann Dairy, Hilarides Dairy, Len Del Chiaro at DC's Extraordinary Cherries and Seko Ranch, José Aguiar, Susie Morrill, Marc and Paul Marchini at A.M. Farms, Mercier Grapevines, Jaime Serrato at Lumberjack Ranch, Monrovia Growers, Adam Englehardt at California Olive Ranch, and Cathy Wolfe at the Wolfe Ranch. Long travels through many properties were more the rule rather than an exception, but Clark Smith, general manager at Monterey Mushrooms, went beyond the call in taking us through the complex world of darkness and refreshing light that is the universe of the *Agaricus* mushroom farmer.

We should recognize, in an honest rounding out of the crops list, an anonymous contact in Northern California whose crop is so illustrative of the Emerald Triangle. A corps of stalwart wine experts we consulted includes winemaker Jessica Boone and her erstwhile colleagues at Armida Winery, Paul Stroth and Sequoia Hall at Stroth-Hall, and Tracy and Jared Brandt at Donkey and Goat, who revealed previously unknown and tasty qualities of the new urban wine movement.

The causes of California agriculture, and the concerns of development at the urban fringe, are ably expressed by a remarkable cadre of people. Conservationist and MALT cofounder Phyllis Faber, of Marin County, is first among them. Upholding the theme is Sibella Kraus, current president of SAGE (Sustainable Agriculture Education) and a founder of the San Francisco Ferry Plaza Farmers Market and its education wing, CUESA (the Center for Urban Education about Sustainable Agriculture). Sibella's continued work with SAGE, now based in Berkeley, is a great confidence-builder for those working with any facet of agriculture. Jessica Prentice at the Ferry Plaza Farmers Market is expanding on Sibella's earlier work, and gave us great facts and perspective, as did Lisa Bush and the singular Paul Michelsen.

Also within the Bay Area, Sheila Barry of Cooperative Extension proved an invaluable source on active agriculture at the urban margin, as did a number of Cooperative Extension experts who answered bemused queries to help us identify localities, tillage systems, and individual crops, although no one ever satisfactorily explained why those peppers (Plate 36) were on sticks. Rick Standiford, an old friend and sometime associate vice president of the University of California's Division of Agriculture and Natural Resources, was invaluable in passing on contacts and advice. Always professional and ready to answer questions were voices—and, later, known contacts—at the Sacramento office of the National Agricultural Statistics Service (USDA), and especially Doug Slohr. The Farmland Mapping Program and Benjamin Sleeter at USGS, each working on aspects of current, past, and future agricultural land use, were great assets, and their work even more so. Books such as these are in part about maps, images, experience, travel, theory, and toil, but they are first and foremost about data, and Ben and Doug (and their respective organizations) put out a steady stream of first-rate information, reliably updated.

On the home front, Dana and Kari Goin, Pamela Henning, Genoa and Carlin Starrs, and Lynn Huntsinger stood firm as only family and friends can. Al Medvitz and Jeannie McCormack showed great forbearance in the face of questions and variously had figs, eggs, olives, and goats on the range always at the ready. Glenda Humiston was a source of mirth, advice, and added contacts. Monica Moore and Sally Fairfax were never-failing advisors, technical and otherwise. Scott Hinton and Donna Macknet, two remarkable former students, and now colleagues, did far more than yeoman effort, and laid groundwork that Peter and Paul could build on—and did. Your labors were hardly for naught—thank you; and, as the quip famously goes, we stand on your shoulders.

The University of Nevada Academy for the Environment (UNAE) and the College of Liberal Arts Scholarly and Creative Activities Committee provided critical fieldwork and laboratory support. As two Regents and Foundation professors at the University of Nevada, we offer our thanks to the board of regents of the Nevada System of Higher Education, and in particular to John Carothers and the University of Nevada Foundation, who recognize the value of work such as ours.

Jenny Wapner of UC Press and Phyllis Faber showed considerable faith in this effort when they could have been skeptics, and we hope additional readers enjoy these results. Early readers—and believers—in this manuscript include reviewers Gerald Haslam, David J. Larson, and Louis Warren, who strongly supported this book's becoming what it is.

Finally, we wish to credit the many field workers who gladly spoke with us and helped us understand the scope of their role in California agriculture. Thank goodness we work well in Spanish—it is eminently clear to us where so much of the labor, knowledge, and diligence in California agriculture come from.

PREFACE

> *"That I can well believe," said the village priest, "for I now know from experience that the backwoods give rise to the eloquent, and the herdsman's hovel shelters philosophers."*
>
> *"At the very least, sir," came the goatherd's retort, "they shelter those chastened by life."*
>
> *DON QUIXOTE*, I, CH. 50

The world of California agriculture is a formidable, buzzing hive of activity whose envoys and practitioners probe every corner of the state's working landscapes. Does agriculture have its own economy and geography? No question. Do labor, climate, technology, and migration affect what is grown and raised—and what is profitable? You bet. Might someone sensibly claim that history, diet, religion, and urbanization are essential elements of the agricultural realm? Scrub even the fleeting doubt from your mind. Agriculture is a boon companion for Californians. Bold in ambition and vast in extent, California agriculture advances in nooks and crannies where growers experiment with novel crops and clever cultivars rarely before seen, yet soon to be on display in markets across the United States. Some innovations are big, others tiny; some agroindustrial, others organic, and at a boutique scale. All matter, and all make agriculture in California what it is.

Just how many distinct ag products exist in California is a roiling argument. Typically, the total number bandied around is approximately 400 items. Most are "specialty" crops and animal-derived foodstuffs: they are not considered staples, nor was production subsidized (beyond inexpensive water and land) in the U.S. Department of Agriculture (USDA) ledgers until the

2008 Farm Bill. We can safely accept that 300 to 400 crops and agricultural products dot California with diversity, add drama to diet, and push the respect accorded some growers and marketers (both big and tiny) beyond superlatives to near beatification. This book reviews the roots of California agriculture and provides tools for understanding farming and ranching, along with the production of food, fuel, and fiber, and then adds fitted lenses that help us grasp why and how California came to possess the most dramatic modern agricultural landscape in the world. The drama is physical and human, cultural and economic—sometimes heroic, and at other times decidedly tragic.

We recognize that readers may come to this discussion with less than a strong sense of mastery. If you are an expert, we hope you may find the occasional pearl of great price, and as veteran teachers, we aspire never to bore. But not many people, even in California, know the state's agriculture. In 1990, the farmer–author Wendell Berry, channeling the spirit of Thomas Jefferson, wrote an essay that he titled, "The Pleasures of Eating." In it, he warned of losing sight of the human relationship to cultivation, stewardship, and animal husbandry. "I begin," he admits, "with the proposition that eating is an agricultural act. Eating ends the annual drama of the food economy that begins with planting and birth. Most eaters, however, are no longer aware that this is true. They think of food as an agricultural product, but they do not think of themselves as participants in agriculture." Although Wendell Berry's quiet, firm message is now a couple of decades old, it is as true today as it was when he first gave it. Looming behind the words is a warning that we had best repair our awareness of the world of animals on the farm and range, of healthy food prepared with time and care, and start where it matters—near us.

California is home to 37 million people (2009) and sees vastly more visitors than that every year. For tourist and natives alike: this is even more your book than ours. Our goal is to provide recognition, facts, and information—to feed you a storyline, adding the occasional drum roll or bass beat of tension to the story of California agriculture—and, in the process, to encourage exploration and discovery.

Posing two questions is eminently fair. First, why now? Collecting information about agriculture in California is no modest activity. Now is the time. The Census of Agriculture, undertaken every half-decade, was conducted in 2007, and once collated and processed, census data began emerging in 2009, with maps and detailed county-level portraits available in Spring 2009. Our fieldwork, across more than two dozen trips and thousands of miles, caught the face and spirit of California agriculture over six years—and as our knowledge crowned, the taps of detailed data opened. Mating fieldwork with information that might be deemed official, we pass along the results. The numbers are fresh, the maps redrawn. Agricultural and geographical analysts in California working for the U.S. Geological Survey, the National Agricultural Statistics Service (NASS), California's Farmland Mapping Program, and the California Department of Forestry and Fire Protection (recently renamed CalFire), along with nongovernmental organizations that watch over land use, planning, and production, add information. Seeing timely fieldwork, fresh federal surveys, new maps, and improved state and county reports, we saw an opening.

The second question is a more complicated one, and is best approached on tiptoes. Why are two people who teach in Nevada the appropriate authorities to write about California agriculture? First, we were willing to put in the time, and not merely conduct this as a library project or an exercise in theory. Also, we know California. Our family roots here date back to the 1850s. We've navigated long enough by feel, sun direction, and two sets of much-abused but detailed DeLorme map books, and recognize north as the side of a tree with moss growing on it, so we have reason to feel sanguine about our results. We've pored over enough statistics and maps to find errors aplenty—and this guide fixes some of them. Furthermore, this book needed to happen, so we pled the case for "agriculture" as a theme suited to the venerable "California Natural History Guide" series.

As the references section of this book makes clear, especially in the listing of essential sources, the past has seen passes made at California agriculture as a prospective theme. Some volumes are chock full of information, with essays by experts whose credentials we deeply respect. Writing aimed at a distinct crop or specific agricultural activities (labor, dissent, organics, ag sociology, food policy, technology, drought, water usurpation) in small parts of California is reasonably done in the literature. But as for grasping agriculture across the whole of the state, only the Berkeley geographer James J. Parsons came close—and he never wrote the book. So many treatments of California agriculture are screeds or polemics; vitriolic arguments garbed in cute academic titles, rather than an honest grappling with the whole of what is there. Even then, this is an empirical study of California agriculture, and another volume of essays on the nature of agriculture in California is emerging from our lenses, laptop trackpads, and keyboards. We've traveled California widely and often—and, we hope, with honest enough eyes—to see and know what is there. You be the judge; if we're wrong, we'll surely hear about it. If you fancy the photographs, every one in this book was taken by Peter Goin, most during field excursions we purposely made together, wanting to challenge and corroborate our observations on the ground.

Please also keep in mind that no book designed as a field guide, and therefore supposed to be pocket-sized, can cover every crop. "What about cherimoya?" a friend in Vermont asks. "Surely it rates its own entry!" Or "Don't forget bananas—they were a great crop in La Conchita, near Santa Barbara along Highway 1." And that last would be true—until a landslide wiped out the sole plantation a few years back. Believe us: we, too, try to share our love for the eccentricity and possibility of California. All those miles, all those conversations (routinely in Spanish, which we both speak with some fluency), have brought agriculture to life.

Finally, let us borrow a few succinct words from Paul Richardson, a Briton by birth but since 1991 a resident of Extremadura, western Spain. In his 2007 book *A Late Dinner*, he offers a suitable thought to close out this preface. True, he writes of Spain: the same Spain that served as a launching pad for the occupation of the western hemisphere by Euroamericans, and that remains only slightly less an agricultural marvel than California. (And little wonder there; Spain was the cultural melting pot where cultures, techniques, abilities, and ambitions came together and launched a knowledge invasion at the Americas.) "Cities represent demand," Richardson writes. "The countryside represents supply." That point may seem obvious—but if so, why does such apparent separation exist? As authors, we hope this book will begin the suturing of the food-to-table world back into an elegant whole—if still one multihued and variously flavored—and in the process again merge city and countryside. Onward.

TOPPY
VEGETABLES
VALLEY OF THE WORLD
NOW GROWING
STRAWBERRIES
Welcome to Salinas
Salad Bowl of the World

A CALIFORNIA CORNUCOPIA

For one who would profess to be a master of this science must have a shrewd insight into the works of nature; he must not be ignorant of the variations of latitude, that he may have ascertained what is suitable to every region and what is incompatible.

LUCIUS COLUMELLA, *DE RE RUSTICA*, FIRST CENTURY AD

California Agriculture: A Panoramic Glimpse

Whether someone drives or flies to California, agriculture is conspicuously what is seen and savored. With fields and livestock, rangelands and stock ponds, tree crops and gene-tweaked forests of pine and redwoods, fully 50 percent of California is privately owned. Half that area is in active agricultural development. From the air, California seems like the empire of a creator bent on constructing the world's most cruelly demented jigsaw puzzle, a vast agriscape cast in shades of green, tan, emerald, and gold. If childlike excitement at the energy and cleverness of this landscape is a normal first reaction, that gradually gives way to awe at the sheer size and diversity of California agriculture. Even someone who prefers food local, community-supported, and

Plate 1. From the air, a patchwork of colors, patterns, property sizes, land uses, aqueducts, canals, and rivers dramatizes the variations of California agriculture, here looking north up the Sacramento Valley on a flight from Reno to Oakland.

organic will find the ambition of California agriculture intriguing, like a super-tall Shanghai skyscraper, an ultra-class off-highway mining hauler, or the huge aqueducts of California's State Water Project. Some enterprises leave the mouth agape.

An initial exposure to California agriculture is just the opening salvo. Mesh agriculture with the thorny word "infrastructure," and all of a sudden the focus changes to a profusion of canals and aqueducts, private roads without public access, and haul yards crowded with trucks and transport bins. With sufficient exposure, worker housing will emerge in plain sight, and the initiated will spot huge steel holding tanks, intensive nurseries, grain elevators and Harvestore dairy silos, working farm yards, and ag equipment dealerships dominated by John Deere green, Kubota orange, and Caterpillar yellow. Flying through the air, or tied down on remote airstrips, is a veritable air force of spray delivery systems for non-organic agriculture—airplanes, helicopters, tractors, workers with spray tanks held in vertical plastic backpacks—that blankets the land on command. The enterprises that sell or depend on agriculture are plentiful, and with farms in the state (2007) on more than 25 million acres, there are customers aplenty.

Agriculture is implicated in features, material and nonmaterial, that the casual traveler may fail to observe: greenhouses, railroad sidings, carefully bordered fields where rice seed was recently flown-on, long chicken houses plastered with biohazard signs, and even seldom visited and unpopulated "grow-houses" stuffed with gro-lites and marijuana. Local folklore and attitudes surface in signs that reflect pride and identity: Zinfandel Lane; Utopia Cattle Co.; Henry Miller Ave.; Welcome to Mendota, Cantaloupe Center of the World. The messages in other cases are less welcoming: stern no-trespassing signs proliferate, and some growers—as the state's farmers are known—resort to private security patrols. That's not to sound unduly menacing, but California agriculture is big business in scale, in profits, and in prominence. Even small producers can go about their business in a big way. The rule, as the wise observer Peirce Lewis once opined, is that "the more you know, the more you see." In that spirit, this is a field guide to California's agricultural landscape—or, better said, to its many landscapes. Once you are attuned, those agriscapes offer a continuing marvel to the eyes and senses.

As a field guide, there are things that this volume can, and cannot, be. It lays out how California grows things and uses its pasture and grazing land, but also what that means for our relationship with nature, including soil, water, clean air, family farming, reasonable quality of life, energy use (and abuse), and the remarkable assortment of people of varied national origins who are growers and ranchers. Keep in mind: crops change, demand decides, and agricultural land goes under asphalt with regularity. So, too, do animal breeds vary with fecundity and fashion: Hereford or Charolais cattle, a kid goat born of a Boer, Thuringian, Nubian, or Appenzell parent. When, for instance, we take up the topic of "grapes," included are raisins and table varieties, the products of premium- and commercial-grade vineyards, the three dozen trellising systems in current use, and wine-tasting facilities varying from hand-hewn humble to bombastically baronial, with every imaginable variation lying in between. Nor can agriculture exist on its own—production is tied into an intricate system of resource movement and monitoring: energy, water, transport, soil, quality control, labor supply, and supervision of government and commodity group.

Agriculture is California's most consistent economy, in part because it is so physically grounded. At various points in history, mining and commerce offered competition, and today, computers, software, and electronics, construction, tourism, and the service economy form slices of the state's economic pie chart. But ag remains huge. When production figures for 2007 were summed, agriculture was better than a $37 billion activity—wholesale—on 75,000 farms and ranches within California, according to latest USDA figures and the California Department of Food and Agriculture. By the time other activities are added in—for someone builds the tractors, sells the deep-well pumps, delivers the irrigation pipe, and flies on the rice seed—the total value of the California's agriculture-linked economy easily exceeds $100 billion, and if ancillary products are added in, might be recalculated to $200 or $300 billion (2007). The wine industry alone, with all that tourism, tasting, and aging, was worth $53 billion in 2007. Is that significant? There isn't a bit of doubt about it: consider that worldwide revenue from Hollywood films in 2006 was $25 billion.

But such numbers threaten to sound like the start of a particularly old-fashioned and potentially tedious textbook in agricultural economics. Just as innovation and glamour are elements in the release of a blockbuster film, agriculture in California, year to year, season to season, is never entirely the same. Straightforward variation in climate, with droughts and deluges, heating days and cooling hours, fog and onset of the rainy season, is a climatological and a regional part of the story. The other chunk is a drama of changing markets, overplanting, pest problems, and an ebb and flow of fashionable crops. Almonds are almighty; pistachios are peaking; grapes threaten to be in glut; apples are old news; kiwifruit crowned—and then caved. Naturally, all that can change with just a bit of time. Almonds, for example, are big water users, and are utterly dependent on bee pollination. When CCD (colony collapse disorder) arrived in 2005, it looked as though yearly almond crop production might plummet because of an absence of pollinators. Instead, at the end of 2008 there are 640,000 acres, with another 86,000 coming on, and production is the highest ever, up 15 percent—at 2,350 pounds—per acre. California currently produces a fifth of a pound of almonds for every single living person on earth—a formidable total.

California agriculture boasts a visceral reality that is a great deal more satisfying than account books, export numbers, or debt ceilings. We may travel in insulated and air-conditioned vehicles, but a pause by the side of the road or the successful identification of a walnut or almond or pistachio grove is a particular pleasure, especially when accomplished thanks to recognition of the signature characteristics of each species. There is even a good bit of humor to be found in agricultural activities. Giant buzz saws that slice the tops off pears and apricots to keep the fruiting

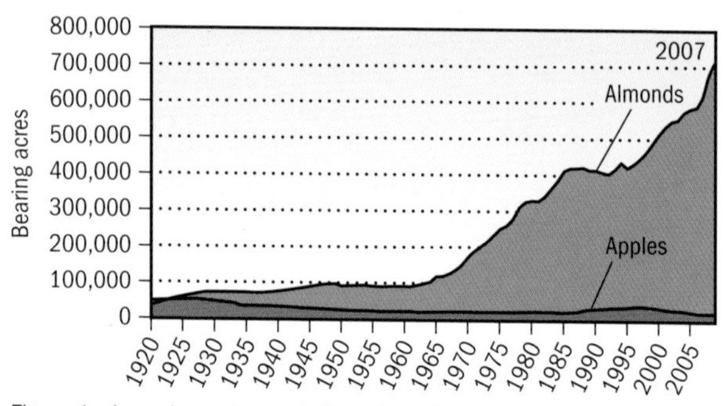

Figure 1. Acres in apples and almonds in California, 1920–2007.

branches at an accessible height look like something bought from ACME in a Road Runner cartoon; outsize monumental statues and signposts laud oranges, olives, lettuce, or loquats. There's a grin to be gained at the solemn commemoration of agriculture among the Cowbelles, or the respectful investiture ceremony for a Dairy Queen and her princesses. Worry not about covering up the smile—growers and producers recognize humor, too.

A fieldside conversation in Spanish with an irrigator or with the grower at a peach orchard, an orange grove, a blackberry field, or a mushroom bin is more than a talk; it's a reconnection to roots, and a recognition that farming feeds us all. In that way, although field borders and the sanctity of crops (don't even think about picking when you're looking!) are not to be argued, California agriculture offers a graduate-level course in pattern recognition, puzzle solving, and memory testing. Granted, the distinguishing characteristics of agriculture in California are, first, an inarguable diversity in products and people involved, and, second, the sheer scale of the state's agriculture, with—by some counts—nearly 400 crops. But an enduring third feature is quality, something not always associated with agribusiness and volume. It's the final distinguishing element. Whether an

Plate 2. A proudly painted dairy cow sculpture at the World Agriculture Exposition broadcasts Tulare County's role as a dairy megaproducer.

agricultural product in question is a unique Marin sheep cheese, a harvest of boutique lettuces, a small-lot production of Adelaida Valley wine, boxes of North County San Diego avocados, or navel oranges from Strathmore, there is attention in California to self-policing in the quality, health, and marketing to a degree that is hardly so conspicuous elsewhere in the United States. That alone merits recognition in this Natural History Guide series, which documents unique features and changes in California's landscape. Agriculture is here to be noticed, occupying more than half the area of California, or exceeding 70 percent, depending on how timber-growing and rangeland grazing on the state's public lands are counted.

What's Seen . . .

No one who travels California with clear eyes and an open mind escapes unscathed from a repeated and gnawing question. We peruse a plowed and check-dammed field, take in a grove of flood-irrigated orchard trees, and mull over long, thin sheds clearly holding livestock of some stripe. Thwacked by windshield bugs while blasting alongside acre after acre of recently planted crops, we pause on the roadside to watch a combine or harvesting rig lumber its way through a field, bright machinery rapidly dulled to the anonymous shade of powdered soil. The inquisitive absorb the idiosyncrasy of an apparent ocean of three-mil black plastic from the top of which only the barest hint of green strawberry leaf emerges. We look, we see, we consider. Each and every one of us has, then, peered, squinted, and scowled at something growing on a farm or ranch and uttered a semi-exasperated question: what on earth *is* that?

Only a few adventurous or obsessive-compulsive souls will consistently take the trouble to get off a limited-access road, stop safely, and walk over and investigate—even then, perhaps to no avail. But there are means of finding answers that this field guide is designed to provide. Curiosity is one thing; satisfying that visual itch with knowledge and certainty is another sort of itch-scratching and satisfaction altogether.

In California, agriculture can summon an untapped well of interest. This "field guide" is about what's where, and why—making sense of a world we perceive but don't necessarily understand in California agriculture. A second volume of essays will chronicle a complicated history, series of land developments, competing uses, ingenious responses to market forces, and novel patterns of resource use and property acquisition. This is beyond the scope of a field guide, which is designed to be on your dashboard, if not clutched in the hand, for a road-based adventure. But getting a sense of the enterprise may be helpful, for along the way California agriculture is both a model for its friends and the despair of its detractors; its huge economy enjoys many complications and a physical presence that no other state (and only few countries) can approach.

Through the years, California agriculture has been the theme of learned books, including several notable tomes from the University of California Press. In an ever more urbanized society, there is quite noticeably a growing attention to agriculture, food, and domesticated animals. It may seem contradictory, but unfamiliarity can breed respect, not obliviousness, spawning a sense of remorse: "I should know more about food and its production than I do." If farther removed from working the land, Californians are nonetheless concerned about urbanization, land trusts and conservation easements, preserving farming, and making sure that food (including chickens, turkeys, and dairy calves) is raised right. Although there is no obvious contradiction in this, concern overlain by ignorance is something striking and very much of our time.

Recognizing the originality of California agriculture requires going back to the *Pacific Rural Press* of the 1870s whose editor, Edward Wickson, a pioneering publisher, a lecturer at UC Berkeley, and an evangelist for things Californian, wrote up the agricultural possibilities of the state's predominantly Mediterranean-type climate (winter wet, summer dry), which provides such dramatically varied farming habitats. It was Wickson, with his advice-for-readers column dispensing horticultural suggestions and his promotion of the state's agricultural land virtues, who truly began the viral marketing of California as an appealing destination for potential migrants. In a sense, he initiated the landslide that would consume and pave over what he so lauded.

Other states specialize in growing a handful of crops, and in value and expertise they may even be powerhouses. With California's variety of environments, the state can grow almost anything—and does. It ships produce across the United States and abroad, with international exports nearing $11 billion in 2008.

At the big end, California agriculture is formidable indeed. Then there are small local producers of carefully raised foodstuffs who distribute through community-supported agriculture (CSA) or farmers markets. The state's "foodies"—ranging from professional chefs to amateur cooks who prefer to eat locally—look toward the locavore movement for special inspiration. And there is no mistaking a movement that has caught on. The local food movement in the San Francisco Bay area in 2005 inspired Jessica Prentice, director of education at the San Francisco Ferry Plaza Farmers Market, to coin the word "locavore," which was chosen by Oxford University Press as its "Word of the Year" a couple of years later. While agribusiness in California has a momentous advantage in scale, there is something to be said for a movement that can orchestrate an enduring vocabulary.

The point, in short, is that agriculture exists at every step along a production scale, from intensive and hand-raised crops with limited outputs to agribusiness growers whose gross production expenses are in the tens of millions of dollars. California agriculture is a notable novelty that provides much of the diversity in American eating through agricultural exports recognized even in film epics such as the 1955 Elia Kazan–directed *East of Eden,* starring James Dean (from the John Steinbeck novel set in the 1910s). Marketing, however, is a crucial part of the California story, and the film-literate will remember that a plot element in *East of Eden* was the crisis triggered when Salinas Valley produce was a glut on the market. Growing is only a start in California; building an organization to research crop improvements, find new markets, and navigate the regulatory landscape is not a simple matter.

Within the hundred million acres of California are more concentrated and diverse agricultural products than at any other site in the world. Twenty distinct plantings can be seen in a dozen miles of driving. Turkey coops, emu corrals, dairy yards, and foothill rangeland hold varied livestock. An orchard might contain eight different olives, or a single vineyard block a half-dozen grape varieties that are not only cherished but fought over by winemakers who want to lock up access to what they might perceive as a crucial component in a Bordeaux- or Rhone-style blend. After all, it was wines from California that featured in the famed 1976 "Judgment of Paris" that pitted Napa-, Santa Cruz-, and Monterey-derived cabernet sauvignon and chardonnay against the best of Bordeaux and Burgundy—and saw the California wines emerge as stars.

In unlikely places in California (but for good reason) are found important crops: nursery stock, flowers, avocados, and heirloom vegetables in rapidly urbanizing North County San Diego. In Monterey County are acre upon acre of lettuce and artichokes, strawberries and arugula. Apricots near Patterson are being displaced by urbanization advancing at rates so fast it could be steroid-stimulated. Or there are dates, still growing at a leisurely pace in Mecca and the Coachella Valley in southeastern Riverside County. Raised around California also is better than $2.1 billion of lettuce (three-quarters of the U.S. total); strawberries earn more than $1.5 billion (2007). In the San Joaquin Valley's hugely fertile counties are almonds, peaches, plums, apricots, melons, oranges, and grapes. Or take in the exquisite medium-grain rice grown on a necklace of counties around the Sutter Buttes, north of Sacramento. There Dr. Hugh Glenn produced a sans pareil California wheat—hard, dry, and white—in the 1880s on the largest agricultural empire the world has ever known, though perhaps almost matched just a few decades later by California's land- and cattle-hungry Miller & Lux Company.

At the other extreme, an organic grower in Marin County, or working the loamy soils of Cache Creek or Capay Valley, may include dozens of crops in a biointensive 20- × 5-foot double-dug raised garden bed. California's residents and visitors travel a great deal. In all that movement, apparent anomalies abound. Fresno is the highest dollar-producing agricultural county in the United States year after year—rivaled only by Tulare, immediately to the south. The other end of the scale holds tiny boutique crops, including wines produced by vintners such as Edmunds St. John, Stroth-Hall, or Berkeley's Donkey and Goat, wines that are marvels of urban agricultural production, drawing on grapes grown in distant parts of the state, brought to city-street warehouses for crushing and fermentation—where, with time, tasting, and careful tending emerge some of California's most intriguing and eccentric wines. Silicon Valley legend portrays the computer industry as originating in the garages and in-law cottages of the San Francisco Peninsula. For a much longer time, the same has been the case with agriculture and ag technology. "Jacuzzi," after all, started on the flats of Berkeley, where a family-owned machinist firm instigated

Plate 3. The historic era when Sacramento Valley and San Joaquin Valley wheat harvesting combines were drawn by 18 horses is commemorated in this mural by Claudia Fletcher that decorates the side of a building in downtown Exeter, California.

the manufacture of novel agricultural pump technology; in the same spirit, Holt Manufacturing in Stockton created a practical caterpillar-type track that offered flotation and kept vehicles from sinking into soft ground in the Delta's peat soils. Then, in 1920, Holt merged with C.L. Best, a San Leandro firm, to form Caterpillar Tractor.

Although small enough in dollar value that they hardly feature in a pie chart of California agricultural production, the state's niche crops, specialty harvests, "boutique" vegetables, and foodstuffs grown to spec for gourmet restaurants are the very model and inspiration for innovative and health-conscious chefs around the globe. For good reason is Alice Waters of Chez Panisse a sacristan in the so-called Slow Food movement, a UC Alumna of the Year, and a model for foodies everywhere. Journalist and UC Berkeley professor Michael Pollan's *The Omnivore's Dilemma* is just one of a now vast assortment of books encouraging Americans to eat better and with more joy, helping them learn about the sources and history of their food. In *Coming Home to Eat*, the biogeographer Gary Nabhan does

Plate 4. Lettuce is a behemoth crop in California, grown in many varieties and to great dollar value. The shades and hues of lettuce are so diverse that they turn a satellite image of the Salinas Valley into a tapestry of greens and blues—and reflect production worth more than $2 billion in 2008.

the same—as, more recently, have novelist Barbara Kingsolver and food maven Marion Nestle. Organic and natural producers of everything from beef to basil are on an influential upswing, in part because they accept cautions about land loss, pesticide use, and water shortages and aim to support profitable family farming. California's largest organic grower outproduces small farmers by more than a million dollars annually. Intensification is a reality across all California agriculture. There are economies of scale to organic food production, and, as the geographer Julie Guthman notes in her work, even the organic sector in California agriculture is industrializing and attaining a global reach.

Although organic and boutique farms of California cannot outproduce huge corporate agribusiness growers, they significantly diversify the commodity mix, providing a degree of variation and inventiveness recognized with prize medals at world competitions, and growers glean premium prices at farmers markets statewide. These facets of California agriculture are part of the innovation cherished from Spanish–Mexican arrival in the late 1700s onward, and they provide a haven for those who seek food and fiber that passes through a rigorous gauntlet of supporters who believe in eating produce from close to home, and who favor paying farmers and laborers a living wage for their work. Such weighing of social conscience represents another way that California may be cutting new ground in the eating and lifestyle cultures of the United States.

A Field Guide

A primary impetus behind the Natural History Guide Series is recognition of unique features and changes in California landscapes. The late Berkeley geographer James J. Parsons observed that "agriculture tops California's uniqueness," and his work shapes our observations. For latter-day Californians, however, ignorance of agriculture's impact upon the land is pretty much pervasive. Tension that rivals strains on the San Andreas and Hayward faults underlies attitudes toward agriculture.

The split state movement in California dates back easily 80 years, and on occasion breaks into active revolt: *San Francisco Chronicle* reporter Stan Delaplane won a Pulitzer prize for coverage of the 1941 "secession" of northern California and southern Oregon into the State of Jefferson. In the energetic 1975 novel *Ecotopia*, Ernest Callenbach splits off northern California and the Pacific Northwest from the United States.

Philosophical and geographical divides in California are no less formidable today. In 2009 the British newspaper the *Economist* reported a fresh split-California movement. Departing from well-aged calls for a north–south division, DownSizeCA.org would partition western, or coastal, California from the rest of the state—severing 45 counties in the interior state from 13 populous counties on the coast (Marin to Los Angeles). The problem, an irascible San Joaquin Valley farmer claims, is that coastal Californians "love fish, [and] hate farmers"; another laments the coast's "agriculturally uneducated city dwellers." Granted, the generalization may reflect more dyspepsia than truth, but it bears remembering that strains form between agriculture and environmental quality, between production and protection—and although such divisions might appear whimsical, they manifest themselves in attitudes about what sort of agricultural production is best (big or small), how much water should be moved around (lots or little), whether cities or farms are more important, and what should happen when cities decide they—or, perhaps, the Delta smelt—need more water. Reducing the gap is needed—or someone will start laying bricks in earnest to wall off the "fifty-first state."

A 2007 study by the American Farmland Trust acutely catches the threat of urbanization and change: 28 percent of land subdivided and developed in California from 1990 to 2004 was high-quality farmland. The urbanization of another two million acres of land is anticipated by 2050. Learning to observe what is going on may help with a moderation of this speeding movement out of agricultural acreage, encouraging other, less destructive forms of development, perhaps even making California's cities more interesting and livable places with higher-density clustering of population instead of sprawl. But gaining better sightlines is important: too many of us see without knowing, glimpse without understanding.

This field guide includes crop and livestock descriptions and finding guides for recognizing agricultural products. An introduction offers a concise view of the past evolution and current manifestations of a highly varied regionalized agriculture. Not every crop is covered in depth—a task that would require three volumes, not one; and besides, a goodly amount of California's agricultural innovation comes at the fringe, with a small number of growers who are more interested in agricultural experimentation than in up-front profit. Growers are foremost among the innovators. Highly trained agricultural scientists may also work for crop commissions, or for the University of California as professors, research scientists, or Cooperative Extension agents, and they are often in the field, exploring new ways to improve a crop. Collectively, from farmer–experimenters to scientists sporting doctorates, these are the people who test exotic varieties, create the hybrids, introduce new breeds, explore agronomic techniques, research rangeland health, and deserve credit for encouraging innovation in California agriculture.

Should you be fortunate enough to run into one of those ag experts, you may see agricultural history ratcheting forward. Learning to recognize the distinctions between current varieties has value, and learning what is where, and why, is important. As various authors have had to grudgingly admit, understanding

Plate 5. A cleaned understory, carefully arranged lines of almond trees, and the emerging drupe (fruit) are much in evidence in this view of an established almond grove in Tehama County.

California agriculture is no simple process. Anyone who sees agriculture in California as an easy story is blind to the world of people and end goals involved. Agriculture can be done up as a triumphal face of California, with an economy, physical influence, a variety of workers and owners, and a sheer range of crops unsurpassed anywhere. But that is not the entire story: what transpires with agriculture is a cautionary tale, too, and that stands as an equal theme.

So Put It to Use

This guide gives casual visitors and longer-term Californians a means for understanding what they see, from asparagus spears rising from deep Delta peat soils to the stark visual contrast of cherished oranges and dark green foliage in Highgrove or Navelencia. The importance of California agriculture is undeniable. But even people who know a good bit about agriculture don't know what to make of the diversity or complexity of California crops. If some are inspired to take action on behalf of sounder growing, working, and marketing practices, then all the better. Although everyone eats, only a fraction of a percent of California's population is actively involved, on the ground, in raising that food. It was sometime Californian Joni Mitchell who sang us the line: "don't it always seem to go / that you don't know what you've got 'til it's gone."

We aim to bring facets of the state's agriculture to a huge urban (94 percent) Californian audience. Better still, however, if you hail from out of town, out of state, or from a foreign shore. The broad coverage of this work encourages a self-empowerment of first-hand observation and field-based discovery for the tens of millions of tourists who visit California and find a state without equal in its agricultural diversity or productivity (and in matters relating to agriculture, many native Californians count as "tourists"). Literally, this is a primer in California agriculture, easing readers from landownership to history, and taking in a production cycle that marches, like a crop itself, from tillage to harvest to market.

Why Is That Raised There? Regions, Climate, and Products

Among the reasons for great agricultural productivity in California is an unparalleled diversity of environments. A quick overview is handy, although 11 distinct agricultural regions and some smaller districts that lie within them are discussed in a final section of this field guide—proceed there for the details. Grasping a couple of broad rules is useful. As you travel from north to south, California dries out; the same is true from west to east. Driest are Southern California's Riverside and Imperial counties, resting alongside Arizona; wettest are Humboldt and Del Norte counties, nearing Oregon in the northwest of the state. Because water is impounded in reservoirs, behind a complex system of dams, water can to some degree be moved to where it is required in an intricate system of aqueducts, canals, and ditches.

Much of the climate of California, from Cape Mendocino south, is classified as "Mediterranean-type," which means that it is dry—essentially without rain—from May onward, often into late September or October. This summer dryness is shared

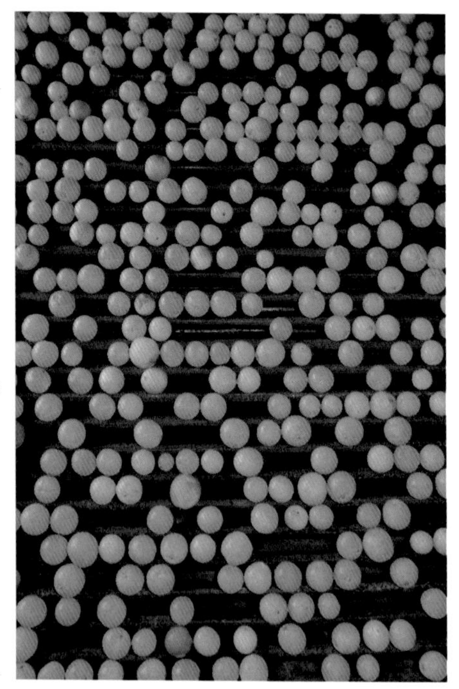

Plate 6. Flowing like digital data, these fruit are emerging from a second wash and are headed toward electronic sorting in the Exeter-Ivanhoe Citrus Sunkist packing house.

by the countries that border the Mediterranean Sea, but is also characteristic of California, western Australia, Chile, parts of South Africa, and select other regions of the world. All those places are significant in their agricultural production; none equals California.

Climate

What makes an area agriculturally rich or not depends initially on latitude, nearness of the coast, and elevation. Coastal California is influenced by cold water of the Pacific Ocean and by the California Current that in summer generates fog banks along the shore. From the San Francisco Bay Area to Southern California, temperatures rise about 1 degree for every mile inland—the mean July high temperature in San Francisco is 68 degrees Fahrenheit, and Walnut Creek, 20 miles east, averages 87 degrees. Neither area has much cropland agriculture anymore, but 60 miles south lie Castroville, Monterey, and the Salinas Valley, where the summer fog is balm to artichokes and berries and, farther inland, helps cool lettuce, spinach, kale, and broccoli. That pattern holds down to the Mexican border, although Southern California is far drier than the Bay Area or North Coast.

Mountain areas are wetter than valleys and in a good year hold a supply of snow that is essential to agriculture and cities alike. The mountain regions are cold and less stocked with highly fertile soil, so an imbalance of water supply between water source areas and agricultural zones is a given, and a common source of tension. A pronounced rain shadow reduces precipitation in areas that lie to the east of several mountain ranges—as in the Carrizo Plain, between San Luis Obispo and Bakersfield, or east of the Sierra Nevada in the climatically arid Owens Valley. Double and triple rain shadows can halve or quarter the rainfall received from coastal mountains to valleys far inland. Yet even an isolated valley can have significantly varied growing environments, depending on aspect, elevation, water table, and soil characteristics, which collectively can creative a different microclimate. Finally, urban expansion is a major influence on agriculture. Areas with the richest and deepest soils were settled early in California's recorded history, and those soils in many cases are now covered with asphalt or green suburban lawns, as in Sacramento, Livermore, San Jose, Los Angeles, Pasadena, and San Diego. In other areas, conversion of ag land to human-occupied space is ongoing, and the demand of cities for water increases with population. In any snapshot of what is raised where, the picture of California changes with time's passage.

Water

Common sense would suggest that desert areas are the least productive, but that's not necessarily so. Sand is actually a suitable growing medium—just add water. Since 1769 and Spanish movement into California, irrigation and a willingness to transport water from place to place were the rule, and that continues unchanged today. The historian Donald Worster accurately describes California as a "hydraulic society," which fairly captures the extent of the state's plumbing. Dams and reservoirs

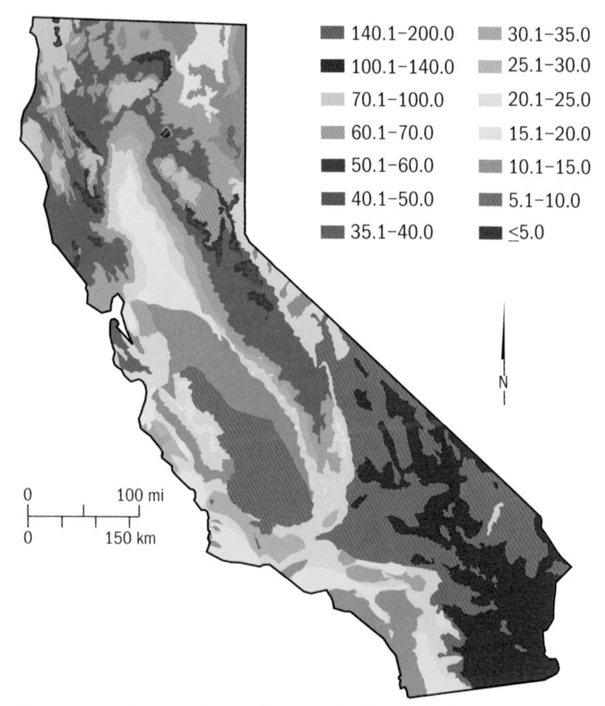

Map 1. Average annual precipitation in California, in inches.

hold not just river water and melt-water, but also topsoil, which accumulates behind dams instead of spreading downstream on fields. Water is moved through the federal Central Valley Project (CVP), by the labyrinthine State Water Project (SWP), or from an alphabet soup of acronym-bearing water systems built by irrigation districts, cities, counties, and entire metropolitan areas. The total extent of the distributaries is in the thousands of miles. Just four main aqueducts controlled by the CVP, the SWP, the L.A. Department of Water and Power, and the Metropolitan Water District stretch better than 2,000 miles, and water in those aqueducts runs deep and fast, sweeping money and influence ahead of it.

When precipitation as rain—and, more often, as snow—is plentiful, water is moved wherever it is needed. But in drought years, as supply dwindles, the stakes grow high as demand increases. In theory, water can be pumped from Oroville Dam to downtown Los Angeles, but in practice, the cost of lifting water over the 4,100 foot rise of Tejon Pass through a chain of pumping stations makes that a rarity—metropolitan Los Angeles has other, cheaper sources. And besides, farmers in the San Joaquin Valley are more than happy to use the water as it wends south in huge canals that roughly parallel Interstate 5. The water is needed. The San Joaquin Valley is for all intents and purposes a desert: Shafter (near Bakersfield) averages 7.34 inches of precipitation a year. Farther east, Riverside County's Mecca has a yearly average of 3.22 inches of precipitation. Yet both are remarkable locations for productive commercial agriculture. By contrast, precipitation in the North Coast towns of Crescent City (75 inches), Orick (60 inches), and Eureka (40 inches) is

Plate 7. Once a signature of the long-established California farm, palms were in the same spirit widely planted as an emblem of urban sophistication. These are agricultural palms within the town of Thermal, eastern Riverside County.

relatively plentiful. Although charming places, they are not known for agricultural output, aside from forest products, a few orchards and pasture, and a profusion of marijuana, untallied in official county agricultural reports.

In the end, California's hugely diverse ecosystems are shaped by the seasonality of rainfall, by the increasing scarcity of precipitation grading less and less from west to east, and by the growing aridity from north to south, which inspired the state's early European-derived occupiers to start an irrigation-based society that continues through this day. That is no criticism: many a location outside of California, including Muslim and Christian societies that rimmed the Mediterranean Basin, were skilled in the management and manipulation of water. But the principle that is defined is of scarcity and need, not plenty of provender. As the first decade of the twenty-first century draws toward a close with threats of a third year of drought, it is well to remember that paleo-ecologists believe that five centuries ago, California went through a 150-year drought cycle. And we may go there again—with what sorts of effects on agriculture and productivity of the land?

California Agriculture through Thick, Thin, and Time

The 10 biggest moneymaking products in California agriculture today are a decidedly different roster than would have topped such a list from a hundred years ago. Changing eras, markets, and technologies inspire the production of distinct animals and crops, and with time, alterations in demand modify the crops and animals grown. What tends to remain the same is a steady rise in agricultural productivity, and the prominence of supremely fertile areas. Even then, prime lands can be submerged under expanding towns, cities, and suburbs.

It does not take full urbanization to drive farms and ranches from a given area: there is a tipping point past which agriculture is no longer feasible, practical, or even enjoyable to the grower or the rancher, and at that point, agriculturalists who remain will flee, selling their holdings and either retiring or moving elsewhere, to less contested and congested terrain. By no means are all agricultural shifts attributable to urbanization: the styles

Plate 8. Two Mexican farm workers are laying in plastic irrigation piping for a strawberry field, near Jensen Road, Monterey County.

and demands of agriculture change with time, and the prized and profitable crop of one decade can in just a few years be yesterday's news. Such is especially the case in California, where agricultural fashions alter direction like a whimsical wind.

The history of California agriculture, though much discussed, is anything but carved in stone. When *Harvest Empire: A History of California Agriculture* was published in 1982, author Lawrence Jelinek rightly noted there was no current—or even suitable older—history of California agriculture. Twenty-eight years later, what continues to surprise is the enduring truth of that remark. The best history—highly compressed—was, and still is in *The California Water Atlas*, edited in 1979 by William Kahrl, with maps by Bill Bowen. An observation about the state's missing agricultural history was renewed in 2004, when in a special report from the Giannini Foundation, Warren Johnston and Alec McCalla, two emeritus professors at UC Davis, produced the far-reaching and insightful "Whither California Agriculture? Up, Down, or Out? Some Thoughts about the Future." A gem of a study, the paper is something only two veteran and deeply published agricultural economics researchers could produce.

Although the short history here takes its own path, Johnston and McCalla's insights into seven constants in California agriculture are worth restating, in somewhat different language, because the points work as well for the past as for today:

1. Entrepreneurial demand drives California agriculture, not subsistence agriculture. California has always been at the edge, if not over it.
2. Resource dependence (land and water) drives California agriculture, and a constant is the search for more and better resources.
3. An absence of water defines agriculture; seldom is water where it is wanted.
4. Agricultural labor, never sufficient, is often imported.
5. Although growth in agriculture is the rule, disasters, economic shocks, and natural catastrophes are regular, costly, and traumatic events.
6. Successful California agriculture since the Gold Rush has required high levels of management skills, as much technical as economic, and the scale of agriculture makes that all the more an issue.
7. Technological skills and innovation are constantly honed and brought to bear on agricultural techniques, production, irrigation, plant varieties, animal handling and resource management, and everything else imaginable.

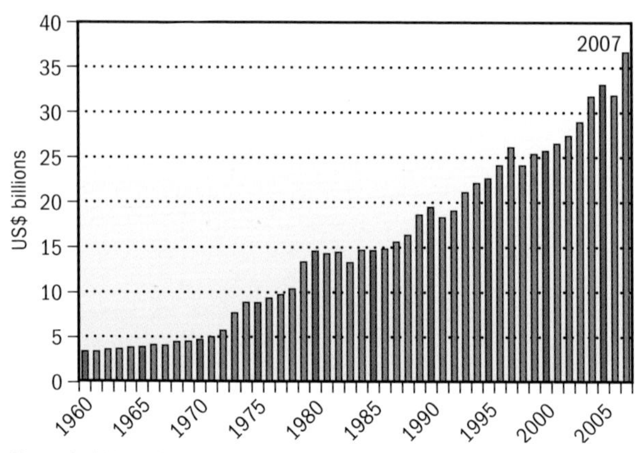

Figure 2. Value of California agriculture, 1960–2007; not indexed for inflation: $1 in 1960 = $7 in 2007 (CPI).

Several additional points are added in the sections that follow. These insights above offer an energetic beginning and are threaded through the history of the state's agricultural ventures.

Native Agriculture and Spanish–Mexican Introductions

Current thinking accepts that California's indigenous population was far larger than anthropologists and archaeologists admitted as little as 50 years ago. The count of native Californians certainly numbered in the hundreds of thousands—perhaps even a million. Many lived by the rules of stored surplus—in a Mediterranean climate, living well through the rain-bereft summer requires preserving extra foods. These were gathered during the wetter fall and winter seasons, with baskets and caches saving smoked fish, acorn meal, grass seeds, or berries. Some Native Americans practiced transhumance, traveling deep into the mountains during summer months to hunt, gather, and stockpile protein- and calorie-rich foods from high meadow environments lately relieved of their snow cover. In the eastern Sierra, some anthropologists even argue, formal Paiute agriculture employed irrigation ditches and cultivation using seeds and tubers, with careful selection of improved plant varieties. In the years lying ahead of us, these arguments will be tested and either verified or cast aside. What is suspected is that cropping and plant selection was understood through much of what is now California, and some Native American tribes were more accomplished at it than others.

Although Europeans visited California before 1769, that date marks the arrival of a delegation from northern Mexico, which established Mission San Diego de Alcalá. With time, a chain of 21 missions roughly paralleling the route of modern-day Hwy. 101 extended north to Sonoma. By design, these were agricultural colonies, incorporating crops and ranching, and the religious and economic conversion of Native Americans living in the vicinity was integral to the missions' purpose. The land selected as mission sites includes some of the deepest soils and most fertile agricultural land in California even today, and the wise choices made then are a tribute to the training and instincts of the mission founders of long ago. Claims to land radiated outward in the vicinity of the missions as soldiers furloughed from military service received parcels of land and governors issued property grants to nobles or favorites, a move that aimed to keep retainers local and loyal.

Extensive livestock ranching flourished, drawing on sheep, cattle, goats, and horses. The details of early ranching are laid out in contemporary narratives, with Richard Henry Dana's *Two Years before the Mast* the best known. This not only forged the start of California's ranch era, but it also marked the birth of a boom economy in cattle hides and tallow. Curiously, meat was a castoff product either dried into *charqui* (jerky), or abandoned to the benefit of scavengers that included the California Condor, a Pleistocene-era survivor that probably reached its peak historic population during the 1840s. Ranch hands were *vaqueros*, known now as buckaroos, and a ranch owner was the *patron*—establishing a split between workers and owners that survives in western ranch culture today. Drought, depredation by predators, and failing markets affected ranch life, and insecure land titles—mapped by hand-drawn *diseños* that were attractive metes-and-bounds documents but rarely precise—guaranteed that a great deal of Spanish–Mexican land was transferred to Anglo hands shortly after Californian statehood, despite contrary assurances in the Treaty of Guadalupe-Hidalgo. Only a small fraction of Spanish–Mexican claims survived in the hands of their original holders. A knowledge base was created that led to the transport of a handful of cattle and ranch hands to the Hawaiian Islands in 1798 as a bequest from the Spanish Crown to King Kamehameha, who unified all the Hawaiian Islands except Kauai. Some ranch hands who traveled to Hawaii to instruct the island natives in cattle-rearing techniques remained, and today's ranch hands on the Islands are still known as *paniolos*—which was as close as Hawaii's residents could come to the Spanish "*español*."

The imprint of Spanish–Mexican land grants exists 160 years later in California. But the huge landholdings that today characterize the largest agricultural operations owe little to the region's Hispanic past. If a few properties such as the Tejon Ranch and the Hearst Ranch were built around a core of Spanish–Mexican holdings, they are the rare survivors. Most great ranches were broken up long ago, and the largest holdings in today's California are traceable to grants made to the railroads that went into private hands and were sold off or were purchased by venturesome petroleum firms, as the geographer Ellen Liebman determined in the 1980s. Still, as a legacy of Spanish–Mexican times, the gridiron township-and-range land division system shows interesting gaps around communities, generally east and west of today's Hwy. 101. In the Salinas Valley, in parts of Los Angeles, and through the Coast Ranges, the imprint of earlier Hispanic land grants took precedence, and land division there lacks that utter regularity of the land office's four-square sections and townships. Ranching in the mid-nineteenth century offered a lifestyle that was no less attractive to Anglo settlers than to their Hispanic predecessors, and the extensive raising of cattle—with a relatively small number of animals grazing a large amount of land, left largely to their own devices—continued into statehood and the Gold Rush era. In fact, livestock, and especially cattle, had the great benefit of being able to walk to market, which made them particularly prized as a food source after the discovery of gold at Coloma began a fevered march of new residents to California, whether from across the continent or abroad. What was available to feed several hundred thousand argonauts deposited on nearly every stream along the west side of the Sierra? Not much, and what there was proved astoundingly expensive. Shortages and a lot of discretionary income forged a respect for foodstuffs that survives to this day in food-devout north-central California.

Gold Rush Agricultural Demands

In the mid-twentieth century, historian Rodman Paul built a case for the significance of mining in California, with observations echoed by his intellectual inheritor Andrew Isenberg in a remarkable study, *Mining California.* Both authors document that the absorption of Gold Rush–era human populations into the Sierra Nevada foothills demanded a massive recasting of California's conventions, resources, and written and informal laws. Mining required settling prospectors, and, soon, also the less glamorous laborers in industrial mining operations. They needed to be close to the mines themselves, because moving unrefined ore was difficult, so a great many people were concentrated in recently rural—and farm-poor—areas. New laws laid out rules for the distribution of scarce resources, and prominently water. Revised rules for the adjudication of claims, the meting out of justice, and the management of property were quick to form, and over just a few years created an entirely new California—part Hispanic, part Anglo, part created out of raw invention. In shortest supply were two things: actual wealth and food. A byword at the mines was that "it takes a mine to run a mine," and fortunes were more often gained by merchants than miners. But food supply and kick-starting California's agriculture merged into a separate problem.

The readiest food was meat on animals that could be moved overland to the mines. Within a year of gold discoveries, California's domesticated livestock supplies were depleted. An era of market hunting—commercial venery—arrived, and, as Scott Stine has written, in no time, wild populations of tule elk, grizzly and black bear, birds of every breed or race, turtles, fish, and oysters were reduced to the point of extinction. Hunters floating the Delta region in boats sported enormous punt guns that could kill a dozen ducks with a single shot, a scene worthy of the Grimm Brothers' valiant little tailor and his seven at one blow. And more than just miners craved meat. Residents throughout California's northern cities—San Francisco, Sacramento, Stockton—prized exotic flesh. The eggs of three dozen species of avifauna on the Farallon Islands, collected by the millions, sated miners' hunger; oysters that provided sustenance for a hundred generations of Native Americans around San Francisco Bay were nearly eliminated in a decade; turtles harvested by the thousands from inland Tulare Lake in the southern San Joaquin Valley were brought by cart, and eventually train, to the mines. Voila!—cauldrons of turtle soup, as documented by Gold Rush era menus deposited in the Bancroft Library at the University of California, Berkeley.

By the 1860s, California could not provide for its swelling mining population. There was no refrigeration, beyond evaporative cooling and stream waters, so fresh food spoiled in heat. The nearest source of produce was a double handful of farms established in the Sacramento Valley, yet once gold was found, ownership of those was almost instantly in question. As functioning farms, these were forlorn. Market drives brought livestock from Arizona and New Mexico—and soon, Texas and farther east—and these movements of animals trailed cattle and sheep to where they were needed, a journey that could also multiply their worth 10- or even 30-fold.

Weather could produce oddities—the devastating Great Flood of 1862 put Sacramento under 12 feet of water, and a three-year drought came in the next couple of years that provided an early warning of the inherent but unpredictable variability of the California climate. Lessons that might have been learned were not. Ranching continued at a muted scale in the Coast Ranges, ranging inland, north, and south from Santa Barbara, but a second branch developed out of Visalia and Porterville, with livestock trailed from foothill range to copious feed in the Sierra Nevada during summer transhumance movements. Sheep operators in the 1870s and 1880s moved flocks of several thousand animals northeast from Bakersfield and the Mojave Desert, up the east side of the Sierra Nevada, and across the Sierra westward through passes that now have roads but didn't then: Tioga, Ebbetts, Sonora, and Monitor. Although himself a sheep herder in the Sierra, the Scottish-born John Muir (later, a prosperous Californian) came to loathe his charges, which he would famously characterize in print in 1901 as "hoofed locusts."

What California faced in the 1850s was a dual culture: urbanization for many, and for the rest, a scattering across rural—but still densely populated—mine sites. San Francisco had its mission-era roots, but Sacramento and Stockton grew out of the Gold Rush as port cities on the Sacramento and San Joaquin rivers, respectively. Irrigation colonies started in both southern and northern California by the 1870s, as did San Leandro and Oakland, two East Bay cities that competed to be named the terminus of the Transcontinental Railroad in a competition that Oakland finally won in 1868. The arrival of heavy iron, in the form of a Central Pacific railroad venture that brought small dollar payments per mile of track laid but yielded fortunes in land, all but guaranteed that California—with its winter-spring rainfall regime that was much more secure than that of the Great Plains, and great access to seaports in secure bays and deep rivers—would rise to success. The hold of propertied landowners during the Mexican era proved more tentative than tenacious, and for many that ended with statehood, as property was wrested away by the California Court of Claims under suspect circumstances involving a lot of lawyering. An era of land speculation seldom equaled elsewhere arrived, and the new big winners in the land claims sweepstakes proved successful beyond the dreams of avarice.

Bonanza Agriculture and Technological Innovation

Throughout American agricultural history, land speculation has loomed hand in glove with agricultural innovation. The last half of the nineteenth century saw many an attempt in California to create a successful blueprint for farming. In the northern Sacramento Valley, a durum wheat, grown hard in California's dry summer, was acknowledged on the East Coast as a prize crop. Durum was enormously popular in Liverpool and elsewhere in Europe and was valued above all other wheat (until Russia discovered it could do the same on its eastern frontier). This boom produced the bonanza wheat era, in which Colusa and Glenn counties were seeded to wheat and huge gang plows, engineered for California's relatively light soils, were pulled by dozens of draft horses. For 30 years, California's wheat traveled the world.

Other innovations were more local. With a fading mining boom, demand for mutton eased in California. Sheep ranchers then trailed animals from the southern San Joaquin Valley across Nevada, Idaho, Utah, and Wyoming to distant railheads, where the sheep, having actually added weight on the trail, were worth far more than in depressed California. By the 1880s, in both Southern California and in the San Joaquin Valley, irrigation colonies formed, drawing on pooled capital and collaborative labor. These groupings of aspiring farmers, often hailing from a single population center in Europe or the American Midwest and Plains, sometimes flourished. More typically, the colonies survived for a time only to disband, with landholders relinquishing their acreage to the heartiest—or the most stubborn.

Such ventures created a conviction that California was an agricultural empire waiting to happen. Technological innovation, new varieties of crops, transportation lines (especially railroads) that could move produce and people cheaply over great distances, and novel delivery systems for water were each a part of the success. Uglier segments of this era are hardly unfamiliar: exploitation of labor, banning of Chinese ownership of land, devastation of the local ecology, fraud in water claims and land ownership in equal measure. Agricultural towns dueled and fought mining ventures, especially as the detritus of hydraulic mining clogged rivers draining the northern Sierra, washing fine debris into the Sacramento Valley and choking the waterways.

Levees were raised to prevent valley towns from flooding in late spring runoff, and court cases proliferated. Agricultural discoveries were made, and entrepreneurial experimentation was widespread, but no single model for agricultural moneymaking was in place. Bit by bit, however, what California farmers realized was the potential for profit, especially if they could secure access to nationwide markets and a reliable water supply.

A Hydraulic Society and Industrialized Agriculture's Rise

The crucial discovery California's farmers and ranchers made was climatic: the California climate was not like that of other places; its summer-dry period of four months could extend to as long as a half-year—from mid-May to late October or even November. In that respect, California was "droughtier" than Spain, Italy, Greece, or other Mediterranean-fringe countries, where rain started reliably in September.

Wet spells were followed by formidable drought, persisting sometimes two or three years, which, in turn, spurred attempts to divert and secure water supplies with weirs, canals, flumes, pipelines, dams, and reservoirs. The arrival of the twentieth century brought an awareness that the control and impounding of water was essential, and California started its way toward what Edwin Cooper memorably called an "aqueduct empire," and California was forged into a hydraulic society, plumbed to the nines. While cities required water—and the San Francisco Bay Area's communities were quick to reach toward the Sierra for their supply—it was the growing population of Los Angeles that most dramatically tapped distant water sources, and did so faster and far more decisively than northern California. But cities use a fraction of the water that is devoted to agriculture; growers stood to benefit, and made sure their interests and claims were heard in the clamor. Through much of the state's history, over 80 percent of impounded water went to agriculture, though with end-of-the-twentieth-century improvements in irrigation efficiency—and skyward-rising urban demand—the total going to agriculture is in descent to 40-some percent.

Once water control began, and monies from the federal government and the state went into massive dam and aqueduct projects from the 1930s through the 1960s, the Sacramento Valley was less fearsome as the scene of massive annual floods. Irrigation in the San Joaquin Valley improved as flows issued from reservoirs in the Sierra or aqueducts brought water from the northern Sacramento Valley. The land market stayed fevered with agriculture on an aggressive rise. And labor, too, would arrive. To the San Joaquin Valley came displaced migrants from the Dust Bowl—not just farmers, but doctors and lawyers: professional people whose livelihood literally blew into planetary orbit as a cloud of aerosolized soil from the speculative suitcase farms of Oklahoma, Colorado, Kansas, and Texas. The migrants abandoned what was left of their homes, and in the 1930s traveled Route 66—the Mother Road—into California. Laborers from Mexico came too, answering demand. The saga of their job seeking and exploitation at the hands of agribusiness owners is a matter of record and careful historical account.

A satellite image of those two valleys, Sacramento to the north and San Joaquin to the south, shows an intricate and varying pattern of landownership. The better ground and older districts, especially along the San Joaquin soil formation, manifest a filigree of small landholdings where settlers took up acreages that they could handle with family or a few workers. The difficult areas, cultivated decades later, are larger holdings in bigger blocks. Reading the geometry of California agricultural property is an art form, with parcels and their borders the runes of our modern age.

A characteristic case is in the Tulare Basin, in the southern San Joaquin, which once included a fully functional lake that provided millions of Pacific pond turtles to make terrapin soup for well-stocked Gold Rush restaurant tables. From the 1870s through the early twentieth century, various attempts to develop the larger Tulare basin were made, with limited success. As William Preston writes, Tulare Lake effectively disappeared, its life ended as the lake bed was sold to agricultural interests who drained the ground, tiled it to improve drainage, and carved what just a few decades before was shown on maps as an active lake into huge dryland fields. Those were soon put to use by J. G. Boswell, the so-called "King of California," and other imperious but indubitably successful growers in cotton's mid-twentieth-century heyday.

By 1930, the "Great Central Valley" was firmly within the focus of large growers, and modern-day California agribusiness was on the march. By the 1950s, Fresno and Tulare counties competed, in good years, with the most productive counties in

Plate 9. A vineyard of inviting Thompson Seedless grapes demonstrates successful trellising technique in Madera County.

Nebraska, Iowa, Colorado, and Texas for bragging rights as the largest agricultural producers in the United States. Since 1990, there has been no competition, except in-state between those two California counties—if Tulare is rarely the outright winner, Fresno County has voracious urbanization, including the city of Fresno with a half-million people, and that may begin to tip the scale's balance arm the other way. In either case, the variety of agricultural production in Fresno and Tulare counties is without equal outside the state of California.

Impacts of Urbanization

Southern California, especially the L.A. Basin, offers perhaps the greatest example of agricultural rise and fall in world history, at least within a 50-year period. Los Angeles became a magnet in the 1890s for those with yeoman farmer ambitions as rail shipment and improving agronomic techniques made navel and Valencia oranges boom crops. Drawn to the Southland were not just farmers from other states but also agricultural ingenues, attracted by a skillful marketing campaign funded by the Southern Pacific Railroad, which had passenger tickets and land to sell. The attraction went beyond advertising, though the colorful orange crate labels shipped around the country (and abroad) on wooden boxes didn't exactly damage the image of Southern California as a paradisiacal landscape. Billions of oranges shipped, and the L.A. Basin became a land of smallholder orange producers drawn (as now to vineyards) to imagery of the Southland as the foremost example of "orange culture." By the end of World War II, the attractions were widely recognized, and a booming postwar population pushed out the groves first from Los Angeles, then from Orange, Riverside, and San Bernardino counties. Large-scale orange growers moved first inland, and then to the San Joaquin Valley, where they remain.

A similar urban shift took place in the southern San Francisco Bay Area's Santa Clara Valley, where apricots, peaches, nectarines, plums, cherries, and other stone fruit reached their apex in the 1940s. Real estate developers were buying up the land of any farmer whose resolve failed even slightly, and before long the cities of San Jose and Campbell changed from agricultural heartland to Silicon Valley. Growers shifted east, into the San Joaquin Valley, in communities such as Lodi and Patterson. Born from the collapse of the Santa Clara Valley fruit industry, Patterson thrived. Then, in the 1990s and 2000s, relocated Bay Area commuters proved willing to trade two-hour-plus one-way journeys to work in return for housing that was cheaper yet apparently more grandiose. Los Angeles and the Santa Clara Valley are only two of the areas where urbanization replaced agriculture; the same can be said of Sacramento, Stockton, eastern Contra Costa and Alameda counties, and, now, even San Benito and Monterey counties. The story of water and land speculation in Los Angeles and the San Fernando Valley is fictionalized in *Chinatown*, still a great and alarming film, in no small measure for its fateful pairing of greedy urban development with damaged human souls. Absent true protection for agricultural enterprises, the cycle will continue to take a toll on cropland agriculture. If livestock ranching is somewhat more resilient, pressures to subdivide are familiar, and with time topple even the most committed ranch clan.

Agribusiness in the Ascent

Since World War II, though some would roll inception back to the 1890s, agribusiness has held the upper hand in California agricultural production. Development of land and water, especially in the San Joaquin Valley, could rarely be undertaken at a small scale. Putting into production big-acreage purchases from the Southern Pacific Railroad and Standard Oil demanded huge equipment, high-end lobbying for subsidized water, and an easy labor supply, all lubricated to assist the economies of scale. At various points, the California Legislature, the University of California, banks, and international markets were brought

Plate 10. Harvesting low-growing crops such as strawberries remains physically taxing agricultural labor, work overwhelmingly done by immigrant Mexican farm workers, as here in Monterey County.

to heel and reminded by agricultural commodity groups just whence came the butter on their bread. These are stories that the 1930s and '40s journalist Carey McWilliams related as well as anyone; certainly he did so better than many of his successors, who can tip toward shrillness.

State officials and the federal government sought to regulate who benefited from subsidized irrigation water. The irrigation delivery systems were designed to aid smaller agricultural producers, and their workers were supposed to be protected; regulations demanding humane housing and working conditions for farm laborers are still on the books, and were even revised and strengthened. There is no doubt about abuses of the past, nor any about the need for vigilance; big agriculture (and many small operations) is run for profit, and the exploitation of laborers is common. It bears mention that Robert Kennedy is remembered by some in California not because he was assassinated here, but because he conspicuously supported César Chávez and the United Farm Workers in their 1960s struggle to gain union representation for agricultural workers. Improvements in the conditions in which agricultural laborers live and work tend to be incremental as legal cases are resolved. Some growers are proactive and make conditions as decent and humane as they can; others, no.

Reliably among the largest agricultural commodities in California are cattle and calves; always in the top five is dairy (milk and cream). Grapes and almonds are relatively recent arrivals at the upper rank of California's top 20 commodities; oranges are typically within the top 10, as are lettuce, strawberries, tomatoes, and the separate categories of nursery products and floriculture, which feed the hunger of builders and homeowners for the "settled" look that comes by planting mature trees around a freshly finished home. California agriculture is quintessentially diverse. The unspoken #1 product is, by any statistical reckoning, marijuana, although it is the Lord Voldemort of products—that which must not be named. Its value may equal all other agricultural products combined. And any history of California agriculture needs to recognize regional specialization and the formidable role of culture in production: definable groups specialize in given products, as with Portuguese (Azorean, in fact) and Dutch specialization in milk, cheese, and dairy in the San Joaquin Valley. There are Punjabi Indians (Sikhs) who loom large in Sacramento Valley stone-fruit production; Bakersfield Basques still involved in the sheep trade, and Italian-Americans who pioneered wine-making in Sonoma and Napa, and who also maintain a presence in Marin dairying. Most conspicuous is a 60-year presence of Spanish-speaking agricultural workers within California. Specialization is not just in crops. It is in the cultural origins of the people who choose to specialize in the raising and marketing of one or another California commodity.

Market Innovation and Alternative Agriculture

Such widespread change over 150 years offers an appropriate introduction to the current state of California agriculture. With legal agriculture recently a nearly $37 billion industry, in the value of its products, and with the growing of marijuana perhaps bringing in easily half again that much in 2009, agriculture in all its manifestations is big business. But this short narrative (a few full-length ones are available, although the story is not revised often) would be entirely too abbreviated were contributors—and enablers—of California agriculture to go unmentioned.

Nothing in California is grown, raised, or harvested in a vacuum. Support structures for agriculture are huge, as evident in the industry, the crop marketing orders and government assistance, and the cooperatives and commodity groups that meet the needs of producers—whether growers, dairy owners, or ranchers. Interestingly, with a few exceptions—rice, grains, dairy—there really isn't much to federal cash subsidies for California agriculture (specialty crops received a first-time $1.2 billion subsidy in the 2008 Farm Bill). California products are diverse and change quickly. In a map of federal subsidy fund payouts, California barely rates compared to the Plains states, the Midwest, and the Southeast. Granted, access to inexpensive irrigation water and public lands used for grazing, and government research monies going to universities and researchers, aren't always counted up front as support. Because there are so many different things raised in California, the amount going to any single crop can't approach what is spent on price subsidies in a state such as Iowa or Nebraska, where there are only three or four main products and ag interests can be highly focused. The producers of individual crops in California, beginning in the early twentieth century, started funding their own research and organizing commodity support to cover research, processing, and marketing, as with Sunkist or the Almond Board. Costs are not small.

Grower expenditures on equipment, land, maintenance, and labor are as big—as outscale—as the size and gross sales of growing, grazing, and dairying operations. In a USDA map showing total production expenses, California farms stand out: agriculture at such a scale is expensive. In short, being a commodity producer in California is no guarantee of big profit. To pursue agriculture at that level requires capital and commitment. Family-owned farms, which are common, embrace an LLC structure, so the family corporation is a frequent presence.

A final anecdote may help convey the outsize scale of California agriculture. We started this project with a return visit to the World Ag Expo in Tulare, which bills itself as the world's largest agricultural exposition; it was inaugurated in 1968. What used to be known as the Tulare Ag Fair is anything but a rustic county event. Set in February, when California agriculture is at its lowest ebb in terms of time demands on growers, the "expo" celebrates technology in agriculture and offers an unsurpassed opportunity for vendors to make sales. Conspicuously absent is anyone (besides fellow growers) with whom

Plate 11. Dairies benefit from feeding milk-producing cows moisture-rich feed such as silage—which is preserved, in characteristic California fashion, under plastic, held in place by split tires held together by cables.

to discuss plant varieties or innovations in agronomic technique or marketing orders. Such enterprises are undertaken in close consultation with seed company representatives, or with Cooperative Extension agents whose specialized knowledge includes those crop varieties best attuned to the requirements of a local landscape. The World Ag Expo is about technology and machinery scaled to a particularly epic Californian canvas. Featured are irrigation pipelines, milking carousels, six-foot irrigation ditch gates, mechanical grape and olive harvesters, 12-foot-tall hardpan rippers, high-tensile strength smooth-wire trellising systems, four-wheel-drive tractors, buzz-saw treetop pruners, 12-bottom plows, and the ever-popular Gophernator (imagine *Caddyshack* with exploding gopher burrows). California agriculture is pursued at many a scale, and the next sections discuss current-day variations and their influential role in the U.S. diet and in California's self-image. But within the Sacramento and San Joaquin valleys, where farmers are growers and crops are "production," agriculture is about scale and volume. Few residents involved in agriculture would apologize for that—quite the contrary, the dominant emotion is pride of place and product.

Toward the end of the first decade of the new millennium, agriculture in California is shifting again in interesting—and not entirely predictable—ways. In California agriculture, the universal is change itself: something is always being modified, tweaked, or tried. Innovation, exploration, and experimentation are the constants, and if something potentially better or paradigm-changing is found, expect to see it where you shop within a year or two. Someone with a sharp pencil will test the financial numbers, and if they work, the product will be available for purchase before long.

Agricultural Variations: Rangelands, Pasture, and Water Use

Crops offer an easily recognizable agricultural category: plants grow from the ground. But they are not all that contributes to agricultural production in California. There is more to agriculture in the state than what is grown from tilled cropland. There is rangeland, pasture, and a significant wealth—and animal-producing capacity—tied up in that land, which has historically been well used.

Driving along a rural road, it is routine to see animals grazing on the other side of a barbed- or woven-wire fence. Grazing is the largest single use of land in the state. Once upon a time in California, rangeland was home to horses, sheep, goats, and cattle—with the last three raised possibly for dairy, or for meat. As open-range operations, commercial sheep and horse ranches are seldom seen now, thanks to high costs and predator issues, and dairies will rarely entrust dairy cows to the wide-open range. But cattle, goats, and in a few settings, even free-range chickens and turkeys, along with the exotic herd of llamas, burros, or emus, graze on rangelands. Properly managed, ranching on extensive range can be a picturesque and, on occasion, profitable activity.

Two forms of livestock raising predominate in California: open-range and confined feeding, with pasturage in between used by both systems. Open-range operations turn animals loose at least daily, and animals may be herded and managed to graze over hundreds or even thousands of acres. Free-range operations with small herbivores such as goats and sheep—or, for that matter, poultry—may return the animals to protected enclosures at night. Large herbivores, such as cattle and the occasional horse herd, can graze unsupervised for days or weeks on end; sheep and goats tend to be watched more closely.

If the breeds and species of livestock involved in open range grazing tend to be distinct from dairy operations, that is not always the case. Feedlot animals constitute a distinct topic; animals eating in close quarters, either in feedlots or in dairy feed yards, are yet another significant and substantial economic force in the livestock economy. In animal raising, there is crossover from grazing to dairy: male dairy calves are a big part of "cattle on feed" in the beef industry. CAFOs (confined animal feeding operations) include about a half-million cattle that are in feedlots at any given time in California, on their way to high per-day weight gain and eventual trips to slaughter. Feeds include a much more diverse mix of foodstuffs in California than in the enormous western Great Plains or Midwestern feedlots, including a variety of agricultural by-products that would otherwise go to landfill.

Confinement dairies are enormous milk producers, earning almost $4.5 billion in 2006—the state's largest crop. More than 1.8 million dairy cows produced an average of 1,860 pounds of milk in December 2008, down somewhat in production from 2007, but by nationwide standards still a formidable number. Goats, raised for milk or for meat, are becoming more and more popular, in particular with California's increasing Hispanic, Islamic, and South Asian (Indian and Pakistani, in particular) population. In the North Bay counties, and, north of that, in the Emerald Triangle, and in limited numbers along the Central Coast, sheep and goats—especially for milking and for weed control—are returning to the scene in California agriculture. Feeding of animals in dairies or feedlots is an intensive form of agriculture, discussed later by type of commodity and livestock.

Although cattle and other livestock are not likely to ever regain the sway they held over the late-eighteenth-century California landscape, they are in something of an upswing in prestige, having found niche markets. Prices for beef, mutton, lamb, and *cabrito*—young goat meat, or kid—are in a slow rise, but market variation from year to year is marked. And innovations

Plate 12. Although the sheep, lamb, and wool industry is well below its peak, young animals are in demand and, as American lamb, bring significant prices; seen here near Firebaugh.

also appear. There is a further boost of interest in grass-fed animals, as several generations of once-vegetarian consumers with baby-boomer pocketbooks decide that they are willing to eat and pay higher prices for smaller portions of meat, so long as it is raised under humane conditions, grazed in a way that minimizes exposure to grain and additives and permits the animals what might, in the aggregate, be considered a good life. Appreciating such a product requires potential consumers to develop a different palate than that of their parents and grandparents, and to front more generous food budgets. Producers tend not to mind meeting market requirements that they observe principles of better stewardship.

In all this, there is a tradeoff: although grass-fed beef takes longer to raise to slaughter weight, it captures a higher per-pound price. If transplanted Midwesterners and high-end restaurateurs roll their eyes at the quaint notion of forgoing heavily marbled prime beef in favor of two- or three-year-old grass-fed animals, modified market preferences are clearly being expressed as a consumer choice U.S. grade. California agriculture is large and varied enough that there is room for the exploration of new market strategies within the beef industry. And although grass-fed beef, for example, was not a huge success when "natural beef" marketing began to be explored in the late 1970s, techniques and the knowledge base of how to produce more palatable beef improved over the last 30 years. An entity such as the multi-product Marin Sun Farms is a highly recognizable brand name in northern California, with a sales reach increasingly far afield from the San Francisco Bay area.

Rangeland

If a single item tends to get lost in a discussion of California agriculture, it is rangeland and its products. For this there is a reason. The term "rangeland" is to a large part of the public a kind of obscure default category: whatever isn't cropland, deep forest, or a subalpine highland is by default considered rangeland. Actually, even forests are grazed by cattle, who sport sturdy bells in the Sierra Nevada or Cascade Range so that they can be tracked in the forest understory. The classification of everything not "cropped" as rangeland carries controversy, since almost exactly half of California is public land controlled by federal government agencies such as the Forest Service, the Bureau of Land Management, the National Park Service, the Fish and Wildlife Service, and the Department of Defense, or by California State agencies, which are equally varied in purpose and constituency. Whether such lands should be grazed or not is a subject of continuing debate, but grazing commonly was and is a long-time use. Likely in the future a good part of California public lands will continue to be browsed by livestock.

The production of beef, calves, sheep, and even goats on the California range is economically formidable, with the category for "cattle and calves" often the largest commodity produced yearly in the state. As decade of inception goes, grazing is the oldest agricultural practice brought to California during the Spanish–Mexican era. In turn, the Spanish experience was itself a hybrid of Peninsular and North African practices that were transferred to Mexico and further adapted before traveling into Alta California in the mid-1700s. Once grazing was a commonplace in the Southwest, rules and laws native to Spain were embraced with livestock, so the "Judge of the Plains" morphed

Plate 13. The widespread insertion of new housing stock into what was historically solely rangeland is a regular phenomenon in parts of California. This land in San Benito County is grazed—around the houses.

into the modern-day brand inspector, and is still an official position in every California county.

Range livestock offers another side of California agriculture, and grazing on rangeland is an extensive use of land, in contrast to the intensive use of cropland; grazing may represent a disturbance—a change—in landscape over time, but done properly, it places a few animals over a relatively large acreage. Well-managed grazing produces nothing like the effects of plowing and laser-leveling, discing and harrowing, building check-rows and seedbeds, and heavy irrigating.

Estimates by FRAP, a division within what used to be known as the California Department of Forestry (now CalFire) hold that the grazed rangeland in California is 34 million acres, and that the area that potentially could be grazed is 57 million acres, or about 57 percent of the state's surface area. The number of animals—cattle or sheep—is much lower in California than in many other states in which grazing can take place on lowland range throughout the year. States in the southeast, such as Florida or the Carolinas, or the Great Plains states, are formidable producers, because their rainfall and grass supply are so much greater, allowing for year-round grazing.

What California offers is a crucial feature of its Mediterranean-type climate: during a wet later winter and spring season, California rangelands grow so much grass that livestock can be placed onto the range in sizable numbers. Stocker cattle are imported, generally by truck but sometimes by rail, and do their best to keep abreast of grazing the fast-growing annual grassland. When the summer dry season arrives, the feed supply plummets, and the animals are moved elsewhere, either to upland range in a modern-day imitation of the cycle of transhumance that was first practiced in California in the late eighteenth century, or to other states—or to feedlots, if their size is large enough to merit that.

Pastureland

Almost any field in California can be grazed. After a crop is harvested, livestock are routinely moved onto the remnants or stubble left on the field. The practice isn't universal, because fencing and controlling the animals can be an issue. But there is an important distinction between rangeland, which is unirrigated and essentially natural, and pasture. Irrigated or not, pasture is a highly significant source of feed and amounts to more than 800,000 acres of land in California. Irrigated pasture is particularly important for forage and as a grazing area for animals that are milked (including sheep, goats, and dairy cows) and for range livestock, especially once the summer dry season hits with full force. However, pasture is a formidable consumer of water, whether delivered by wheel-move line, impulse sprinkler, or flood irrigation. It offers lush feed and sanctuary from confined feeding, although the cost in water and the labor demands of sustaining a good pasture, whether in improved breeds such as Johnson grass or other imported natural pasture grasses, will seem increasingly high in drought years to come.

Hay Crops

A common variation on pasturage is hay lands, which are planted to various species of introduced grasses or legumes, allowed to grow to maturity (instead of being plowed back into the soil), cut with a swather into windrows, cured dry in the field, and baled for transport and feeding. Hay crops include rye grass, timothy, orchard grass, clover, brome, and fescues, and when a high-protein hay is considered risky for animals (notably, horses, which lack a rumen to aid digestion), a grain field of oats, wheat, or barley also can be made into hay. Most feed stores offer several different varieties of hay, and a relationship of trust between consumer and vendor is crucial: bad hay not only is a waste of money, but can imperil the animals to which it is fed. The crucial ingredient in good hay is an appropriate mix of seeds and leafy material, so the grower will attempt to harvest when seed heads will remain attached to the stalk as the grass is mowed and then baled.

The archetypal hay is alfalfa, a perennial legume that sets deep roots and that can produce multiple crops throughout a growing season (as many as seven to 12 cuttings in the Imperial Valley, where alfalfa produces essentially year-round). The nitrogen-fixing alfalfa is conventionally harvested as hay, although it can also be chopped into a high-protein silage for feeding to dairy animals. Pollinated by bees, alfalfa honey is light-colored and prized by consumers who may not like the darker and heavier honey (sage, oak, chemise, wildflower) produced from wildland hives. Hay is cured, or dried, before it is baled, and the packaging is tight enough that further curing will take place slowly when the hay is stacked properly. Wet hay is a danger; a tight haystack, however, whether built by hand or by a cruise-stacker—which builds layers of a haystack as the bales are picked up and then pivots them from horizontal to vertical layers at the stackyard—can sluice off a good bit of water and weather and retain nutritive qualities.

Domestication and Its Aftermath

There is a joint irony in animal agriculture that is born of a contradiction. First, animals raised on the range tend to be a substantially forgotten element in agriculture. They are not particularly visible in day-in, day-out life, unless someone is traveling in foothill environments. Obviously, the long sheds of turkey or chicken farms, or the concentrated ammoniac smell of an imperfectly maintained dairy or feedlot operation, are readily recognizable—but those animals are not in an open-range situation. Although crop fields are evident on a satellite image, grazing cattle or sheep generally are not. Yet, as California voters made clear by approving a 2008 ballot initiative, the public does demand that animals be raised in humane conditions (even if the execution of 2008's Prop. 2 was highly imperfect). That has not spread to ranchers who graze rangeland or pastureland, but ranchers would argue that their animals must be raised in suitable conditions lest they cease to thrive, and thus claim the mantle of good stewardship as well. That is often true, and few would argue that ranchers are deliberately unkind to livestock. The perception of what is appropriate depends on who perceives what is happening. Refreshingly, at this point in California there is negligible discussion of cruelty to crops—but caring for animals is clearly a matter squarely in the public eye and awareness.

Rangeland, pasture, and hay lands represent an additional side of California agriculture, a part that serves the needs and interests of livestock, which puts animals on the land. Use of land in an extensive (contrasted with "intensive") fashion is approaching a 250-year history in California. Rangeland use, in particular grazing, is the largest single agricultural activity in California. Long ago in California's history, land claimants bought up control of bottom lands and access to water sources and managed these often-large tracts of land for fodder—grazing animals during the months when California can support a substantial population of cattle, sheep, horses, and goats. Not all the animals helped are domesticated; wildlife benefit from a number of improvements on rangeland or pasture that can provide added water sources and feed in the dry season.

Open space has value, too. Rangeland kept as grassland or woodland is not given over to city streets and suburban lawns, which are poor habitat for wildlife. So as long as rangeland is maintained, habitat use and improvement is a possibility. That use continues and has an added benefit: ranchers are custodians on the land whose interests are clearly aligned with maintaining the fertility and good condition of the ground, so future years will be at least as productive as now. The concept of stewardship, if not universal, is a principle supported by every rancher we have visited through the years. Ranch families tend to be made up of people who appreciate wild—at least, less developed—country, and they are far less enthused about the sorts of schemes that cities play with to encourage forms of "development" within the

Plate 14. This golf course near Coachella is a suitable reminder that according to statutes of many California cities and counties, golf courses are tabulated as "open space" and taxed as agricultural land.

urban limit line, such as sponsoring the creation of golf courses only to then classify them as "agricultural" land.

A mantra recited by ranchers happens to be true: livestock take vegetation unsuited to other uses and convert solar energy, which with water nourishes the feed into flesh that in time becomes meat. Nothing else can complete this transformation, and the grass-fed end product is healthy and—unless someone is a scrupulously careful vegetarian—even necessary, in small amounts, to human health. Livestock partisans might add that the presence of livestock further serves to enliven our existence, making the world a more interesting and diverse place, since animals were domesticated in the Neolithic by our distant ancestors and goats, cattle, horses, and sheep are therefore in some sense our boon companions—which ought to make our lives more interesting. This quiet satisfaction in the results of domestication may fade as an urbanized population roams cross-country less and less often, but awareness should not be allowed to disappear: rangeland is important, and its presence unmistakable in its significance for California agriculture.

Farming at the Urban Fringe: Innovations and Challenges

The growth of cities, in time coalescing from town to city to megalopolis, inevitably pushes away profitable agriculture at the urban fringe. For a couple of thousand years, one of the few absolute geographical laws is that the most valuable, labor-intensive, and delicate agriculture lies close by the city, and radiates outward as the city grows. Transportation inevitably affects the process. Freeways open up land to development, bumping out agriculture with new office parks, as along northern California's I-80 corridor. But then, easier access can open up fresh ground for vast orchards. Such is the story of the great recent upswing in almond, orange, and pistachio acreage along I-5 in San Joaquin County. If the symphony title is "Growth and Development," then for this song California knows not just a few words of the chorus, but every variation on all the verses: Agriculture on the spectacular soils near Sacramento disappeared in the 1880s, farming was waning from San Francisco,

San Leandro, San Jose, and the Santa Clara Valley by the 1930s, and the Los Angeles Basin was fading to a handful of orange trees and scattered dairies in the decade after World War II. The important thing to understand is that urbanization is a potent lever exercising force in numerous parts of California. The challenges and possibilities deserve mention.

Valuing Local Products

Strikingly, in an era when globalization is a byword in debates about economics, public policy, and labor, a diet grounded in locally grown foods can attract sizable interest. Such writers as Wendell Berry and Verlyn Klinkenborg, and, more recently, Gary Nabhan and Michael Pollan, emphasize the pleasures of eating what you know and what you may have helped grow through community-supported agriculture or by buying a share of a

Plate 15. Evocative white caps protect and shape frisée, a form of curly endive sometimes known as chicory, growing in Bolinas, California.

cow, lamb, or goat. At a more abstract level, the theory of food miles, first bandied about in the 1980s, holds that the ideal foodstuffs travel a minimal distance from their place of production. Because California (like Australia and New Zealand) ships up to 80 percent of selected agricultural produce as international exports, this idea's gaining sway could be a decidedly mixed bag. But for those eating in California, and potentially setting food fashions, the notion of buying as much locally produced food as possible resonates, with sizable implications for pricing and what is raised.

If interest in local produce and cuisine rooted in tradition is longstanding, family preference and regional habit has shifted from a casual consideration to conscious practice. Two widely separated Mediterranean-type ecosystems, Italy and California, feed and irrigate the Slow Food philosophy at the heart of this movement. The Slow Food effort commenced in Italy in the mid-1980s and, as the movement's name suggests, sought to distance itself from the American embrace of "fast food." With some 800 *convivia* chapters worldwide in 2008, Slow Food believers attempt to generate a culture of cuisine that pays attention to local animals, seeds, and foodstuffs. The movement supports farming that acknowledges local products and a distinctive regional "ark of taste" that promotes sustainably produced foods with outstanding taste and local provenance.

"California Cuisine" is a political and agricultural endeavor that began fully a decade earlier in the late 1970s. Although sometimes attributed solely to Chez Panisse owner Alice Waters, credit to the effort's progenitors is owed to a larger circle of chefs and farmers who emphasized shared themes. From its inception, emphasis in California Cuisine was on the fusion of distinct cooking styles. More important for agriculture was an emphasis on the use of ingredients raised locally and, at times, acquired at farmers markets. At Chez Panisse and other restaurants in the movement, menus change often—sometimes daily, observing the rules of what is sometimes called market dining: cooking to take advantage of what is freshest and finest in the market. (Interestingly, in 2008, when the first Michelin Red Guide to the San Francisco Bay Area came out, Chez Panisse was awarded one star, and consensus suggests that that was because the restaurant offers no fixed menu.) Other emphases include paying agricultural workers a living wage, and giving grower–producers a reasonable return on their investment. Such ideas put the movement into elite territory, supporting more expensive food—with distinct advantages in nutrition and freshness. The shared consciousness of Slow Food and California Cuisine took and held. In 2009, these two ideas represent a strong alternative to mass, industrial-scale production of food products. This is part of the interesting dual nature of California agriculture: there is an outsize commercialized production of high-quality and relatively inexpensive foods that may circle the globe, but also a contrasting intimate and much pricier yield of foodstuffs from small-scale producers who sell to a local high-visibility market that supports farm-to-table dining.

For all the chichi attention, the "locavore" movement doesn't have everyone convinced. As author–restaurateur Anthony Bourdain writes, even among elite chefs there is a division between those who want to serve up only locally produced fare and the chefs who prefer to draw on an entire world economy of top-flight raw materials that can be moved from place to place, making anything available, if at a price. Great chefs buy into one camp or another—and to be fair, it's a lot easier to be a happy locavore in California, with hundreds of foods and farmers markets, than it is to embrace such ideas in Kyrgyzstan or Iceland—or Minnesota. Local California producers of top-flight food radiate star quality, and their products are watched for, purchased—often at premium prices—and cherished in the realm of urban food fanciers. Connections are maintained and cemented between city people and farmers beyond the urban fringe, and chefs, growers, and consumers routinely meet, as John McPhee and Adam Gopnik note, at the specialty produce store and the farmers market.

Distribution

Tractor-trailer bins of tomatoes, garlic, and boxes of berries, kale, or lettuce routinely move along Hwy. 99 and Hwy. 101, delivering thousand upon thousands of pounds of "product" to storage. Like a constant flow of trucks and tractors on farm fields, warehousing and storage allow California agriculture to cater to national and global markets. Truck yards along the major freeways hold literally hundreds of trailers topped by massive plastic produce bins, awaiting harvest season. Those massed trucks are just one added tribute to the upper end of the scale of California agriculture, which is big-big-big. On the road, produce trucks are a moving ballistic menace, and the impact of a tomato dislodged at 65 mph and lofted into oncoming traffic would incite any motorist to fervent religious observation.

For all the value, size, and heft of big agriculture, it's worth remembering that California agriculture exists at another size, one that deliberately tries to stay small, local, and adaptable, despite ongoing struggles with major producers in matters relating to standards and licensing. If embracing efficiencies of scale and a business plan that may extend to national or even international markets is one model in California agriculture, another way to approach production exists. Beginning in tandem with Earth Day and the intentional community movement of the 1970s, an interest in organic and alternative agriculture reared its head in California. This became a bellwether movement that today is still spreading resolutely around the United States.

Farmers markets were common in California during World War II, and they faded without USDA support in the 1950s and '60s. But as outlets for farmers willing to bring meat, eggs, or produce to a central point whence they could offer direct sales to consumers, farmers markets returned with energy in the 1970s. They were a means for farmers to market seasonal produce and avoid the entanglements of marketing orders, middlemen, and the grocery chains. The results of a 2008 survey show that there are over 500 Certified Farmers Markets in California; the informal count is larger by far. By no means do all small-scale farmers and ranchers sell through CFMs—but they provide a licensed supervision with fairly strict production rules, and the number of participating growers gives an idea of the rising consumer affection for finding fresh food by alternative means.

Community Supported Agriculture (CSA), U-pick operations, and Farm Trail excursions are just a few of the added venues through which local populations may elect to do non–grocery store forms of food purchasing. If a world apart from the scale of production normal to the Sacramento or San Joaquin Valley, alternative markets represent the embrace of a different philosophy and approach. They really are not a separate universe of cropland. In fact, organic growers may farm next to commercial producers —and some growers of organic

Plate 16. Because family livestock raising is rarely lucrative, ranchers sometimes branch out, as Sally and Mike Gale (pictured) have at their Chileno Valley Ranch in Marin County by adding heritage apples for U-pick to their grass-fed beef operation.

produce are now multi-million-dollar operators. That said, however, standardizing the rules for *organic*, *natural*, *pesticide-free*, *biodynamic*, and other agricultural labels is a complicated matter (as cautious consumers are aware)—but the variations in production norms is something for a future discussion.

Changing Agricultural Tolerances

Opening any discussion of ethics and practices of food-raising risks edging toward the boundary of incivility; it's like debating religion in a barber shop, or singing loudly in a packed elevator. Some people, for reasons of budget, culture, or lack of concern, purchase whatever food is cheapest wherever they can find it.

Within California, this may not pose a problem, since produce and animal products move to distribution quickly, and the quality, as overseen by marketing associations, tends to be good. But as shipping distances grow, so might doubts about freshness and quality—a function of the tyranny of distance. And just as some folks worry about imbibing tap water or breathing air that may be tainted, so are there anxieties about what form of agriculture is best for the earth, and best for the consumer. Predictably, there are choices.

Today's swirl of local food issues are refreshingly varied—and concerned with the agriscape. Should purchasers plan on eating locally grown food? Should they buy seasonally, and not buy food when it is from afar, or out of season? Is it best to buy organic food—and if so, at what level of organic certification? Should eggs and meat be free-range? Is it acceptable to buy produce, or should it be grown at home, raised in place, planting beds in the biointensive style (optimum yield from a given space, as suggested by French intensive gardening guru Alan Chadwick), or using biodynamic techniques, following "the method" prescribed by 1920s Austrian educational philosopher Rudolph Steiner? What are the ethics of cruelty-free eating, or is there in fact an implicit duty to eat meat, so that the animals domesticated in the Neolithic may continue to occupy a meaningful place on earth? There is little consensus in how these questions are answered, but each reply has partisans, with the debates heartfelt.

There are books written about every one of these themes, and bioethicists will produce many more. "Food studies" experts proliferate much faster than new cultivars. Even in the hyper-competitive world of California wine, better than three dozen wineries are certified biodynamic: they treat farms as unified and individual organisms, emphasize the interrelationship of soils, animals, and plants as a single system, and aspire to sustainable agriculture. If certified, they assiduously make and apply biodynamic preparations #501–508 according to the lunar, solar, and astrological calendar, adding appropriate additives to compost and including oak bark "preparations" for humus according to a schedule that has to be vouched for by Demeter International, the supervising association of biodynamic farmers.

Let it be agreed: food production, whether backyard or commercial, is a realm of believers.

Urbanization

In a moment of budding imagination, the geographer John Fraser Hart crystalized developing work on urban sprawl into an evocative phrase in 1991, couching his model of the spread of cities as a "perimetropolitan bow wave." Describing the wave formed as the prow of a ship cuts through the water, he transported that image to the blunter effect of city-spreading, and the overturning of agricultural land into suburban developments or edge cities. Concern in California about the effects of urbanization is pronounced, and has been so for some time. At the west side of UC Berkeley, a notable phrase is chiseled onto Hilgard Hall, the original agricultural building on that pioneering land-grant university campus. Lettering across the architrave declares that the building's purpose is "To Rescue for Human Society the Native Values of Rural Life." What is striking is that the building was dedicated in 1917, which suggests that nearly a hundred years ago, concern was already widespread about challenges threatening the disappearance of agriculture in California.

In situations such as the Pauma Valley, in North County San Diego, the legacy of a hundred years of careful cultivation of avocado orchards and orange groves is being picked away. Lying close to burgeoning suburbs in not-distant Riverside County, with Native American casinos arriving along the Hwy. 76 corridor, Pauma Valley's days as an agricultural haven are numbered, in a story just as easily told elsewhere in California.

The American Farmland Trust released a late 2007 report on farmland loss in California with the apt title "Paving Paradise." It notes that from 1990 to 2004, more than half a million acres of California farmland were converted to urban uses. The accompanying warning in the report is that the state's cities are by and large built on what used to be the most productive California farmland. There, valleys and coastal plain soils are deep, and the climate is right for raising crops that are essentially unique to California, and that are highly sought after in out-of-state markets and abroad. Drawing on public data from the State of California, from the USGS, and from the Bureau of the Census, the report looks at the top 10 counties in agricultural production, and at distinct regions within California. Much growth comes at city edges. That has the virtue of orderliness, but because so much of California's urban development is

Plate 17. Several citrus varieties and avocados in Pauma Valley lie at the foot of Sam's Mountain in North County San Diego, an area under active development pressure.

on top of what was once prime farmland, suburbs are concentrated where the problem is already severe. On other sites, the growth is hodgepodge, with ranchette development injecting big-footprint suburban-style homes on rural 1- to 10-acre sites, fragmenting agricultural holdings.

Whether examined by satellite images showing city lights, by close analysis of census reports, by data collected from County Cooperative Extension agents, or by nongovernmental organizations such as the American Farmland Trust, the warning is constant: once-productive ag land is being turned to nonagricultural purposes, and there are limits to what can be made up or repaired. Three-quarters of the San Francisco Bay Area is urbanized, and a third of that was once high-quality farmland. In the East Bay's Contra Costa County, the number of farms is down by 30 percent since 1987, and Cooperative Extension specialist Al Sokolow was recently quoted as saying, "It's a new aspect of agriculture: the survivability of farming on the edge of an urban area." As farms disappear, so do farm-related businesses, making survival all the more difficult for the remaining growers. There can be benefits—in East Contra Costa's Brentwood and Oakley, U-pick farms proliferate, enabling local urbanites to gather fresh fruit without a prolonged trip to the Sierra, and allowing farmers savings on labor. But the relief is likely short-lived; only an entire collapse of semirural real estate will bury demand for suburban acreage.

Nor are all the problems in urbanization with prime farmland and ranchettes. In the North Bay counties of Napa and Sonoma, and to some degree in the coastal hills of San Luis Obispo and Monterey counties, fervor for the cultural and agricultural

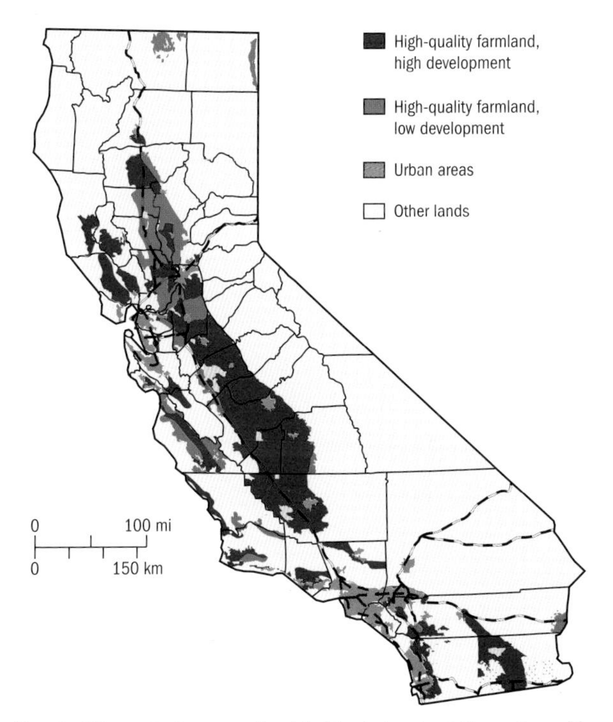

Map 2. Urban-suburban sprawl and California farmland. The red area is exaggerated, according to J.F. Hart, but more detailed maps since this American Farmland Trust effort reinforce the scale of development plans.

possibilities of vineyard establishment drive expansion of agricultural land up the hillsides into what was historically rangeland. This happens partly because demand for vineyard land is pronounced, but also because some growers are looking for unusual sites where the quality of difficult soils may be expressed in *terroir*, the French expression that holds that grapes take on the characteristics of their parent soils. Finally, hillslope expansion comes because urbanization has taken prime bottom land soils out of production in the last 20 years. From 1990 to 2004, 11 percent of the newly urbanized land in Napa was taken from optimal farmland; 14 percent in Sonoma, 9 percent of San Luis Obispo, and 44 percent of Monterey County came at farmland's expense. If the numbers are telling, that's because they trace the boundaries of a continued reshaping of California agriculture.

Working Landscapes

The term "working landscapes" is a relatively recent addition to agricultural language, although the intention goes well back in time. A working landscape balances social, ecological, and

Plate 18. A sizable house under construction rises here from what used to be pear orchards, west of Snodgrass Slough and Locke, Sacramento County.

economic needs—while recognizing the value of ecosystem services that may be difficult to quantify. Often, that is where various forms of productive work coalesce. Support for working landscapes recognizes that there are public policy issues involved in conserving land, preserving and sustaining agriculture, all the while adding to landowner values and accruing benefits to the public. Environmental benefits are often part of a working landscape—wildlife habitat, wetlands, improved water and air quality—but food, fiber, and timber are products, too. A working landscape combines ecological and economic benefits in a way that is considered advantageous to society as a whole.

Agricultural landscapes are not automatically "working landscapes"—a severely overgrazed dairy pasture, or a cotton field assailed a dozen times a year with various pesticides might offer two hard cases to make. But a rice field in the northern Sacramento Valley, managed for rice production and then fallowed and reflooded to maximize feed and habitat for birds migrating along the Pacific Flyway could easily fit the bill. So, too, can well-managed organic farms, on which beneficial insects are encouraged and water quality is carefully guarded. A working landscape can have room for recreation, and multiple uses are frequently possible. Some regional parks in California are prime visitor habitat for walkers, bikers, and hikers. Livestock—generally cattle, but sometimes sheep—can be grazed at the same time that recreational hikers tromp the hills, and so long as dogs are kept under control, coexistence is often possible. Because ranchers are willing to maintain water sources and fences, they invest in a public landscape, and the advantages are mutual. It is when parties are unable or unwilling to coexist that problems arise, and the difficulties can be ugly.

A significant problem with working landscapes is incompatible uses, which can lead a landowner, whether public or private, to close off all but a single use. Even then, wildlife may benefit, but there can be a singular importance to sharing in a working landscape. For agriculture, demonstrating the benefits to a public at large is particularly important. The 50 percent of the area of California that is in cropland or rangeland production is a conspicuous part of the state's public life, lived out in the open.

Making working landscapes into a successful part of daily life requires effort and forbearance from all concerned. Growers or ranchers must become used to an interested public, which in turn cannot threaten lawsuits every time someone snags a pants leg on a strand of barbed wire, slips in an irrigation ditch, or steps on a cow pie. Recognizing that well-managed agriculture provides ecosystem services and significant benefits to the public and to nature itself demands a little self-education and awareness of how agriculture works.

Agriculture's Demands on California

Problems challenge the future of California agriculture, guaranteeing that the state's growers and producers will be kept on their toes. None of these is unique, or even necessarily new, to California, but in light of the issues at stake, meeting the requirements needed to maintain California agriculture will keep the next generation of purchasers, interested graduate students, Cooperative Extension agents, producers, and policymakers more than busy. Each of these themes has at least a doctoral dissertation or two written about it, but for this guide, a few paragraphs on each will have to suffice.

Labor Supply

Pressure on labor supply rises and falls with public and government sentiment toward immigrant labor. If there is a simple reality, it is that California—and the United States—has been unable to meet the accelerated demands of the state's agricultural production and labor since the 1920s. The Okie migration during the Dust Bowl era was a question of both supply (there were lots of unemployed and homeless people on the westernmost Great Plains) and demand (California needed farm labor). California has enormous ongoing needs for agricultural labor, some of it seasonal and some year-round. Whether Congress permits a porous border with Mexico will make the difference: in recent years, considerable produce has gone unharvested because of agricultural labor shortages, though financial and legal strictures and consumer preferences also influence trade.

As we did field research for this book, it became abundantly clear that the language of agricultural workers in California is Mexican Spanish, not American English, and that trend is unlikely to reverse. Whether those workers can be placed on a legal footing, or whether illegal immigration and support for the Mexican remittance economy will be the future depends on U.S. and Mexican government action and public recognition that there are jobs that Californians will not or cannot take, even in a recession. Estimates of the absolute number of workers required to sustain California agriculture varies, but it is substantial, especially because after a half- or whole decade of work, many Mexican laborers turn to other jobs, especially the lucrative construction trades, as an alternative to the uncertain housing, health, and salary support in agricultural work.

New Markets and Changes to Old Ones

The supermarket or corner grocery is no longer a sole distributor of agricultural products in California, nor to the rest of the United States. Wholesalers and even producers themselves recognize varied outlets, some of which provide better return, lower overhead, and quicker turnaround from field to consumer than would ever prove possible through conventional channels. The number of farmers markets in California has shot up in

Plate 19. What were once alternative markets for farm and ranch produce increasingly feature regularly in family shopping, as with the Ferry Plaza Farmers Market, visited by 10,000 to 15,000 shoppers a week, in San Francisco.

the last decade, and farmers markets have penetrated throughout the state, from inner-city venues to mini-mall parking lots to the plazas in front of city hall and onto streets closed off to allow direct sales tents to rise. In San Francisco, the renovated Ferry Building hosts the Ferry Plaza Farmers Market under the auspices of the Center for Urban Education about Sustainable Agriculture (CUESA), with up to 15,000 shoppers attending weekly to make their purchases of regionally produced jams, vegetables, fruits, cheeses, eggs, bread, and meat. These outlets, big or small, provide price flexibility for vendors, a quick turnaround of fresh produce for consumers, and, in prospect, better and more healthful nutrition for all concerned, and they admirably reconnect city residents with the surrounding area growers and ranchers.

Another program that has wrought successful change in urban diets, with sizable benefits for producers, is the formalized CSA—Community Supported Agriculture—in which subscribers pay a monthly fee and receive weekly or biweekly shares of diverse seasonal foods. CSAs guarantee a market for agricultural products, and since the makeup of each delivery is rarely known in advance, there is plenty of opportunity for CSA participants to learn new recipes and how to deal with crops they may not have encountered before—a learning experience. Some CSAs allow community members to pay for their share with labor, contributed through turns at irrigation, tilling and weeding, planting and harvest, or picking and packing, which makes for a still better deal and solves labor questions. For meat-eaters, purchasing a share of a steer or heifer is doable, too: write a check, wait for a few months, and an e-mail or letter arrives, alerting you that your quarter-, half-, or whole animal has been dispatched and sent to a meat cutter for butchering and packaging, and now awaits you in a freezer for pickup. Fanciers of grass-fed beef, lamb, or goat have made this a particularly popular way to support local agriculture while having more confidence in the source of meat.

Environmental Issues

The ties between agriculture and the physical environment are weighty. Agriculture in California is like everywhere else, only more so: climate, soils, pollution, and pest control are all tied to creating a healthy product. There is a tyranny to distance, as well, since products not only have to be harvested safely, but also must be transported. When the smooth execution of these processes goes awry, as with the 2007 spinach scare, a multi-million-dollar industry drops to its knees as public fear and skepticism rise and government investigators swarm over dispersed fields trying to ascertain cause and effect. If diversity and quality are two bywords by which California growers and producers try to live, when problems develop, questions arrive about the food supply's health. Were maintaining food health a simple matter, there would be no concern. But issues in environmental and food health are not simple at all. There is legitimate disagreement as to just what a thoroughly safe and sustainable agricultural ecosystem might look like. What to one grower seems safe, and in keeping with best practices, might seem wayward, even slipshod, to a certified organic farmer—the standards are not fixed. Accidental incursions can take a toll. In the spinach crisis, a group of wild boars breaking through fences and contaminating spinach fields east of Hollister is generally accepted to have started an *E. coli* O157:H7 outbreak. But it took almost four months to settle on when, where, and what caused the outbreak—hardly a quick and efficient result.

Longer-term problems, such as climate change, will trouble California agriculture, and a worst-case scenario, experts warn, might see Sierran snowpack reduced by up to 90 percent. Predicted temperature increases by the end of the century could reduce production of dairy between 7 and 22 percent, while grape varieties currently growing may disappear from the scene, unsuited to a climate warmer by 2 or 3 degrees—on average—than today's. Add to that warnings from UC Berkeley researchers who in 2008 estimated that California has $2.5 trillion in real estate assets, including agriculture, that would be endangered by global climate change. In time scale, geographical range, and quick incursion of problems and pests, California agriculture faces frequent challenges, some of which can only be solved at a global level.

Water Supply

Of perhaps 400 crops produced in California, only a handful could grow without irrigation water. The federal-built Central

Plate 20. The wording of the welcoming arch almost says it all for Modesto, a community whose substance and wealth through the last 150 years was strongly tied to agriculture, including marketing, finance, processing—and the control of water.

Valley Project (20 dams, 11 hydroelectric plants, 500 miles of canals and aqueducts) was designed to supply water to relatively small producers, something of a rarity in an agribusiness-dense state, and arguments about who is entitled to what water source make for bracing debate. The federal and state subsidies that went into building this hydraulic network constitute the single biggest subsidy to California agriculture, and although the argument can be made that repayment has come through inexpensive food and fiber, that case is hardly a foregone conclusion. And the durability of the system is suspect. No one who lives in California fails to understand that wet and dry cycles come and go; the sticking point is, at what timetable?

Careful work in paleoecology suggests that five centuries ago, what is now California underwent a 150-year dry cycle. Were that to reappear, the effects would devastate agriculture and drastically redistribute the West Coast population. As it stands, droughts of two or three years produce a series of bumps that threaten to topple an intricate design of standing dominoes. Agriculture claims more than 40 percent of California's surface water supply—a proportion that has dropped from 85 percent 20 years ago. But some crops are still water-intensive and relatively low-value, and with summer being the prime production season for many of California's crops, the greatest pressure on the water supply comes when rainfall in a state with predominantly summer-dry Mediterranean-type climate is at its lowest ebb. The role of fresh and unimpeded water in natural ecosystems is better understood now, and the effects of water diversions to agriculture are seen in the decreasing health of fish populations in northern California's rivers, from the Delta and San Francisco Bay to the Oregon border. Salmon fisheries are in free fall, and judges threaten to halt the pumping of water from the Sacramento–San Joaquin River system, which not only will dry up agriculture water supplies to the San Joaquin Valley, but may also put cities on water rationing.

The human plumbing of California exists at different levels—federal, state, county, municipal, irrigation district, grower. The uses often compete, especially because water supplies are routinely oversubscribed—famously so in the case of the Colorado River Compact, in which the yearly flow used to determine water allocation was, in fact, at least 20 percent higher that the historical average. As a report by the Pacific Institute in 2007 puts it, agriculture "is important to our economy, culture, and environment, but is subject to mounting pressure from uncontrolled urbanization, global market pressures, and threats to the reliability and availability of fresh water." Water shortages, and arguments about the best and most beneficial use, will continue as part of the story of California agriculture; water markets will reallocate water to those able to pay and make their case for the "best and highest use" of a regular water supply.

Product Prices

The prices of agricultural products in California do not rise and fall entirely at the whim of consumer demand; more than half of California's crops are regulated by marketing orders, laid down under federal charter by producer associations linked to specific crops. There are cotton, rice, almond, orange, avocado, lettuce, spinach, and myriad other marketing agreements that limit the amount sent to market and prescribe the quality of produce leaving farms and packing sheds. Costs also move

Plate 21. Especially for wine grapes, rootstock is crucial. High-quality varietal grapes are seen here grafted onto disease-resistant rootstock at a French-owned greenhouse complex near Vacaville that offers a front-end product to the lucrative wine grape industry.

upward, and growers find their profit margins influenced by a familiar pair of effects: rising raw material costs and decreasing prices for products (prices can rise, too, but that is less worrisome to producers). The costs for raw materials can include fertilizer and additives to soil and investment in irrigation water and equipment for distributing it. Fuel costs may rise, and with them the price of purchasing and operating machinery; labor costs can ratchet upward, too, whenever shortages in worker supply arrive.

An episodic, if difficult to time, rise and fall in prices is by no means uncommon for harvested agricultural products. Particularly at issue is increasing production in California. When a crop takes hold and becomes popular, growers routinely stop planting an older and less profitable crop and turn to a newcomer or to an up-and-coming variety of an older crop. The degree of unsettling can be considerable; tree-crop producers will on occasion rip out orchards or vineyards to plant something novel. The earliest adopters who find that a new product is dear in the public eye may make a tidy profit for a few years, but inevitably, imitators will come and plant the same variety or crop, and as acreage swells, great profit is replaced by glut. This happened with peaches, and later with kiwifruit, and it threatens to be the case with almonds; innovators are switching to pricier almond varieties (which may become common), or removing almonds and substituting pistachios. Agriculture is a waltz: a drought in one growing area, a pest outbreak, the suddenly rising popularity of a particular cultivar can produce a boom year or two. A particular grafted grapevine may prove the difference between indifferent and spectacular yields: uncertainty rules.

Pests

Precisely what constitutes an agricultural "pest" is semantically complicated. The short answer is that a pest is anything that is not wanted in a growing or livestock-raising situation, a category that can include rodents, birds, bees and wasps, nuisance animals, a variety of soilborne pathogens and soil-dwelling creatures, and quite a broad array of rusts, molds, mites, fungi, wilts, and viruses that individually can affect crops in undesirable ways. At some point in a crop's life, as with cotton, even healthy leaves might be considered "undesirable," and until a few years ago, in a late summer ritual, cotton plants would be defoliated by air or ground treatment to remove a plant's leaves so that cotton bolls might be harvested more easily. Just what to do with pests similarly involves a multibranched decision tree: supporters of organic agriculture use hoes and human labor to restrict weeds; biocontrol advocates add beneficial insects that feast on weedy species but not on the crop in question; other farmers may opt for chemical control, which can be costly, and not always healthful for the broader environment. But when a pest appears, a vast scientific and research engine shifts into gear, trying to determine what the pest is, and how it is best treated, according to the tolerances of each school of agricultural production.

Pests can pose great risk to California agriculture, and how to respond is always a crucial question. An illustrative case is the history and ongoing risk of an outbreak of grape phylloxera, an aphidlike insect native to eastern North America that when transported accidentally in the late nineteenth century to Europe devastated French vineyards, slashing production by 75 percent and infecting 90 percent of vines. Grafting of traditional varieties onto resistant rootstock was the only salvation, and the "Great Wine Blight" was followed by the "Reconstitution" of French and other afflicted European vineyards. Use of grafted grape vines is much preferred, although some of the grafts planted in California proved similarly susceptible in the 1960s to phylloxera, and a number of vineyards subsequently had to be torn out and replanted to resistant varieties. The expenses are huge—not only because resistant varieties add cost, but even more because old, established, and high-producing fields have to be destroyed, which, even if they are replanted immediately, means another four or five years before production returns to significant levels. One hundred years later, the glassy-winged sharpshooter, a leafhopper native to northeastern Mexico, began its move into California in the 1990s, feeding on almonds, grapes, citrus, stone-fruit, and oleander and spreading Pierce's Disease, a particularly feared scourge of California vintners. With the example of grape phylloxera in mind, growers and the USDA are responding with broad action.

Diseases and pests among California crops are as varied as the crops and animals that are raised. California's summer-dry climate provides protection from some pests that require a wetter summer, and the degree of seriousness in an infestation depends on the pest involved. Some affect appearance, posing no risk to humans but reducing marketability (as with lettuce mosaic virus or citrus canker). Other plants, including grapes, oranges, and almonds, are subject to a remarkable range of bacterial diseases, fungi, rots, stipples, speckles, parasitic nematodes, viruses, and viruslike diseases, and animal pests may proliferate in equally diverse assortment. Crop description sheets, pulled up online from California Cooperative Extension Services, frequently end with a multipage listing of pests that afflict a given crop, and the various means of responding to the malady in question, and offer a mind-expanding, if sometimes grim, read.

Organizations such as the Pesticide Action Network act assiduously to try and find alternatives to a full-bore aerial and ground spray assault on pests, and IPM (integrated pest management) attempts to limit action to a proportionate response.

Energy Costs

Rapid shifts in energy costs, generally on an upward trend, make the ag producer's life complicated. It's important to understand that energy costs affect many different aspects of agriculture, sometimes from angles not readily seen: deep-well irrigation pumps are driven by diesel or electricity; pesticides are commonly petroleum-based; tractors and farm equipment run on diesel, which in California means on a special CARB formula that runs cleaner but that is costlier; transportation of harvested crops involves fuel for trucks and delivery vehicles; propane and natural gas are used to dry crops at grain elevators and cotton bales at the gin. For dairy alone, costs are way up for heating and cooling involved in processing cheese and butter and evaporating water from milk to produce powdered or evaporated milk. All these are occurring as agricultural production loans have failed to see the sorts of interest rate drops that have offered benefits to homeowners. But particularly for petroleum-based energy, California is susceptible to rises in world prices, which in 2008 reached levels never seen before on the global market.

Shifting Global Demand in Food

Crops developed in the Mediterranean form a robust and significant portion of the world's favored foodstuffs, and California is arguably the most successful producer of diverse food crops in global agriculture. Take almonds: California produces more than 80 percent of the world supply, and three-quarters is exported. Fruit grown in California, especially table grapes, oranges, peaches, and nectarines, cross the Pacific to Japan, South Korea, Taiwan, and China. Wine produced from California grapes is a worldwide commodity. The total value of exports was a shade under $11 billion, nearly 30 percent of the state's agricultural production. But demand rarely rises and falls in straight lines, and shifting global demand, especially associated with health concerns and food politics, looms as a perennial question.

Simple trade restraint—and an attempt to boost local products at an importer's expense—shape some resistance to imports. And within local markets, consumers have more than once expressed reservations about modifications introduced into the food chain—unease about Alar, worry about bovine growth hormone (rBGH and rBST), or fretting about other injectables or subtherapeutic additives to feed or animals. A technological tussle is ongoing; the savants of agricultural wizardry can come up with means of producing more per animal, or of "improving" crops so they yield more. But the public responds with skepticism: With a glut of milk, why is more production desirable? With transgenic crops, will the food taste better—or even nearly as good as it used to?

There are larger issues that weigh into the marketing of California food. Increasingly, European Union buyers are reluctant to purchase foods grown outside the EU that are produced using certain pesticides, or that are GMOs or GEOs (genetically modified/engineered organisms). The concerns are various, and trade negotiators work hard to find means of opening those markets. Even in the United States, a significant proportion of the population is skeptical about the use of recombinant DNA and transgenic techniques in foods.

The distaste may be aesthetic or religious, or it may be driven by doubt about overproduction that can ensue, or the insertion of genes that mean that farmers cannot collect seeds from the field, warehouse them, and plant those seeds the next year without losing the transgenic traits—effectively meaning that seeds must be purchased each year from the seed supplier. Some of the most prominent GMOs are in corn, cotton, and soybeans: more than 80 percent of each of these crops are GMO varieties, although much of their use is in livestock feeding, or clothing. None of these three crops is a major factor in California agriculture—but one of the very first field tests of GMOs came in 1986, with a bacterium designed to protect strawberry plants from frost damage, the so-called ice-minus bacteria. The controversy was predictable. Many of the modifications introduced to GMOs involve the use of herbicides and pesticides, creating a recipient plant tolerant of low levels of a pesticide, which makes weed control easier. Other modifications extend shelf life, reduce spoilage, facilitate crop growth, produce chemicals toxic to pests, and allow plants to survive harsh environmental conditions, including those of high salinity or cold. So-called "Golden Rice" is rarely grown in California, but its supporters in the International Rice Research Institute argue that transgenic rice might cure vitamin A deficiency. Benefits accrue, but controversy arrives in the assessment of costs, and in consumer resistance.

Concern about the genetic modification of crops—and animals—is based on reasonable worries and on a decidedly less than perfect record of seed companies in testing and marketing their wares. Pollen from GMO corn has drifted to nearby fields and produced immense difficulty—and intransigence from the GMOs corporate parents. No one is interested in seeing the food chain dented or broken. At the same time, it is true that almost all improvements in crops, through the ages, have come through genetic manipulation: think of Gregor Mendel's famous pea plants, which, combined with Charles Darwin's work, led to the vaunted "modern synthesis" of evolutionary biology. Brother Mendel was himself experimenting with the fundamental structure of plants—his way of doing so is simply seen as more "natural," but the differences are, in fact, not so great.

❁ ❁ ❁

Finally, a note before this guide shifts to an inventory of crops and animal products and moves away from essays on agriculture as a larger theme. Agriculture is a remarkable feature of the Golden State, and no place in the United States—perhaps, in the world—is quite like it. California produces much of the diversity in the North American diet, in ways that extend farther afield in the global food chain. Innovations travel elsewhere and are imitated or rejected, and because of them, diet changes. In diversity, scale, and food quality and health, California leads the nation; conspicuously, it forges ahead as both a large producer and in small fine lots. Production of boutique foodstuffs has changed how we eat, and in the ways that California Cuisine and its offshoots are viewed worldwide. In short, what happens in California not only doesn't stay put, it rockets elsewhere, in food as in other aspects of life dear to the world economy. And many Californians would have it no other way.

THE PARADOX AND POETICS OF AGRICULTURE

A Gallery

GAMBONI
FARMING.CO.

HOME GROWN
TOMATOES

Gallery Plates

Plate 22. Long tables of sliced fruit await citrus growers at the Annual Growers' Citrus Tasting Event, held yearly at the University of California's Lindcove Research and Extension Center in Lindcove, Tulare County.

Plate 23. A modest, yet iconic building owned by Gamboni Farming in Dos Palos, Merced County, hardly evokes an agribusiness venture that received $2.4 million dollars in federal subsidy payments from 1995 to 2005, largely for cotton production.

Plate 24. Johann Dairy in Fresno provides large huts, outdoor space, and pen-to-pen bottle service for dairy calf rearing.

Plate 25. Boer goats, a breed originally from South Africa, sprint to fresh ground at McCormack Sheep and Grain in the Montezuma Hills of Solano County.

Plate 26. A species of agricultural tourism is indicated by rows of pie pumpkins that are carefully fitted into CD cases at the Uesugi Farms Pumpkin Park, an annual event in Morgan Hill, Santa Clara County.

Plate 27. Signature crops sometimes get a public tip of the hat, as with this ornamental apple next to a bank in Sebastapol, California—a town historically best known for its apple harvest.

Plate 28. The model of a large ripe (black) olive watches over parking at the Olive Garden Restaurant in Lindsay, California, longtime holder of the title "Olive Capital of the World"—an honor likewise claimed by Corning, and perhaps a few other sites worldwide.

Plate 29. The Moro blood orange is particularly flamboyant, with subtle gradational shadings to its internal color that make it sought after for both juice and eating; seen here at Lindcove Citrus Food Tasting and Display Field Day, Tulare County.

Plate 30. Although not an obvious agricultural crop, oysters are produced in several locations in California. Backed by Tomales Bay, in Marin County, this striking Hog Island Oyster Company exemplar is a *Crassostrea gigas*, the Pacific oyster.

Plate 31. Sweet potatoes fill a substantial fresh market niche and are a crop grown in vast majority in Merced County, where they are distinctive in look and feel.

Plate 32. Agricultural technology can sometimes have an obscure look and feel, as with this insect strip in an asparagus field in Imperial County.

Plate 33. If tomatoes were not important enough in the home garden landscape, the "homegrown" theme at this Dairyville fruit stand has special significance ever since Guy Clark's 1983 song *Homegrown Tomatoes.*

Plate 34. A young artichoke harvester stands tall, braced against a full *canasta* of artichokes on his back during the March peak of harvest near Castroville, Monterey County.

Plate 35. Delon and Eva take a postharvest break from picking cherries at an eastern Contra Costa County U-pick farm.

Plate 36. Bell peppers can be placed on wooden stakes, as here near Camulos, in Ventura County. Although the piked-pepper action was repeated through this field, the meaning is unknown.

Plate 37. California agriculture will take a crop through many ephemeral phases, including topping and tree-trimming after harvest, in this view from Stanislaus County along Hwy. 33.

KEYING ON CALIFORNIA AGRICULTURE

ORCHARD
RULES
1. You must buy what you pick
2. No Climbing Trees
3. NO Throwing Fruit
3. NO Running
5. Watch your kids
6. NO Eating Fruit in Orchard

AGRICULTURAL PRODUCT IDENTIFICATION: A HEIGHT-BASED GUIDE

The products covered in this field guide raise interesting issues in terms of their classification into a key and the creation of a finding guide. A single system needs to include animals and their products, manufactured goods such as cheese or honey, and crops that grow in sizes ranging from 100-foot-tall trees to subterranean plants—think sugar beets, carrots, and potatoes—which have an aboveground expression that can be deceptive. Consider it said: products coming out the California farm gate are nothing if not diverse.

Reluctantly, we omit a few crops listed in the 2008 County Agricultural Commissioners' Report, published late in 2009. The omissions include 27 products ranging from anise to watercress, and there will be readers who miss them. Crops such as Christmas trees we passed over because they are so seasonal and specialized; others, including guava and hazelnuts, are exotics that can be grown in California but are not big players in value. True, anise is worth $8 million in California farm-gate value. But consider, by way of contrast, the likes of bees. Packaged bees (sold to bee fanciers) are worth $2.6 million (2008), queen bees $7.5 million, bees (unspecified) $17 million, beeswax $3.1 million, honey $50 million, and pollination fees $224 million (2008), for a total of $294 million. At some point, value speaks: bees we include, anise we do not. How nice it would be to offer entries on mohair and rabbits, okra and mint, jojoba and emu, manure and biomass (the last three worth respectively $1.4, $18, and $30 million in 2008)—and perhaps in a future guide, with an unlimited page limit, we will. Crops may be tucked into other categories: Asian greens are listed with bok choy, and endive, escarole, frisée, and radicchio appear with lettuce (a category that also includes salad greens). Consulting the index will help in finding a particular favorite. Crops are in some cases such local specialties, or grown in such small lots, that they never make it into the Ag Commissioners' data.

The level of detail varies by crop. There are kinds of cows and sheep and goats that we ourselves likely cannot identify on the fly—to do so would require a photograph, setting aside some time, maybe having an expert at hand, and likely working on the Internet. This guide will not help any but the most sophisticated user in the apricot world differentiate between Pattersons, Blenheims, and Tiltons. But what we can and will get you to are entries for "cattle" and "apricots."

We hope nothing excluded is too much missed. Novelty is always tricky: crops are constantly introduced to California. San Diego County and the Sacramento–San Joaquin valleys are laboratories for agricultural experimentation and introduction. The Agricultural Experiment Stations maintained by the University of California and the extremely active research wings of the individual crop research boards (supported with grower funds) are especially strong innovators. In our research travels, we met a number of clever and curious growers who were experimenting with new varieties—crops that could burst out into commercial success at any moment. Dignified, if devoted, magazines such as the *Fruit Gardener*, from California Rare Fruit Growers, discuss the introduction and cultivation of crops that few of us have ever heard of—and we applaud their work and recognize the supreme skills of the dedicated enthusiasts and specialists out there. Recognizing limits, we have used discussions with Cooperative Extension agents and data on the value of agricultural production in the 58 California counties where crops are grown as our guide to what should be included. The data on California crops appears in the annual County Agricultural Commissioners' report, available online, and a minimum of detective work can track down the producers, should you be sufficiently interested.

The key system to be used for crops and products was a subject of discussion. This is not a typical field guide, in the sense that users stroll a tree-lined trail sighting birds, gently lift brush or rocks looking for beetles, or reach for a weather guide while sitting on a porch observing clouds forming on the horizon. Our conclusion was that most users will be traveling, generally by car, pickup, or van, and will be sighting and identifying features that they are traveling past at some speed. Our scheme for initial classification, therefore, is not based on vein patterns of leaves, or incipient growth form, or hair coloration. Even features essential to the taxonomist, such as the color of flower or seed shape, are put aside, for the most part.

We use a two-part key. First, the key turns on whether the agricultural product is animal or not. All animal products are covered in the first part of the key, which includes derivative products such as eggs, honey, jerky, and cheese. Oddities include okra, ostrich, and oysters, a singular California product listed in reports from the Marin and Humboldt county ag commissioners. Items such as chickens and turkeys represent one of the few failures in this guide: despite repeated contact with poultry producers, not once were we permitted to visit and photograph a commercial turkey or chicken producer. Various allusions to avian flu were aired, but in light of the bad press such industries have received in the last five years, refusals are more likely tied to concern about activists inclined to protest loudly on themes of animal cruelty. Of course, the net effect of such refusals is further doubt and skepticism among the public, and exasperation for field guide authors. The interesting by-product is that we can and do tell you what you are likely to see: long, skinny barns inside which poultry are confined. With luck, the observant motorist will find corners of Stanislaus and Merced counties where great white birds range free outside during daylight—a turkey farm—but access is rarely, if ever, permitted, and the view is from a distance.

The animal key, like all crops, is maintained alphabetically within the category; we assume that readers can work between cattle, goats, and chickens, or cheese and eggs. The grown crops are a more complicated matter, and we settled on a separate visual key—by height. There are four categories for crops.

Tree crops are first, a relatively self-explanatory category that includes trees capable of growing quite tall (chestnuts, avocados), more shrublike forms (which include pomegranates or kumquats), and a few outliers, including dates (which, strictly speaking are found on a palm, not a tree—but readers will grasp the idea). Some trees are modified in cultivation to new growth forms, which may make identifying them a slightly trickier process: apples have been espaliered (trellised) for generations and are familiar in concept, but in California, plantings exist of olives, trained and trellised in relatively low height growth forms (super-high-density plantings) so that they can be harvested

Plate 38. Hand-held cotton bolls, supported above a defoliated cotton field in Merced County, speak to the scale of cotton production in the San Joaquin Valley, where cotton (both upland and American-Pima) is a sizable economic contributor to farm-gate value.

mechanically. If you run onto these (near Oroville, Artois, and Corning in the Sacramento Valley), sussing out what you are looking at may require a roadside stop.

Vine, bush, and trellis crops include crops that grow to a height of about six feet, although some, such as kiwifruit, can be taller in certain cultivation schemes, and marijuana—should you see some—can grow another three feet taller than that. Crops in this category range from artichokes (a tall thistle-like plant) and pumpkins (low, ground-covering crops) to grapes—a huge category in California. Good eyes and careful driving are required when sorting out this part of the key.

The field, root, and row crops category displays a shorter growth form and includes a variety of crops: brussels sprouts and celery, chili peppers and cilantro, lettuce and strawberries, spinach and onions. Peculiar items such as nursery crops intrude

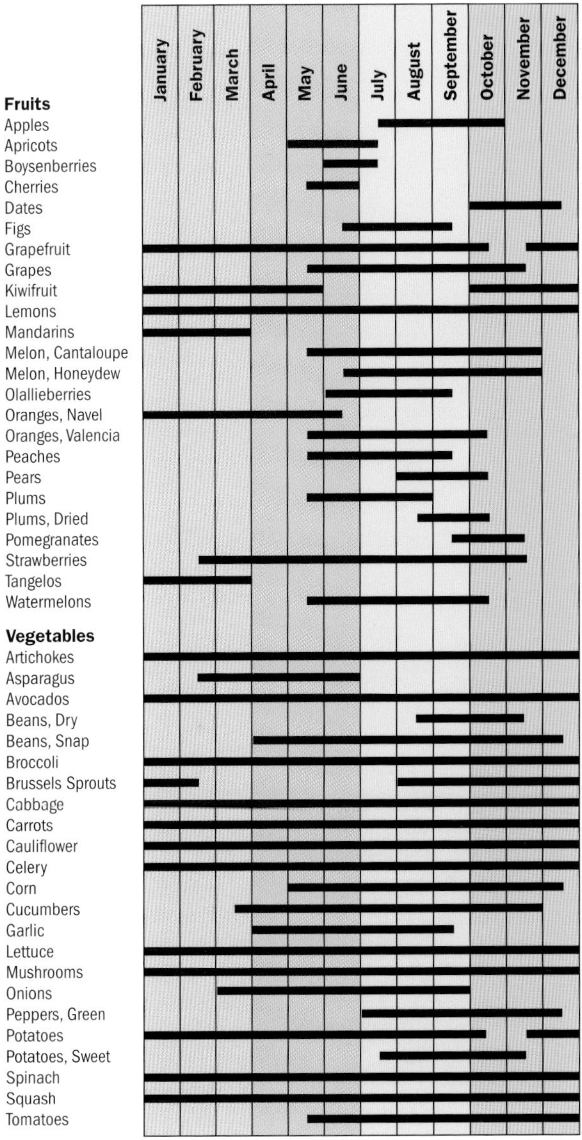

January
February
March
April
May
June
July
August
September
October
November
December

Fruits

Apples
Apricots
Boysenberries
Cherries
Dates
Figs
Grapefruit
Grapes
Kiwifruit
Lemons
Mandarins
Melon, Cantaloupe
Melon, Honeydew
Olallieberries
Oranges, Navel
Oranges, Valencia
Peaches
Pears
Plums
Plums, Dried
Pomegranates
Strawberries
Tangelos
Watermelons

Vegetables

Artichokes
Asparagus
Avocados
Beans, Dry
Beans, Snap
Broccoli
Brussels Sprouts
Cabbage
Carrots
Cauliflower
Celery
Corn
Cucumbers
Garlic
Lettuce
Mushrooms
Onions
Peppers, Green
Potatoes
Potatoes, Sweet
Spinach
Squash
Tomatoes

Figure 3. Crops produced through the seasons in California agriculture.

TABLE 1 Top 20 Agricultural Commodities in California, 2007

Rank	Commodity (2007)	Dollar Value	Top 3 Producing Counties
1	Milk and Cream	$7,328,474,000	Tulare, Merced, Stanislaus
2	Grapes (all)	3,077,769,000	Fresno, Kern, Tulare
3	Nursery/Greenhouse	3,065,995,000	San Diego, Ventura, Monterey
4	Lettuce (all)	2,178,041,000	Monterey, Fresno, Imperial
5	Almonds	2,127,375,000	Fresno, Stanislaus, Kern
6	Cattle and Calves	1,785,101,000	Tulare, Fresno, Imperial
7	Hay (all)	1,434,850,000	Kern, Tulare, Imperial
8	Strawberries (all)	1,338,585,000	Monterey, Ventura, S. Barbara
9	Tomatoes (all)	1,241,735,000	Fresno, Merced, San Diego
10	Floriculture	1,002,571,000	San Diego, S. Barbara, S. Cruz
11	Walnuts	754,000,000	San Joaquin, Tulare, Stanislaus
12	Chickens	713,000,000	Fresno, Merced, Stanislaus
13	Broccoli	669,405,000	Monterey, S. Barbara, S.L.O.
14	Cotton Lint	599,352,000	Fresno, Kings, Kern
15	Rice	583,833,000	Colusa, Butte, Sutter
16	Pistachios	561,600,000	Kern, Madera, Tulare
17	Oranges (all)	518,496,000	Tulare, Kern, Fresno
18	Lemons	512,550,000	Ventura, Kern, Tulare
19	Carrots	496,916,000	Kern, Imperial, Monterey
20	Celery	401,206,000	Ventura, Monterey, S. Barbara

here (a massive value in California, largely thanks to influential commercial and homeowner landscaping). Landscaping can range from ornamental plants to plantations of palms, purchased by roaming agents from homeowner front yards and transplanted into huge wooden pots, awaiting an impatient buyer with deep pockets willing to pay for some instant mature landscaping. But nursery crops, flowers, and ornamentals are treated as a single—and highly valuable—aggregate category, and sit here.

Finally, there are grain or pasture crops: corn, barley, oats, potatoes, and alfalfa hay—the last a billion-dollar crop in California, and essential to the dairy and cattle-feed industries. These are familiar densely planted crops, although some, such as rice, have a highly distinctive form.

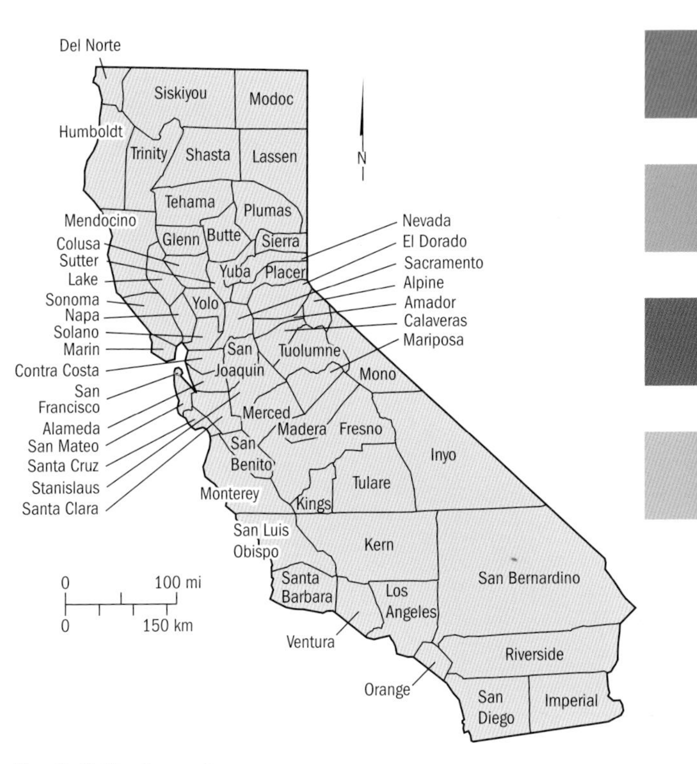

Map 3. California counties.

There is another way to recognize crops in California, and that is by the season in which you see them. Some crops are astonishingly difficult to recognize at particular times of year—which is to say, at certain points in their growth cycle. Cotton, for example, is a magnificent deep-green plant, growing two to four feet tall, with spectacular pink and off-white flowers. Sighting immature cotton is not easy; the "cotton" that most of us would recognize is defoliated, and displaying bright-white cotton bolls—in short, a far cry from the youthful form. This guide includes a crop seasons chart (Figure 3) that will aid in identifying what crop is likely to be active at what season. Some agricultural crops, such as artichokes, lettuce, squash, and cruciferous vegetables (broccoli, cauliflower, kohlrabi, and more), are essentially year-round producers, but others such as pears, cherries, boysenberries, garlic, and dry beans grow and produce in limited periods. In this chart, the latter part of any plant's growth cycle will involve maturing fruit or vegetables, which will make identification easier.

Two warnings to keep in mind: first, data is taken from the most recent, most complete year, which is generally 2008, 2007, or 2006. Sometimes, USDA–Census, USDA–NASS, and California County Ag Commissioners' data are at odds, in which case we used the most detailed, generally NASS. Second, when historical patterns of crop production are referred to using National Agricultural Statistics Service (NASS) data, the dollars used are for then and now, without any correction for inflation (as in Figure 2). For indexing and comparative purposes, here are two comparisons: $1 in 1920 would equal $10.35 in 2007—about a 10-fold increase. And $1 in 1960 would equal about $7 in 2007, so $3.3 billion in total agricultural production in 1960 California would be the equivalent of about $23 billion in 2007. This can be useful. Because the actual farm-gate value of production in 2007 was $36.6 billion, an upswing in agricultural production is not just imagined, in dollar value, it's real—in inflation-adjusted dollars, agriculture in California is worth nearly 170 percent of what crops and agricultural products were worth 50 years ago, in 1960.

ANIMAL PRODUCTS, CROPS, AND COMMODITIES

Livestock and Livestock Products

Livestock have a particular fascination for some; others among us are unmoved. But the process of domestication is a particularly important reality for humans. The domination, the taming, and the befriending of animals is not easily done, and was something our distant ancestors undertook long ago. First came the dog, which experts believe may have domesticated itself, volunteering into human company. The goat was next, perhaps 15,000 years ago, then the sheep; chickens were early on, pigs not too much later. Bringing cattle, the donkey, and the horse—to say little of the alpaca, the guanaco, the duck, the llama, the rabbit, and the turkey—into the mix was momentous. By 5,000 years ago, more or less, the deed was done, and our bestiary was complete. After that, humans did not want for companions of other species. The geographer Yi-Fu Tuan, in the wise 1984 book *Dominance and Affection: The Making of Pets*, makes a simple yet startling point: we domesticate many things, including landscapes—think of water fountains, or ornate topiary gardens—and treat them as our minions, even as friends. It is entirely normal for humans to surround ourselves with "natural" things turned to our purposes—but we have to appreciate the responsibilities that come with that act.

In an agricultural society, people are accustomed to slaughtering animals and harvesting animal products. But many Americans, and certainly most Californians, are urban people who have not grown up eating their own animals or hunting game. The family farm, where animals are almost a part of the extended family, is increasingly rare. People separated from domesticated animals except dogs and cats are uneasy about animals and their contribution to food stocks. Debates about cruelty to animals, about the "right" conditions under which animals should be raised—and slaughtered, since that too is an element of the cycle—are part of the political and economic scene. Agribusiness, farming at an industrial scale, is a player in dairy, in cheesemaking, in feedlots (confined animal feeding operations, or CAFOs), in the raising of chickens, and in egg production. Today's consumers, however, may—and do—choose to purchase animal products based on the specific form of husbandry used. Industrial farming can apply to almost any crop or animal product, keying on scale, efficiency, mass production, and distribution, and taking advantage of resulting cost-effectiveness.

For all the sizable producers, there are small ones, too, experimenting to see if they can find a market, contacting locavores who are committed to buying, eating, and supporting locally raised products, and going to old breeds or new ones to see whether they can thrive. Experimentation often takes place at the fringe of agriculture, something without doubt true of livestock and animal products. It's not just big producers who innovate. An enormous amount of knowledge comes from farms, ranches, ranchettes, fields, and backyards. The instinct for tinkering with plants and animals is where many an eccentric and ingenious idea resides, presumably to the eventual benefit of us all.

BEES

European or western honey bee ***Apis mellifera***

Pls. 39, 98

FAMILY: APIDAE.

RELATED PRODUCTS AND SERVICES: APIARY PRODUCTS, HONEY, AND POLLINATION.

RANK: U.S. #2; CALIFORNIA SHARE 13%.

SEE ALSO: HONEY.

Bees are not what comes to mind when most people think of livestock, but that they are, and their role in California agriculture goes far beyond the production of honey. Bees are the essential pollinator for no less than 53 of California's crops, including major players such as avocado, plum, strawberry, walnut, pomegranates, safflower, chili pepper, apple, apricot, and almonds; various other crops produce higher yields when bee-pollinated. Although bees themselves are not prominent from a distance, evidence of their hives definitely is, especially because the bees upon which most of these tree and field crops depend—the European, or western, honey bee—is gregarious, and hives are preferably stacked in groupings of 10 or more. Native bees are important pollinators, too, and recent research suggests that European honey bees actually travel faster and work harder when native bees are in a field. There are more than 1,600 species of native bee in California, but the native bee's limitation is habitat: when local vegetation is cleared from around a field, especially by herbicide use, the native bees find no food, and they disappear. As a pollinator, the honey bee is preferred for a major added attribute: hives of well-domesticated honey bees can be, and are, rented out to farmers, and the fees paid are significant—$224 million, in 2008. Within the race of bees, the role of sexual surrogate is a loftily paying business.

Some crops (with almonds and kiwifruit at the extreme end of the scale) require a high density of bees for flowers to bear fruit. The mechanism is familiar to any amateur gardener: the bee visits a flower and picks up pollen grains from the male anther; the pollen is transported with the bee to another flower, where the body brushes up against the stigma, or female organ, of the flower, depositing grains of pollen. The pollen grains grow, delivering a pair of sperm cells from the previously visited flower to the female ovule, a journey down the pollen tube that may be a prolonged one. When an unfertilized ovule is found, the sperm cells are released; one sperm cell becomes the new plant's embryo, the other develops into the endosperm, which nourishes the embryo and later becomes the seed leaves of the new plant. By this mechanism, almost all flowering plants are pollinated. Some plants are wind-pollinated, some are autogamous (self-pollinating), and still others are self-sterile (unreceptive to their own pollen) and may have to be pollinated from another cultivar or variety of plant. Even then, nectar and pollen must be transported, and bees are a favored vessel. But biotic pollination, quite common on Earth, requires animal or insect intervention. And for that there are bees.

Although beeswax and honey are products that have a long history of human use, recognizing the contribution of bees in the pollination of fruiting trees and other plants is a more recent discovery. Honey-hunting goes back 10,000 years in human history to Spain, Egypt, India, and Africa, and most religions have a place for honey in their rituals. In terms of sweetness, honey is on a par with granulated sugar, and is preferred in many baking recipes. Even in spoken language, the term "hon" is an affectionate one—and is spun straight from the bee's work. The United

Plate 39. Backed by a sizable cherry orchard are beehives, set in place to act as pollinators in this Santa Clara County orchard. Bees are the essential pollinators for more than 50 California crops.

States produces a great deal of honey, but still imports more than half the honey that is consumed annually.

The bee life cycle is complex; the queen bee is crucial to hive health, the organizing figure of a hive. Loss of a queen can be devastating, and a main preoccupation of beekeepers is to keep a queen in place and prevent swarming, which depopulates a hive as the workers follow the prime queen when she departs. The maintenance involved can be considerable. One narration of an annual saga of a Washington State beekeeper notes that to get bees to California for almond pollination, he annually transports 13,000 hives on 32 trucks, each loaded with 432 of the familiar white hive boxes of bees. The scale of planning is nothing if not logistically complex—and the hives must arrive in time for the honey bees to acclimate and get to work right as the trees they are to pollinate begin to bloom.

The varroa mite was broadly blamed for a massive decline in honey production numbers and plummeting hive health in 2005, although the mite had arrived in the United States in the late 1980s. Current thinking suggests that so-called Colony Collapse Disorder and the varroa mite infestation was a symptom, rather than the cause—the most likely culprit, still identified only tentatively several years later, is thought to have been malnutrition: with hives in less than optimal health, partly from stress and less than ideal feed, the mite could attain an upper hand, with resultant damage to bee health. Active programs attempt to improve bee nutrition, including, in some cases, supplements, and in 2009 the bee industry is on a marked upswing in health and profitability, with honey production topping 15 million pounds.

Products derived from bees are various: the big-dollar revenue is from pollination fees ($224 million in 2008). Honey in California—with two dozen flavors, including alfalfa, buckwheat, eucalyptus, lavender, safflower, and pumpkin—was worth $50 million, beeswax $3.1 million, packaged bees (sold to beekeepers) $2.6 million, and queen bees $7.5 million. Royal jelly, produced in the hive for young bees, is sometimes sold; a variety of health-giving qualities are attributed to it. Granted, with all of California agriculture generating a value above $36 billion, the $294 million brought in by bees is hardly momentous, but consider that there would be no almond industry (worth more than $2.6 billion in 2008)—and that all those other commercial agricultural products would be nonstarters—without bees to pollinate them. Honey is in its own right a remarkable product, so sweet that bacteria cannot grow in it. It may be of some interest that despite the crucial role that bees play in almond pollination, almond honey is generally considered to be unpalatable, so the honey gleaned from all that work is used in baking (if harvested) or left in the hive as feed for the bees. Bees remain a constant source of fascination for scientists, some of whom have noted with acerbity that, structurally, something shaped as bees are should not be able to fly. Yet fly they do, and the dance of the bee, which directs fellow workers to pollen-dense sites, is widely studied.

In short, the bee provides a variety of products, but also an outsized benefit in the culture of California agriculture. Honey takes on tastes peculiar to the surrounding plants where pollen is gathered, and the tastes and shades of honey are remarkably diverse—and sometimes prized. Honey, it might be said, has its own terroir—the quality of taking on the qualities of local plants and soil that is much appreciated by some vintners. Little wonder, then, that the bee is a creature of particularly rich symbolism. The beehive is a widely accepted symbol among Mormons, who admire the industry, hard work, and what might perhaps be considered family values of *Apis mellifera*. Those familiar white beehives are to be respected as the engine of agricultural fertility.

CATTLE AND CALVES *Bos primigenius*

Pl. 40

FAMILY: BOVIDAE.

NAMES: OFTEN PREFERRED NOW AS *BOS PRIMIGENIUS TAURUS*; UNDER LINNAEUS, SPLIT BETWEEN EUROPEAN CATTLE (*BOS TAURUS*) AND ZEBU CATTLE (*BOS INDICUS*).

RANK: U.S. #7; CALIFORNIA SHARE 4%.

SEE ALSO: CATTLE, DAIRY; CHEESE; HAY (ALFALFA AND OTHER); PASTURE.

Over about 25 million privately owned acres of California rangeland, cattle—mostly cows—are the conspicuous residents: great creatures, weighing 500 (for a big calf) to 1,000-plus pounds, horned and not, cud-chewing until bothered, curious or skeptical in mien. As an activity, ranching is both an economy and a commitment to stewardship, drawing on cattle—and sometimes other livestock—as partners. In summer, when upper elevation range is snow-free and patterns of transhumance (seasonal grazing) are in effect, cattle are scattered throughout the state, from lowland fields to high country (which is often public land).

Some observers view cattle as intrinsically attractive. Breeds selected vary according to what uses ranchers have for them, and according to prevailing weather—zebu-type cattle do better in hot climates, and European cattle in the northern states with their colder climes. Seen on the range, cattle offer a visual feast. They bear polychrome tints ranging from the alternating red whiteface of a Hereford and the off-cream or blond Charolais to the sternly black Angus and the bricklike red of the Brangus. There are middling colors caused by the dilution gene, which shades animals to deep grays, duns, the brindled Beefmaster, spotted black and white Holsteins, and even the charming and characteristic hues of the Shorthorn breed: blue and red roan. Although cattle are now rolled into one species, Linnaeus, the father of modern taxonomy, thought they derived from two lines: European breeds (*Bos taurus*) and the Zebu line (*Bos indicus*)—which has longer ears, a hump on the back above the shoulders, a dished face, and loose skin at the throatlatch and dewlap. Under the ruling Integrated Taxonomic Information System (ITIS), the lines were merged as *Bos primigenius taurus*. There is considerable variation in the breeds and hues of cattle in California. They interbreed readily, and when seasonally used public range is included, cattle graze half the area of California. Grazing is rarely without controversy; all some suburban naysayers need to become enraged is to see a cow or calf grazing public or park lands.

Only with difficulty can the breeds, well-being, and behavior of cattle be separated from the habitats where they eat and grow. In California, the rangeland grazed by cattle (and other livestock) often appears essentially natural or minimally changed—or so it may seem. In fact, capable exotic invasive species of grass, thistle, brome, and sedge came from southern Europe and North Africa, beginning in the 1700s with the Spanish–Mexican arrival. Given time and a distinct genetic advantage in migrating from a

Plate 40. Cows and calves are substantial producers in California's agricultural economy and a reminder of the time-tested relationship of humans with large herbivores. These cattle grazing an Imperial County irrigated pasture are of varied types and breeds.

weed-intensive and grazing-heavy Mediterranean fringe, invasive plants are a visual dominant, and constitute much of what the modern untutored eye sees on the open range. The early Anglo-period history of California was full of small homesteads and ranches that carved up what was previously rangeland or wildlife habitat. Hardscrabble farming, pursued without advance knowledge of what should not or could not be grown, broke the ground and eliminated any advantage that native grass and shrubs might have had. When farms failed, the open soil was a waiting seedbed for invasive plants adapted to occupying damaged land. Cattle, sheep, goats, and horses, along with people and trade, helped spread alien seeds from the coast and Southland throughout California in a matter of a few decades. And grazing could be abusively heavy. At various points during early California history, up to three times as many domestic livestock (counted altogether) grazed rangelands as do today. Public lands were in those days regarded as open for free use, and through the 1880s, fences on range were relatively few and far between.

The history of California for the last 200 years is inescapably tied to cattle. Among domesticated animals, they carried the highest prestige, and even as the Anglo era arrived in California during Gold Rush days, the life of the *patron*, the cattle owner, was something that many old-time Anglo "Californios" aspired to. Cattle being raised went through a change of purpose with the development of a burgeoning market for meat during the Gold Rush. They were converted from biological factories for hides (for leather) and tallow (rendered fat, destined for candle factories on the East Coast) into widely wandering meat. The English term "cattle" is derived from the Old French word *catel*, which meant movable property and, generally, any form of livestock. Milking animals were held locally but were not widespread in the state's economy until the late nineteenth century. Because fence technology remained primitive until the arrival of barbed wire in the late 1870s, much of the California range was "open," and the cattle and calves historically were sorted out in rodeos, or roundups, supervised in a reunion of area ranchers who would claim their calves after watching mothers and offspring together. This complicated process was adjudicated by the Judge of the Plains, a traditional position (dating back to distant Arab–North African precedent) for a respected member of the community, one of the many traditions persisting from the Spanish–Mexican period. That job was recast and renamed, in subsequent years, to that of brand inspector—still an official position in every county in California.

All these antecedents have changed, and so has range management as a professional and academic field. Although "food and fiber" are still emphasized as products, there is ever more attention to using livestock, cattle included, as management tools to better rangeland quality, and the timing and intensity of grazing can be a potent force in weed control and range improvement. Yet the origins of modern California still owe much to "the cattle on a thousand hills," as Robert Glass Cleland named his history of Southern California ranching (its title borrowed from Psalm 50:10), which chronicled the transition from Spanish–Mexican to Anglo times. To see cattle grazing on new lush spring growth, dodging sprinkler water on an irrigated pasture, or traveling quickly through desert pathways, is to witness a reminder of those origins. Even a cow or steer chewing complacently on a mouthful of agricultural aftermath (stubble, or grain dropped in threshing), last year's residual dry matter, or good grain gummed from a feeder has a certain bucolic quality. Not everyone is enthusiastic about the outcome, but who would expect otherwise?

Cattle are kept on rangeland literally throughout the state. They will graze into dense forest; there they are often belled to help cowhands find them later. In arid lands, cattle will work through desert scrub or high desert sagebrush, shadscale, saltbrush, or rabbitbrush range; animals may also be turned onto agricultural fields to work off the stubble or remnant material once crops are harvested. Some form or another of cattle (calves, culls, steers, bulls) is sold in every one of California's 58 counties except one (San Francisco), although fewer than 1,000 animals are sold from tiny Alpine, Santa Cruz, and Orange counties. Cattle are ubiquitous—even Los Angeles saw nearly 3,000 beeves sold in 2007.

Exclusively vegetarian feeders that prefer grass to browse (shrubs and larger plants), cattle will nonetheless thrive on varied vegetal feed, if that is all that presents itself (surplus pumpkins are sometimes fed to cattle). Cattle are ruminants, with divided digestive systems that allow them to break down plant-based feed, improving the nutritive value of roughage that would otherwise be considered of no worth. They pass food through the rumen and reticulum, with solids worked, sometimes repeatedly, by regurgitation and cud chewing. As solids are broken down by repeated treatment, they then move into the omasum, finally continuing to the abomasum, the true, or monogastric, stomach, equivalent in function to that of a human or pig. The ability to process marginal feed is shared with sheep and goats (and other ruminants). The remarkable digestion of ruminants is one reason there are objections raised to feeding high-quality feeds, such as soybeans, feed corn, or other grains to ruminants in confinement: they do well on feed of far lower quality, although average daily gain may be smaller. Objection to the aesthetics and food politics of feedlot confinement has led to a conspicuous upswing in grass-fed beef, with animals raised to slaughter weight on forage and hay. Grass-fed beef, often sold by contract in shares, costs a good bit more per pound, but intermediaries are sometimes eliminated, which can give a rancher a considerably higher price per pound. A main issue is distance to a small number of slaughter facilities.

Especially at certain times of the year, cattle inhabit irrigated and dry pastures. The largest impediment to cattle ranching in California is climatic. The dearth of rain from May to October dries out rangeland, reducing forage quality. If irrigated or high-elevation pasture is not used, careful owners must leave enough feed for summer. Supplements—including hay—may be needed. Pastures, especially when flood- or sprinkler-irrigated, offer a grazing ground for summer and into fall. If agricultural fields are fenced, cattle can be turned loose after crops are harvested. Others go to feedlots. Animals are collected, located in one place, and fed to put on weight before slaughter: these include cull cows (dairy cows past their prime, and cows that turn out not to be pregnant), steers (male animals, castrated to mollify their temperament and bring faster weight and height gain), and heifers (a female that has not calved). They are concentrated on feedlots (CAFOs, or confined animal feeding operations) that are scattered around central and southern California, including the western San Joaquin Valley and the Imperial Valley, near Mexico and the Arizona border.

Dairy cows in the San Joaquin Valley, in particular, are held in the country's largest milking operations. Dairy stock inject complicated variations into any simplified "cattle and calves" story, mostly having to do with animals leaving a dairy herd and going to meat production. Male dairy calves, since they produce no milk, go straight into meat production. Some calves are consumed as veal, before the young shift from milk (or milk substitute) to solid food. In an odd middle ground, dairy steers at that point stop featuring in dairy statistics and shift to the cattle and calf data tables; so do culled dairy cows, which are removed from the milking herd when they are no longer able to produce at a sufficient rate. Cull dairy cows go to feed or straight to slaughter and are part of the "cattle and calves" total production. Cattle, especially oxen (castrated older males), were once used for traction (pulling), but they are singularly rare in today's California. Nor are great-horned and aged steers anything common; most steers are consigned to slaughter before they are two years old.

Although California's share of the United States cattle market is relatively small, the dollar value is not unimpressive—almost $2 billion in 2008. And the details tell an interesting, if intricate, tale. California range cattle offered the bulk of income in 2007, in the form of undifferentiated cattle and calves ($1.8 billion), but dairy replacement heifers ($235 million), culled milk cows ($113 million), veal calves ($40 million), stocker and feeder cattle ($94 million), and cattle on feed ($610 million) all feature in the overall figures, each with its share. "Complicated" is the word. Tulare, Fresno, Imperial, Merced, and Kern counties occupy the first five spots; not far behind in calf production are Stanislaus, Kings, and San Joaquin, with San Luis Obispo County, and its huge rangeland count of cattle, checking in at $88 million in value. Big business, indeed.

A few myths deserve note. First, in terms of recognizing cattle and calves, gender distinction is not difficult, although ranchers invariably are amused when a well-meaning parent explains to an awestruck child that "the bulls have horns; cows don't." The error is significant: there are both polled (non-horned) and horned cattle. Most cattle are naturally graced with horns (male and females, and calves above a certain age if they are not dehorned), and those horns are used for defense, to display physical prowess (for males), to scratch, and to maintain status within the herd (bulls, steers, cows, calves). Although a great deal of money and effort was invested at one juncture in developing naturally polled (non-horned) cattle, for animals grazing in the open, removing horns is not necessarily a kindness. Polled cattle sometimes bring higher prices at a sales yard, because they are less likely to damage one another, and more polled cattle can fit side by side at the feed trough. Any mature *Bos* can have horns; to distinguish females from males, look for the tuft of hair two thirds of the way between front and back legs that marks the sheath for the male's penis—cows, of course, have nothing of the sort. Bulls are easily recognized in range livestock herds and, to do their job best, are kept in a ratio of one bull for every 15 to 35 cows, depending on the ruggedness of the terrain and the age of the bull. But dairy cows may never see a bull—artificial insemination is particularly common within dairy herds.

Second, cattle burps (not flatulence) are sometimes singled out as a contributor to greenhouse gases. The internal fermentation tank called the rumen allows cattle (and other ruminants: sheep, goats, deer, bison, and others) to make use of forage that cannot be consumed by non-ruminants. High-fiber, low-protein grasses and shrubs are digested with great efficiency, broken down by bacteria and protozoa living in the rumen that can use otherwise indigestible cellulose. This has served human populations well, allowing the grazing of crop by-products and aftermath, dry and climatically extreme rangelands and tundra, and scraps of uncultivated land. However, such digestive efficiency comes at a price, as methane is burped out as a product of fermentation. Because every source of methane today is a concern, studies are proceeding on ways to reduce or mitigate gas release by ruminants.

Finally, cattle maintain an interesting, if inconsistent, relationship with the general public in parkland and other public land situations. While ranchers will take on paid permits to graze public land, if those are offered, hikers and mountain bikers are not always prepared to come over a ridge into a group of cows and calves, especially if the cyclists don't respect the diverse uses of the landscape and bring dogs. Cattle have good reason to mistrust domesticated dogs, which turned loose to roam during the day, are a familiar threat to livestock. Balancing cattle and humans takes some doing. Little wonder that many parks (and some national forests) are proving more open to the grazing of sheep and goats than cattle. Recent studies suggest that many people admire cattle and feel close to them; certainly that is true of goats used to reduce fire hazard on public and private lands.

At the end of 2008, "cattle and calves" constituted the fifth-largest agricultural product in California, according to the state's agricultural resource directory. Although there will always be discussion and dissent about using animals to feed humans ("milk and cream," the #1 value product, is in the same category), the human relationship with cattle offers a fascinating glimpse into domestication, a process that happened long ago and that somehow brought a large and powerful animal into the human circle of acquaintances. Some laud that; others not.

CATTLE, DAIRY *Bos primigenius*

Pls. 2, 11, 24, 114, 124, Map 4

FAMILY: BOVIDAE.

NAMES: OFTEN *BOS PRIMIGENIUS TAURUS.*

RELATED PRODUCTS AND SERVICES: MILK, CREAM, BUTTER, AND DAIRY CALVES.

RANK: U.S. #1; CALIFORNIA SHARE 19%.

SEE ALSO: CHEESE; CATTLE AND CALVES; HAY (ALFALFA AND OTHER); PASTURE.

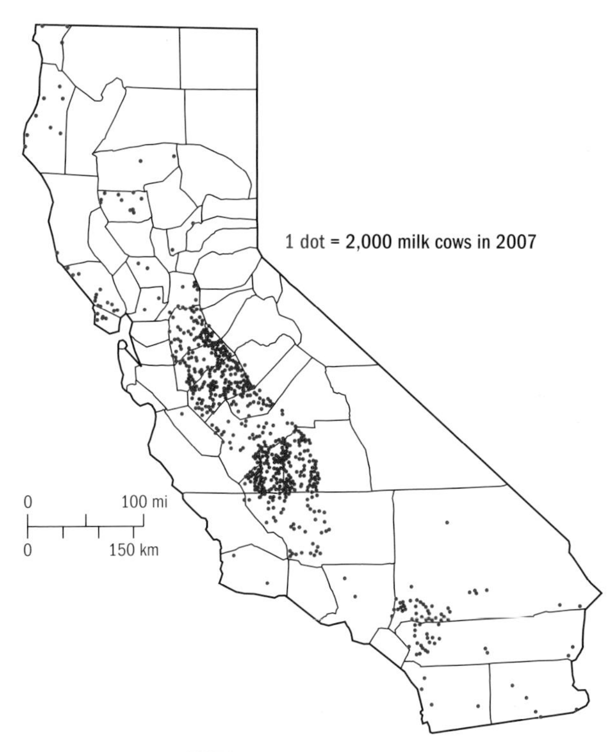

Map 4. Dairy cows, 2007.

Dairy farming of milk cows in California brings a genuine physical presence, partly in landscape, partly in cows, and partly in people. Distinctive surnames mark the landscape, with the names of dairies generally reflecting the Portuguese (Azorean), Italian, or Dutch origin of their operators, whose ethnic specializations in dairy farming have roots better than a hundred years old. The presence of a dairy can be olfactory—even the best-run dairy farm at night gives off a whiff of concentrated animals and their byproducts. Or the recognition may be visual: California dairies are striking things, recognizable by their low, open-sided barns, which shield cows from rain and sun. Every dairy farm has chopped-hay silage, either contained in blue Harvestore silos or, more generally, covered with white high-mil plastic and an intricate network of automobile tires that are cut in two and cabled together to keep protective plastic in place. Dairy yards can extend for several acres, and should you be allowed on-site to mingle with the well-kept cows—usually leggy black-and-white Holsteins, but perhaps Jerseys, Guernseys, Brown Swiss, or Ayrshires—they will eye you like skeptical debutantes assessing potential dance partners at a high school formal. Dairy cows are an appreciative audience. From the outside, the California dairy presence is formal and scrubbed; these are professional operations. They are big, too. Dairy is a giant in California agriculture.

Although climate and concentrated urban populations distinguish California from other states, in dairy farming it is the statistics that set California apart, as in so many other crops. The contrast is illuminating, and deserves a quick run-through. With 1,960 dairies, California has roughly one-seventh Wisconsin's total number of 14,170 active dairy farms. In 2007, 78 Wisconsin dairies owned more than 1,000 milking animals. But in California, 593 dairies milked more than 1,000 animals, and 1,100 California dairy herds topped 500 animals. In scale, no other state comes close. And dairy revenue kept pace: California produced half again as much milk and cream in 2007 as Wisconsin (CA: $7.3 billion, WI: $4.5 billion). Production per cow is somewhat higher in California than Wisconsin or other dairy states (New York is third), but total cow numbers are about proportional: 1.8 million dairy animals in California, 1.3 million in Wisconsin. In 2007, California produced 22 percent of the U.S. milk supply, 23 percent of the cheese (nearly 2.3 billion pounds), 31 percent of the butter (500 million pounds), and 15 percent of the ice cream (129 million pounds). Dairy activity in California is concentrated in (but not exclusive to) the San Joaquin Valley, where more than 70 percent of state milk production takes place: Tulare, Merced, Stanislaus, Kings, and Kern counties are the top five. Tulare County alone pours forth more than a quarter of California's milk production.

In the field, California dairies are diverse. The most concentrated operations are the quite significant urban-fringe dairies that are on the edge of San Bernardino County in the Inland Empire's Chino Valley; dairies extend into adjoining parts of Riverside County. Dairy cows on these urban milkshed operations began meeting the milk needs of Los Angeles–area residents in the early nineteenth century. Dairies were once common in Southern California, but dairies shifted north to the San Joaquin Valley as urbanization crowded in and massing city populations proved unwilling to share space with dairy animals. While the count of Southland cows is not huge, there is something to be said for reducing the transportation distance that milk must travel, given a ready local market for milk. Another outlier in California dairy production is the North Coast region, particularly in Marin and Sonoma counties. The North Coast dairies are small, especially compared to the San Joaquin Valley behemoths, but they produce milk and cheese that is sought after not only in-state but throughout the United States.

Dairies in the North Bay are a story unto themselves. Marin and Sonoma counties, dating back to the early 1900s, were centers of an Italian–American rural populace that specialized in dairy, tree crops, and vineyard production. As urban populations swelled in the 1970s, dairy farms along the western edge of both counties faced compression and potential eradication. Instead of watching that sucking tide sweep farms from the land, agricultural and land-use activists in Marin and Sonoma set about finding a means of protecting farm acreage. Ultimately, a series of land trusts and conservation easements were created that made it possible for farming to continue with a minimal amount of urban sprawl infringing on the dairies and other ranch operations. During the chartering of Point Reyes National Seashore, permission was gained to permit grazing leases on the Seashore, for continued grazing was needed to keep enough dairies in Marin to support a creamery. In fact, Marin County is widely credited, along with a land trust movement that had strong roots in the Northeast, with spawning national interest in preservation of working farmlands, and the dairies in Marin survive to this day. And they do more than survive; they are producers of some of the highest-priced, and most prized, dairy products in the state, with Cowgirl Creamery, Straus Family Creamery, Bellwether Farms, and other dairies and artisan cheesemakers resolutely active in farm conservation politics. Small dairies such as these constitute a second front of dairying, at another scale, as is so often the case in California. Organic producers with a limited number of cows milk the herd, manage animals for a longer lifespan, and emphasize varied products such as yogurt, ice cream, and butter, beyond merely traditional milk and cheese.

The story of dairying in California is inevitably an account of dairy animals, and a parable about scale. Milk and cream is the largest commodity in California agriculture, as has been the case for better than a decade. Occasionally, the cattle and calves category takes the #1 spot, but seldom. Dairying in California has undergone change at a fantastic rate through the last 50 years. In 1950, California dairies averaged 40 animals per operation, and there were almost 20,000 dairies. By 1970, three of four dairies had disappeared, but the count of cows per farm was up to 150, and yield per cow doubled in 20 years to 13,000 pounds. After 30 more years had passed, in 2000, the figures were similar to the numbers of today: 2,100-plus dairy farms had an average of some 850 cows each; now, each cow produces almost 22,000 pounds per year. In 1993, California eclipsed Wisconsin as the leading dairy-producing state and since has followed its own course, although California still lags fractionally behind Wisconsin in cheese production.

Improvements in breeding certainly help make California the milk giant that it is. There is no implication that giant dairies in California are inhumane—they absolutely cannot afford to be, with watchful eyes always turned toward the industry. And producers are sensitive to criticism; they are toeing a line drawn by consumers. Like dairies elsewhere in the United States, California dairies were for a time using bovine growth hormone (rBGH or rBST) to boost yields, a process that generated consumer skepticism. As public sentiment turned against rBGH, first in northern California, and then in the Southland, the California Dairy Council chose to instruct its members to no longer use the hormone in their herds. The same shift—recognizing controversy and moving away—can be noted in the production of veal from dairy calves, a $40 million business in 2008. Male dairy calves, fed to slaughter weight, are a component in the California feedlot, range, and meat market.

Dairy animals are fed silage, made of chopped and slightly fermented alfalfa, corn, or other grains. Silage is fed to dairy cows because, as a moist and readily digestible feed, the chopped and processed mix increases milk yield. Because cows have to consume nutritious forage, even when confined in a dairy yard production setting, they depend for survival on the harvest and delivery of off-dairy crops. Breeding better Holstein cows also boosts yield; cows are able to produce at a rate unthinkable four decades ago. There is a toll—dairy cows last about four years, as producers, and then go to slaughter—perhaps a half or a third of the lifespan of a cow on the range. Dairy cows are highly refined milk-producing organisms, specialized to a degree that fully matches any advance in crop genetics, automotive design, or computer technology.

But there are other parts of the story. Although there is an unusually large size to the dairy industry, as is common in California, there is a smaller, more intimate—and, supporters would say, personal—side to the dairy industry. Of California's dairy operators, 500 keep fewer than 50 milk cows apiece. Several dozen organic dairies have to meet stricter USDA organic standards, and produce about 15 percent less milk per cow, in herds that average 200 cows. Some organic dairy producers go one step further and produce raw milk—subject to the greatest vigilance and sanitary standards, because the milk is unpasteurized. There is a relatively upscale market among raw food believers in health food stores. California dairies produce more than two million gallons of organic milk each month. Feed in organic dairies costs a quarter to half again as much as in conventional dairies, but the price per gallon of milk sold is higher, too—commonly twice the price per hundredweight of milk.

Finally, technology reaches a contemporary pinnacle in adding carousels to the milking parlors. The largest dairy farms in the San Joaquin Valley turned in the 1990s to technology developed in New Zealand, and brought carousel technology across the Pacific to the United States. Traditional dairies sport a long line of stanchions. Cows walk up, put their heads through a locking gate, and are milked by workers who walk behind them. What would happen, though, some soul reasoned, if the cows moved, and workers were stationary? A milking carousel is a simple concept but a complicated bit of machinery, a circular platform that rotates around a central well with its radii divided into slots for individual cows: up to 120 "stalls" at a time. Cows queue up to get onto the carousel, one to each slot, with their backsides facing outward. A decade ago, capacity was 25 to 50 cows at a time; that rose to 60, then 80; now, a new large carousel can hold 100 or more animals at a time, to be milked twice a day. Dairies may run double carousels, side by side, to milk 240 animals at a time. When the cow steps into the stanchion, a worker cleans the udders and teats and attaches the milking machine. As milk flow stops, the milking machine disconnects automatically. Each cow has its own identifying code, typically held in an RFID chip, and statistics are kept on each animal, so the dairy farm manager can keep track of high and low producers and make note of any issues. The technology changes almost constantly, with dueling technology from Germany, New Zealand, and the United States. What ties all the animals together in industrial-scale dairying are computers, sophisticated monitoring, and assiduous herd record keeping. This is a world apart from the small, organic producer—but as in so many other areas, the cow dairy in California introduces a novel concept of scale.

A dairy farm in the California of 2009 is at a far remove from the operations of just a few decades ago. The degree of sophistication is startling, extending all the way to a dairy cow's genetics and ability to produce up to 80 pounds of milk each day—10 gallons—through each lactation cycle.

CHEESE

Pl. 41

SOURCES: MILK TO MAKE CHEESE CAN COME FROM GOATS, COWS, SHEEP, BUFFALO, OR FROM A MIX OF THESE.

RANK: U.S. #2; CALIFORNIA SHARE 25%.

SEE ALSO: CATTLE, DAIRY; GOATS; SHEEP AND LAMBS.

The cheese buzz in early 2008 was all about volume: output from cheesemakers in California was threatening to eclipse Wisconsin's timeless status as the largest cheese-producing state in the United States. Later reports backed down from claiming the record, but with time, California will prevail. In 2008, California produced 2.1 billion pounds of cheese (yes, billion), an 80 percent increase over the 1998 total. This means one of every four pounds of cheese sold in the United States hails from California—surely good news for the California Milk Advisory Board and its Happy Cows advertising campaign. In 2007, the California dairy industry produced $5.4 billion in milk and milk products. The economic reach is deep and far: a 2004 study suggested a total impact of dairy of $47.4 billion in the state, including the provision of 434,000 full-time jobs. Other statistics are no less impressive: 50 cheesemakers in California produce 250 varieties of cheese. Those numbers come from the big-cheese side of the industry. Including the artisanal—small-batch—producers would up the total of both producers and cheeses considerably, but an exact number is difficult to come by, and artisan cheesemakers are constantly generating cheeses rarely, if ever, before seen in California.

The virtues of cheese are too many to recite, including benefits to producers, to purchasers, and to government. Assuredly, "cheese" is a product, rather than a landscape trait or an observable feature, but cheese is born of a long acquisitive process: getting land, adding dairy animals and milking them, and, from close study or long knowledge, learning how to convert milk to cheese, with curing and marketing. Cheese is produced in close quarters and usually with strict sanitary standards to keep the FDA at bay. The advantages of cheesemaking are pronounced. Consumers get a product with a far longer shelf life than liquid milk. In fact, for 2007, 47 percent of California's milk supply went to making cheese.

In cheese, producers have one of the few agricultural products in which modest processing can substantially increase the value of a consumable and lift skyward the income derived from it. Wine is the other prominent example—little wonder that wine and cheese are often consumed together. Cheese travels well, reflects local terrain, and, up to a point, improves with age. It does not hurt that since the 1990s a remarkably impressive range of European cheeses has been widely imported into the United States, giving American artisan cheesemakers models and inspiration. Although some cheeses are produced for uniformity and standardization, the artisan cheesemaker tends to follow just the

Plate 41. Cheeses come in varied shapes, sizes, and milk types (and from various tiny artisanal cheesemakers), unsurprising since 47 percent of California milk went to cheesemaking in 2007. These Serena cheese rounds, from Three Sisters Farmstead Cheese, are made at Hilarides Dairy in Lindsay.

opposite path, aiming for originality and a distinctive blend. The variation is formidable, as befits a food whose origins are lodged deep in human prehistory; in the written records, Columella's *De Re Rustica* describes using rennet to coagulate milk and lays out in detail the subsequent steps of cheesemaking—and his account dates from AD 65. The official Cheese Board in Great Britain claims to have denoted 700 distinct local cheeses, and across the English Channel, France avers that it has more than 1,000 cheeses.

Although such cheesemaking innovation abroad sets an elusively high benchmark, cheese production at the mass-produced end amply demonstrates California's cheese brawn.

California's production, in types and percentages, includes mozzarella (50 percent), cheddar (21 percent), Monterey Jack (14 percent), Hispanic cheese (aka: queso fresco) (5 percent), parmesan (3 percent), and provolone (3 percent); the remainder falls within the ever-popular "other types."

The cheese production process, in its essentials, is straightforward: milk from whatever animal or mix of animals involved is treated (the most famed cheeses in the world are often blends of different milks), and the milk protein, casein, is coagulated. Solids are separated, and pressed into a final shape or a hard form. At that point, molds or bacteria may or may not be added; herbs or spices may be included for flavor; cheese may be pierced with needles; the resulting cheeses may be aged, or not. Cleverness comes in the details, and in the environment where cheeses are aged—there is cheese greatness made.

Although commercial cheeses are pasteurized, raw-milk cheese remain popular in select circles, if also a subject of controversy. FDA regulations dating from 1951 hold that all cheeses, imported or domestic, must either be pasteurized, or must be aged at least 60 days. This essentially rules out an entire class of soft, raw-milk cheeses that include some of the most famous European varieties, and more than a smattering of the artisan cheeses of California. How strictly these FDA standards are adhered to varies by maker. Although it is rare to be busted by the FDA for holding immature cheese, the possibility cannot be ruled out.

Among the "other types" of cheeses, which constitute the most limited part of California cheese production, is virtually the entire world of California artisan cheese. The names are noteworthy for their embrace of a sense of place: from Bellwether Farms (Marin: sheep cheese): carmody, fromage blanc, San Andreas; from Cowgirl Creamery (Marin: goat, sheep, cow, blends): Pierce Point, Red Hawk, St. Pat, Mt. Tam; from Cypress Grove (Humboldt: sheep, goat): Humboldt fog, truffle tremor, Bermuda Triangle, Lamb Chopper; from Meyenberg Goat Milk Products (Stanislaus): goat cheddar, smoked goat jack; from Barinaga Ranch (Marin: goat): Pyrenees-style tomme; from Redwood Hill Farm and Creamery (Sonoma: goat): Gravenstein gold, bucheret, camellia, feta; and from Farmstead/Three Sisters Cheese (Tulare: Hilarides Dairy, cow): serena, serenita. These cheeses run the gamut: goat, sheep, cow, buffalo mozzarella, milk blends. Some are eaten fresh, others are ripened and aged. There are fresh cheeses, soft and soft-ripened cheeses, semi-hard and hard cheeses, very hard cheeses (aged dry Monterey Jack), spiced and flavored cheeses. Prices range from $5 to $65 a pound. That remnant one-twentieth of California cheese output may be as memorable as the large-scale production of 95 percent of California cheese.

CHICKENS *Gallus gallus*

Pls. 42, 43

FAMILY: PHASIANIDAE.

NAMES: SOMETIMES *G. GALLUS DOMESTICUS.*

RANK: U.S. #12; CALIFORNIA SHARE 3%.

SEE ALSO: EGGS.

Chickens are a force in California agriculture, and they must fight off a difficult press, in part brought on by the less-than-gracious relationship that elements within the poultry industry have with the larger public. In California, chickens are substantial contributors to the agricultural economy, coming in at tenth among agricultural producers (chicken eggs come in at twenty-first). There are two crucial components to the production of chickens in California, as elsewhere in the United States. First is food supply: feed for chickens is strongly based on corn production, and as a feed corn producer, California is relatively insignificant. The second contributor is demand for chicken—and there, California looms large, with 11 percent of the U.S. population. It pays, then, to have chickens in significant supply, and that California does, as both broilers (for cooking), and layers (for eggs). The industry is highly concentrated, with 229 producers in California generating a bit less than 10 percent of the total national broiler production. The San Joaquin Valley counties, especially Fresno, San Joaquin, Madera, Stanislaus, and Tulare, dominate production and sales of chickens for eating, with San Bernardino and Riverside respectable contenders. In 2007, more than 280 million broilers were produced, and if turkey and chicken production are grouped together, they are California's seventh-ranking commodity in farm-gate value, at $1.2 billion.

Spotting chicken-raising operations in the field is not difficult. The long, thin sheds of metal-roofed poultry barns are

Plate 42. With health-conscious eaters turning less often to red meat, chicken is an increasingly important element of the California diet. However, doubts concerning the conditions in which fryers, broilers, and roasters are raised and processed have led to an upsurge in the purchasing and the price of "free-range" chicken, a designation that suggests humane conditions but still fails to match guarantees of mobility and outdoor access provided by European Union agricultural doctrine.

immediately recognizable and characteristic; should chickens be allowed outside, they appear entirely chicken-like—far smaller than turkeys, and black or red or white in color. The only problem is differentiating turkey-raising operations from chicken facilities. Turkeys are far larger animals, and are less commonly raised in strict confinement. From satellite images or a roadside vantage point, the bays of poultry barns are easily picked out.

Be wary of moving any closer: chicken operations are militant in their desire to exclude the general public from the grounds of chicken-raising operations. Large biohazard signs will often warn off any potential interloper. In fieldwork for this guidebook, we contacted a number of chicken producers, requesting permission to visit and photograph their operations, and in each case were turned down—the only time during the research for this book that access was denied, except at one unpleasant roadside produce stand in northern San Diego County. Wariness about access to chicken-raising operations may be for good reason: chicken-raisers nationwide are concerned about public image, which has taken a hit when workers or imposters have smuggled cameras into poultry barns and recorded images of densely packed battery cages where egg-laying hens were confined. Although there are chicken and egg-raising operations that are run in humane and appropriate fashion, separating the humane producers from the less so was quite impossible without site visits; we withhold judgment, beyond recognizing that it is hard to believe that animal cruelty benefits anyone in the long run.

Nor has the public at large grasped the finer distinctions: a response to cruelty concerns came in the form of California's Proposition 2, appearing on the November 2008 general election ballot. Prop. 2 passed with a 63 percent "yes" vote in support of the Prevention of Farm Animal Cruelty Act, which many voters believed would make egg production in California cage-free. In the section of the proposition that affected chicken-raising operations, standards were laid down to require that living spaces for hens be large enough to allow egg-laying chickens to turn around, lie down, and fully extend their legs and/or wings. There is no provision requiring that chickens be allowed to range more widely, or go outdoors. Suffice it to say, the measure passed and will become operative, although not until January 1, 2015. Prop. 2 is directed only at egg-laying chickens, not at meat-producers, but the larger issue in question is the conditions under which chickens are raised in California, and what the public is willing to put up with.

Chickens raised for meat represent a preferred diet alternative for many U.S. residents attempting to cut back on red meat consumption. The market is ready-made and beckons to the public, as total sales that approached half a billion dollars in 2008 strongly suggest. Entities such as the California Poultry Federation (representing chicken, turkey, and squab producers) must tread a thin line of public opinion. Among domesticated livestock, chickens are the animals most commonly raised in backyards or open fields, and the chicken can be said to have a strong constituency. Hens, in particular, were common backyard pets into the 1960s, and enjoyed a resurgent constituency in the back-to-the-land movement of the 1970s. Chickens manifest a characteristic goofiness that is both curious and something that human flock-tenders find endearing, and statistics cannot capture the more interesting, if eccentric, side of a chicken's character. As a caution, it might be added that a continuing fascination with the aggressiveness of roosters, some of which are illegally bred to fight, has kept cock-fighting alive in California, as in many parts of the world, and the distinctive cock-fighting A-frames are easily recognized, some of them roadside along major approaches to the San Francisco Bay Area.

The big chicken producers in California are either in the San Joaquin Valley (Merced, Stanislaus, San Joaquin counties), or in the Southland (San Bernardino, San Diego counties), with Sonoma County the only other county having over a million laying hens in 2007. Production in the San Joaquin Valley is enormously larger than the Southland—nearly 70 percent of California's total chicken production.

California does have a second front, at a scale smaller than the agroindustrial, for commercial chicken broiler and roaster production. Following the lead of hobbyists who long insisted that free range chickens grow more consistently and are better-tasting than confined broilers and hens, attention in the early 1980s turned to the possibility of commercial-scale free range operations. Sonoma County's Petaluma Poultry earned recognition as a large-scale pioneer when in 1986 they began raising chickens with room to move, though still held in close quarters, and dubbed the resulting product "Rocky the Range Chicken." With time, various attributes were added—Petaluma built a proprietary feed mill to guarantee quality, abandoned antibiotic use in feed, and in 1989 added a second line of chickens raised organically. Unsurprisingly, exactly what legally constitutes "free range" is a complicated issue; the USDA-designated industry standard holds that any access to the outside allows a chicken to be labeled as free range. "Rocky" chickens are raised with one square foot per bird; Petaluma Poultry notes that that is about 25 percent more than in conventional poultry operations. But when push comes to shove, European Union regulations require four square meters, or 43 square feet per bird, a good bit (40-plus times) more than Petaluma Poultry's online definition of "free range." European Union labeling requirements further require the stamping of each egg or packaged chicken with a designation of how the food was raised. Providing those statistics is not intended to pick on any single producer; the reality is that poultry in the United States is raised under standards that markedly depart from those of other developed nations.

California commercial chicken (and turkey) producers may go further, allowing pastured poultry access to grass, range, and dirt, and in some cases creating movable chicken houses that can be wheeled to new sites to avoid constant use of the same site, reducing the risk of coccidiosis and other diseases (in Great Britain, movable coops are referred to as "chicken arks"). Buyer preference may weigh in yet again to promote a shift in current practices within the industry; avid consumers will pay significantly higher prices for birds raised under circumstances found more worthy of approval. These sorts of debates take place in a good many circumstances in which animals are involved in agribusiness-scale production. Because in California producers are accustomed to economies of scale—and there is a strongly positioned alternative agriculture version of the "best practices" in chicken raising—it's safe to assume the discussions are not over.

EGGS, CHICKEN *Gallus gallus*

Pl. 43

FAMILY: PHASIANIDAE.

SOURCES: EGGS OF *GALLUS GALLUS*; SOMETIMES, EGGS OF *G. GALLUS DOMESTICUS*.

RANK: U.S. #5; CALIFORNIA SHARE 6%.

SEE ALSO: CHICKENS.

Everyone knows what eggs look like, although few people have actually eased a hand underneath a warm hen to pull out a freshly laid egg. In fact, that experience is entirely possible in a farm visit, but less likely in visiting a major egg production facility, where the Leghorn chicken, with origins in Italy, has become a mainstay of the egg trade, laying around 280 eggs a year in densely packed battery cages, efficiently converting feed to hard-shelled white eggs. All hens lay eggs, of course, but some do it better than others. Additional options for an egg laying operation include barn-laying and free-range hens, or with special feed, eggs may be advertised as organic (but not necessarily, then, free range). Many variations apply.

Into the 1970s, a variety of breeds were used in California laying operations, but specialization has taken place and the regularity and cost-effectiveness of the white leghorn (known within the trade as the single-comb white leghorn) makes it a favorite of the industry. Almost never used for meat production, and somewhat skittish and unsociable, especially when compared to other chicken breeds that interact well with humans, the white leghorn is a specialist's specialist. And it rolls out the eggs.

Market eggs in California work to meet the demand of California's 37 million person population; California produces about five billion eggs per year from about 20 million laying hens, for a 2008 value of about $440 million. Not since the early 1970s, when California was producing over nine billion eggs, was the state a net exporter. Most years, California produces about 6 percent of national production, well behind internal demand (California consumes a total of 12 percent), so the state is a substantial net importer of eggs. That deficit has become still more problematic recently, since with passage of Prop. 2 in 2008, eggs from outside the state can come from caged chickens, but eggs produced in California (as of 2015) cannot be kept in small battery cages that prevent movement—a complicated state of affairs that may be dealt with by changes in law already being proposed in 2009. If there are no changes to require the same standards in- and out-of-state, some experts predict an exodus of the egg production industry from California. Estimates suggest that cage-free egg production costs about one-third more than standard egg production. In grocery stores, eggs are regularly featured as a "loss leader" to draw in customers; that increase may become a larger issue in the future. Eggs for reproduction of broilers, and the sales of eggs from the more exotic breeds such as ducks, turkeys, emus, ostriches, and other birds, is a separate matter not covered here for reasons of length.

Production of chicken eggs in California goes back to the eighteenth century. Demand for eggs was a constant through the history of California during territorial and statehood days and dates earlier still, into the Spanish–Mexican period. With the Gold Rush, eggs were regarded as the miner's perfect food: resistant to spoilage, well packaged, and—subject to some care—easy to transport, although moving several hundred eggs on a pack mule or horse is not for the timid wrangler. If chicken eggs were rare in mid-nineteenth-century California, demand from mines and cities was high. The Farallon Islands, 25 miles outside the Golden Gate and a beneficiary of upwelling from the California Current, had deserved fame as a bird rookery, which set the Islands up for devastation by the market collection of wild bird eggs, up to 500,000 a month. The eggs were destined for the mines and for fine dining in Sacramento, Stockton, and San Francisco. Egg collection actually led to a violent confrontation, including two fatalities, in the 1863 Egg War between lighthouse keepers and egg collectors. Lighthouse keepers, who wanted the eggs for themselves, won out. The egg industry today is less

Plate 43. An enormous contributor to the poultry economy, eggs are produced in conditions approaching lockdown, supposedly for sanitary reasons. Fortunately, a cottage industry in egg production is alive and well, supplying local purchasers, as these eggs of many colors attest.

colorful, perhaps, but still fills a crucial niche in protein provision, especially as the reputation of chicken eggs as a healthy food has improved in the diet literature. The widely scattered counties of Merced, Riverside, San Diego, San Bernardino, and Stanislaus are major producers within California, and laying operations tend to be in nondescript buildings with efficient air conditioning and relatively limited access to the open air.

Eggs can be purchased from many a source, including health food and specialty stores, and at times from the yards of farms or ranches where chickens are kept. Roadside signs will often advertise fresh organic farm eggs. The results, kept in refrigerators with "trust baskets" nearby for the placement of payment, can be a cornucopia of color: shades of tan, dark brown, cream-colored, even blue eggs from the Araucana, a Chilean chicken bred by the Mapuche people and widely introduced to the United States in the 1970s. Laying chickens come in two basic sizes, the standard and the bantam; the standard is larger, and lays larger eggs. Although it is difficult to offer a fixed figure, there are something on the order of 120 distinct breeds of chickens, so for the home fancier, there are choices to be had, from the Vietnamese Ac to the Yokohama of Japan.

In the Americas, the genetics and background of chickens remain controversial. Some experts believe chickens arrived in the Americas before Europeans, and were introduced by Polynesian or Chinese travelers. That is one for the experts, although recent dates tend to support a fourteenth-century arrival of chickens in the New World—well before Columbus. The eighth book by Roman author Columella offers an extensive discussion of chicken breeds, and of appropriate practices for raising chickens for egg production. Although the detailed suggestions might certainly change, in light of Columella's first-century AD date, the advice indicates the depth of time in recorded relationships of long-ago humans with the rooster and hen.

GOATS — *Capra aegagrus hircus*

Pl. 25

FAMILY: BOVIDAE, SUBFAMILY CAPRINAE.

NOTES: MEAT, HAIR, AND MILK GOATS.

RANK: U.S. #2 (INVENTORY); CALIFORNIA SHARE 5% OF ALL GOATS; 12% MILK GOATS; 2% HAIR GOATS; 3% MEAT GOATS.

Goats are thought to be the second-oldest domesticated animals (after dogs, which many experts believe actually domesticated themselves, volunteering to go into human companionship). Traces of domesticated goat bones in the Zagros Mountains of Iran date from 10,000 years ago. In Norse mythology, the god of thunder—Thor—drove a chariot pulled by goats. Indeed, there are milk goats and hair goats and meat goats, and goats may be hooked to a cart and used for driving, or loaded with pack saddles to haul goods into the backcountry. Goats integrate with remarkable success into human society and defend themselves better against dogs or other predators than sheep do. Little wonder that goats display a rare understanding of human sentiment and foibles, and can be fascinating companions.

Goats are fastidious browsers, preferring to eat shrubbery or weeds, rather than grass, and in grazing style they are more like deer than they are like sheep (which closely crop what they eat). Goats will sample food they are unsure of; using nearly prehensile upper lips, they nibble unfamiliar foods, retaining an unerring memory for what they have eaten before. Goats will not again eat something that unsettled their digestion, and seem able to pass on their knowledge. Like cows and sheep, goats are ruminants, which means they are quite effective as free-range animals. They do well on rough grass and forage that would be utterly unpalatable to horses or pigs, which are not ruminants.

In California and other western states, a blossoming industry leases goats for use in vegetation management and type conversion—ridding entire areas of noxious plants. Goats can eat their way through steep hillsides, blackberry bushes, and poison oak, removing unwanted brush without chemicals. Goats are a common sight in urban locales, grazing several acres enclosed by a temporary electric fence, generally near roads, with a tender living next to the herd in a modest trailer. Used for weed control and to ease fire hazard, goats are often considered a part of the neighborhood scenery during their stay.

Goats can grow to a substantial stature: the sizable Boer goat, a meat breed, can grow to over 200 pounds—a large buck to 300 pounds. Uses vary. In California, not many goats are raised for their hair, either angora or mohair, with about 4,000 of each in official inventories; the long-haired breeds remain popular in sections of Texas and Oklahoma. The milking breeds range from small to extremely tall and leggy, and although they are unlikely to challenge the meat breeds in overall stature, they offer a substantial presence, complete with horns, irreverence, inquisitiveness, and attitude.

It is widely believed that goats were dropped off on Southern California's Channel Islands by Spanish vessels even before incursions into Alta California culminated in the Spanish establishment of Mission San Diego de Alcalá in 1769. Goats would be deposited on the islands so they could reproduce and offer a food source to returning navigators. Archeological evidence of this is uncertain, although goats were known early inhabitants of Santa Cruz Island. But there is no doubt that goats arrived with the earliest overland Spanish settlements, and ever since goats have been an on-again, off-again feature of California life.

Goats come in three variant forms, though only two are popular in California. Hair goats, such as the Angora or Cashmere, are bred for their luxuriant coats: there are variations on management and harvest of the hair breeds, though as a rule the goats are clipped yearly for their downy undercoats and angora hair. Meat goats—heavy set, and raised for flesh—are increasingly popular, especially among ethnic communities (Sikh, the Islamic *halal*, and Hispanic) that prize goat meat (sometimes known as *chevon*) or kid—the meat of young goats—as food for important events. *Cabrito*, or kid meat, has had a meteoric rise in popularity in California, and the 2009 estimate has 95,000 meat goats in California—admittedly, just 10 percent of the total in Texas. The other prominent use of goats in California is for dairy, either as fresh liquid milk or for the making of cheese or yogurt. In fact, in 2008 there were more milk goats in California—36,000—than in any other state, although Wisconsin was mounting a challenge, presumably seeking restored supremacy in all things dairy. A good dairy doe will give six to eight pounds of milk per day through much of a 305-day lactation period; sheep are less constant. Goat milk may be consumed as fresh milk, made into cheese on its own, or blended with other milks.

The value of goats in California agriculture is varied. Meat goats are increasingly popular and are raised in the San Joaquin Valley (Fresno, Kern, Kings, San Joaquin, Stanislaus, Tulare). But they are common throughout the Sacramento Valley and the Delta, where they find a ready market among Hispanic and ethnic Indian populations that especially prize kid and intact (uncastrated) older male goats for feasts and festivals. The count of California meat goats is hardly formidable—88,000 in 2007. But that represents nearly a 50 percent increase over the 2002 agricultural census and suggests a steady shift of interest and demand. Milk goats are another matter. Demand has held more or less steady, with just a fractional increase in milk goat numbers over the last decade. Merced, Fresno, and San Joaquin counties dominate the milk goat numbers, but Humboldt and Sonoma are players also. As with meat goats, there is an upswing of interest in goats in the western counties of the Sierra Nevada, likely because goats are efficient removers of unwanted vegetation, and because there is concern in the foothills about clearing vegetation to provide a defensible area around living space. Goats are a prime resource in brush eradication.

The total value of goats to the California economy is difficult to estimate. In the annual agricultural resource directory summary, goats provide too small a total to estimate. Separating the component of "goat cheese" from all of California's cheese production is impossible. But according to the quite incomplete ag commissioners' data in 2008, goats and kids were worth $2 million, and milk goats another $5 million. But that is for the animals and not, in the case of milk goats, for the milk that they produced. In short, a statistical anomaly . . . but goats are a most interesting element of the California agricultural landscape, and one with numbers on the upswing.

HOGS — *Sus domestica*

FAMILY: SUIDAE.

NOTES: *SUS DOMESTICA* INCLUDES DIVERSE RACES OF SWINE, INCLUDING THE WILD BOAR.

NAMES: PIGS, SWINE.

RANK: U.S. #30; CALIFORNIA SHARE LESS THAN 1%.

Although pigs (hogs) are consumed in quantity in California, they are far down on the ranked list of California agricultural products, standing in at sixty-third in farm gate value. Pigs are a notch or three down from sheep and lambs (sixtieth), and a few spots above goats. As omnivores, pigs live a life remarkably parallel to the human presence in California; they eat broadly, they adapt to changing diet, and they are astonishingly bright animals that are resourceful, capable, and social. It is interesting to ask why California does not have a larger quotient of pigs. The answer is that given an emphasis in the United States on feeding pigs a corn and soybean diet, pigs are largely excluded from the California landscape—not because soy and feed corn cannot be grown, but because other uses yield higher profits than feeding prime feed to swine. Other states can do so more cheaply, and with less controversy.

Pigs are concentrated at the margin where the Sacramento and San Joaquin valleys meet, not in the Delta, but at its edges: Tulare, Fresno, Stanislaus, and Sacramento counties. The only interloper in the top-five list is San Bernardino, with a small but intense concentration of pigs worth about $1 million. Hogs are slaughtered in California at an average live weight of 245 pounds, under the national average. Domesticated pigs are raised primarily for meat (pork), but also for leather, and in some cases, the bristles of their hair go to manufacture brushes. Pigs may also be kept as pets, although full-sized pigs would constitute a very substantial addition to the average household—the small-statured Asian pot-bellied pig is a more usual companion animal.

When returned to the wild, pigs are able to survive as feral animals, and wild swine or boars pose a formidable destructive force and are regarded as a difficult, even formidable, pest in parts of the state, especially through the Coast Ranges. Rooting for grubs, bulbs, roots, or insects earns wild pigs a reputation as biological rototillers—a phrase not considered a compliment. In circumstances in which wild boars or feral pigs are common, their main fan base is hunters, who welcome the opportunity to stalk a wily and difficult prey, while offering something that might be considered a public service. Eliminating pigs—sometimes known by their Spanish name, *jabali*—is a marked difficulty for the conscientious landowner who wishes to clear a predatory and disruptive species from land. There can be crossover effects: wild boars are credited with introducing *E. coli* 0157:H7 to an agricultural field in San Benito County, in the process initiating the spinach quarantine of 2007 that cost producers tens, if not hundreds, of millions of dollars.

In a field guide setting, it is ritual to mention that pigs are raised in confinement, in large barns or bays. The interesting innovation in California is in the feeds that they are given, which can be more diverse than in their Midwestern and Great Plains settings. In truth, the diet of pigs is as unnerving and eclectic as that of humans. If given free rein, pigs will take advantage of considerable variability in what is put in front of them, including agricultural aftermath or animal by-products. This counts as healthy fare for pigs. They are animals of singular charm, brought to California by the padres and early migrants, and they became nativized readily, and with vigor. In a natural, free-range setting, pigs are omnivorous, but their preferred food is acorns and mast (the duff and decomposing leaves and litter found in deciduous woodlands). There are accounts from early residents of pig drovers herding a long line of swine from the upper Sacramento Valley to livestock auction yards north of Sacramento. That would have been a sight to see: nearly something out of Larry McMurtry's *Lonesome Dove.*

California hog farmers marketed 300,000 animals in 2008 and earned $33 million; in an indication of the survival of a rural economy, another 7,000 pigs were consumed on-farm; there remains a domestic economy on the farm for pigs and pork products, and a willingness to do slaughters on-site. The cash receipts of California for swine production were one-tenth those of South Dakota or Mississippi, and California's total hog production was less than 1 percent of the income of Iowa farmers for factory swine. In fact, California in 2008 accounted for $46 million out of a total $16 billion in U.S. pig receipts—three-tenths of 1 percent. Clearly, this is a business that California is willing to see pass by, at least for now.

HONEY

Pls. 39, 98

FAMILY: APIDAE; INCLUDES *APIS MELLIFERA LINGUISTICA; A. M. CARNICA; A. M. CAUCASICA; A. M. REMIPES; A. M. MELLIFERA;* ALSO, MANY HYBRIDS.

SOURCES: *APIS MELLIFERA* MANY SUBSPECIES, BUT ALSO ADDITIONAL SPECIES, EUROPEAN OR WESTERN HONEY BEE.

RANK: U.S. #2; CALIFORNIA SHARE 13%.

SEE ALSO: BEES.

The flavor depth and variety of honey is amply familiar to anyone who has spent time in a grocery store or for those who spot beehives in a field setting. The link, of course, is to bees (*see* Bees), and the crucial role that honey bees (the European or western bee) play in the pollination of California crops. It is pollination that is most profitable for commercial-scale beekeepers, but honey provides its own benefits, and was a $50 million contributor to the California agricultural economy (2008).

Honey is a product of the nectar and pollen collection of bees, which is deposited in hives. Bees will use honey as a food source in cold weather, but it is palatable to all sorts of creatures, which savor the sweetness. Bees partially digest the flower nectar that they collect, regurgitate it until it reaches the right quality, and then store the honey in cells of honeycomb. Beeswax can, in fact, be another product of bees, with incredibly diverse uses ranging from mustache wax and a safe coating for cheese to lubricants or jewelry-making. Many beekeepers will carefully centrifuge honeycomb so that the honey is removed, but the comb remains. Since producing the honeycomb is a major energetic expense for the worker (female) bees, leaving it in the hive allows bees to devote more energy to nectar collection, and less to hive-building.

Honey is one of the world's oldest agricultural products, and old traditions are still sometimes followed, whether the hives are wild ones harvested by beekeepers who track bees to their source, or from two half-circles of bark from the cork oak that provide a hive location for bees. In California, beekeepers use hives made up of a series of supers, or sections, that are added one atop the other; each super contains eight to 10 frames. Beehive construction makes it possible for the keeper to harvest frames of honeycomb, which when full are capped by bees with beeswax. The caps and extra combs are used for beeswax products. The beekeeper can take full supers and extract the honey. When the honey flow (nectar supply) is strong, a beekeeper will add extra supers to hold the added honey.

There is a significant split in California honey production between commercial producers and large-scale hobbyists, who may keep a several dozen hives and produce a few hundred jars of honey. Because of honey's notable qualities—it is antibacterial and rarely spoils—purchasers can feel reasonably confident about the quality of the flavor or type of honey they might buy. But the real drama in California honey involves the variety of places and the variety of honeys produced throughout the state.

Commercial production of honey is reported by 22 counties that generate amounts over 10,000 pounds. Unsurprisingly, the largest producers include sizable acreage of crops that are attractive to bees: Tulare, Kern, San Diego, Merced, Riverside, and Fresno; each with more than one million pounds of honey production, and better than $1 million worth of honey sold (2008). The varieties are a waltz through the vegetation, natural and cultivated, of California. More than two dozen honey types are commonly marketed: alfalfa, buckwheat, eucalyptus, lavender, six distinct sage honeys, orange blossom, pumpkin, avocado, star- and mint-thistle, mesquite, cotton, safflower, chestnut, tupelo, lemon, and vetch—to name about half.

Honey production in 2008 was up 35 percent from 2007, indicating the honey industry's continuing recovery from Colony Collapse Disorder (*see* Bees), and the number of colonies was well up. In production, California lags behind North Dakota, but state honey yield will pick up as the effects of CCD and the varroa mite presumably fade. The effects of CCD are evident in the record: nearly two-thirds of one million bee colonies in California produced 15 million pounds of honey in 2007; in North Dakota, 390,000 colonies (just 60 percent of California's total) produced 30 million pounds. North Dakota is a prodigious producer—but what really shows is a California industry still in toddler steps toward recovery.

OYSTERS — ***Crassostrea*, var. species, *Ostrea*, var. species, and *Saccostrea*, var. species.**

Pl. 30

FAMILY: OSTREIDAE.

SPECIES: INCLUDES *CRASSOSTREA* VAR. SPECIES, INCLUDING *C. GIGAS* (PACIFIC OR JAPANESE OYSTER), *C. SIKAMEA* (PACIFIC KUMAMOTO OYSTER), *C. VIRGINICA* (EASTERN OR ATLANTIC OYSTER); *OSTREA CONCHAPHILA*, (ALSO KNOWN AS *O. LURIDA*, OLYMPIA OYSTER); OR *SACCOSTREA* VAR. SPECIES, INCLUDING *S. GIGAS* (JAPANESE OYSTER).

UNRANKED—BUT CALIFORNIA IS A MODERATE PRODUCER.

To think of oysters as an agricultural product may seem a defiance of reason, but oysters are counted in county production estimates and fit into federal statistics on aquaculture. Tomales Bay and Drakes Estero, in Marin County, and the area around Humboldt Bay (Humboldt County) are the two main California producing regions, with small yields from other districts. In the nineteenth century, oysters were collected and raised in San Francisco Bay, near river mouths and sloughs throughout Southern California, and all along the North Coast, where visiting an oyster farm is today a favored weekend activity. Oyster farms, unlike many ag sites in California agriculture, tend to welcome the public. Signs along the road note the business name, and if an oyster farm is open to visits, will advertise that fact. Equipped with picnic tables and a motley selection of iron barbeque grills, the sites favored by locals provide an oyster shucker's knife and thick rubber gloves for those dexterous souls disposed to shuck their own.

The bon mot is true: oysters are the only animal food that is eaten alive, and the very thought can cause the uninitiated to succumb to queasiness. But that said, raw oysters have been a culinary mainstay in the United States and abroad through much of history. Although it is true that there are dozens of time-tested and famous recipes for cooking oysters, purists take theirs fresh and raw. Oysters provide high-quality protein, zinc, iron, and omega-3 fatty acids, but oysters are small; it takes a lot of oysters to complete a meal, if that's all someone is eating. Native Americans in California consumed so many oysters that entire seashores are middens, made up of the shells of oysters harvested as meals through prehistory, and more than a few modern cities established their foundations on shellmounds. Oysters are regarded as one of the very few sustainably raised seafoods, a "Best Choice" from Seafood Watch.

A robust bivalve mollusk, oysters in past centuries constituted an essential human food, especially for working-class populations, and William Rorabaugh reports that a working man's breakfast on the way to work in the 1830s could typically include two dozen oysters and eight ounces or more of whiskey. In the context of such a high-octane breakfast, massive deaths by mechanical equipment were inevitable. All this makes today's return of oysters as a prized and pricey food somewhat ironic. But, tasked by overcollection and by declining environmental quality associated with hydraulic mining, logging and, more recently, suburban development, California coast and bay oyster populations have declined. Oysters remain a small, but colorful and significant, element in California's agricultural production.

Oysters are worth about $12 million per year in California, with better than three-quarters of that total issuing from Humboldt County. There are small oystering operations in Morro Bay, and the rest of the state's commercial oysters come from fewer than a half-dozen producers in Marin County. Oyster farms totaled 15 statewide in 2005. It is remarkably difficult to offer a firm figure for oyster production in California: reporting is variously by number of oysters raised, by the pound (which includes the shells), by dollar value, or by meat taken out of the oyster—but in Marin County, for example, only Drakes Bay Oyster Company actually retails oysters in a jar. Most operators sell oysters in the shell only. Estimates for the "Census of Aquaculture" suggest total oyster revenue for 2005 in California was $12 million. Although that figure has certainly been eclipsed, the dollar amount of sales was also 10 times the total of 1998, and suggests something about the industry's upward trajectory.

Viewing oysters as a crop is less difficult than it might seem. At Marin County's Drakes Estero, the Drakes Bay Oyster Company operates in an estuary of the Point Reyes National Seashore, and produces some 450,000 pounds of shucked oysters a year, about 85 percent of the Marin County total. Just to the east of Drakes Estero sits Tomales Bay, produced by down-dropping along the San Andreas Fault, where several additional oyster companies operate: Tomales Bay Oyster Company, Marin Oyster Company, and the ever-popular Hog Island Oyster Company, which pulls up three million oysters a year in mesh sacks, where they are checked, sorted, and, if ready, harvested (otherwise, they are returned to the water). The ambiance at an oyster farm open to the public parallels that of a winery: a good bit of admiration and jubilation, and serious attention to oysters, as the subject at hand.

Known as filter-feeders, oysters live on phytoplankton and nutrient-rich detritus by pumping water over their gills, which filter material and pass it to the oyster's mouth. Oyster reproduction can be complicated, but in California commercial oysters are almost always spawned elsewhere and brought as cultured spat to grow to size in Pacific waters. At each site, the oyster producing experience is slightly different, but California oysters (Humboldt County operations, included) are cultivated and carefully managed; the harvest has not been of naturally occurring oysters, which all but disappeared in the early twentieth century. Practices embraced include bottom culture (Humboldt Bay), rack and bag culture systems, stake culture, bag and longline, and Stanway cylinder culture. Which is used where depends on preference, currents, water depth, and the oysters involved. Oysters take one to two-and-a-half years to grow to harvest size and, even on-site, bring in a good bit more than a dollar per oyster.

Oysters are harvested in 18 of the 21 contiguous coastal states of the United States, and production varies depending on each state's fishery. Louisiana was a strong up-and-coming producer, for example, but Hurricane Katrina thwacked the industry hard; Maryland and Virginia, historically, were large producers, but problems with Chesapeake Bay health have downsized the oyster catch. The New York City vicinity was so great a consumer and producer of oysters that when Mark Kurlansky wrote a history of the city's appetite for oysters, he titled it *The Big Oyster*. Washington and Oregon are active oyster cultivators but use a native oyster (*Ostrea conchaphila*) as their standard. Native oysters also grow much more slowly.

Through two centuries of oyster-eating, Californians focused on native oysters and on introduced Atlantic oysters. About 50 years ago, the Pacific or Japanese oyster started its rise in popularity. As a hybrid, known as a triploid, the Pacific oyster is less susceptible to the taste changes that come with spawning, and so can be harvested as a year-round crop. Today, the rather large Pacific oyster remains dominant, but in Marin, the Kumamoto oyster and some Atlantic oysters are raised, too, and the small size but intense flavor of the Kumamoto makes it a favorite. Efforts to restore an oyster fishery within the San Francisco and San Pablo bays are underway; sedimentation from the Delta is much reduced by upstream inland dams, but a major concern is Delta health and the quality of water and sediment. Since oysters feed off the tidal flux of water, polluted water or contaminants such as heavy metals can be concentrated and kill off a susceptible oyster population. Nonetheless, historical documentation of the oyster industry makes for interesting reading. And demand is strong for oysters in the restaurants of the San Francisco Bay area and elsewhere within the state, where a dozen shucked oysters with a spritely mignonette sauce is regarded as an encounter with paradise. Since they are raw, the oysters have hardly changed a bit since they left the sea.

SHEEP AND LAMBS *Ovis aries*

Pls. 12, 44

FAMILY: BOVIDAE, SUBFAMILY CAPRINAE.

SPECIES: DOMESTICATED; INCLUDES WOOL.

RANK: U.S. #2; CALIFORNIA SHARE 10%.

SEE ALSO: CHEESE.

Sheep and lambs are remarkably cheerful animals when spotted in a pasture setting, and they come in a charming variety of colors, with white faces and black—and simply nothing is cuter than a pair of frolicking lambs. Strongly gregarious, sheep are bound together by mutual interest, and often are herded with sheepdogs, especially border collies, which can display a preternatural understanding of sheep psychology and instinctual behavior. Sheep prefer to graze, rather than browsing as goats and deer are inclined to do. That is sometimes used, however, as a management tool to remove noxious vegetation, though that role is more commonly relegated to goats, which will eat rougher fare.

Most sheep breeds common to California lack horns, which limits their ability to defend themselves. Sheep are multipurpose animals. The total world count of breeds varies, depending on who is doing the counting, between 200 and 1,000; distinct traits are emphasized in selecting sheep breeds, and in Europe, sheep are often designated as "upland" or "lowland" breeds. But alas, in California predator concerns and a general lack of profitability have through time reduced the sheep population. An exception to this is in raising sheep for milk; the production of sheep cheese is a business that has done well in the last decade, with varieties such as feta and blue cheeses increasingly popular (*see also* Cheese). The versatility of products from sheep—meat (mutton or lamb), wool, and milk—have always made sheep favored animals in a farm or ranch setting; they retain strong support in some quarters. And sheep can contribute to a bucolic backyard setting: in 2008, income from sheep and lambs was reported in 40 of California's 58 counties, though it was notably absent from all of the more urbanized counties. There are some notable patterns—the North Bay counties (Marin, Sonoma) are significant producers, owing much to the strong pro-farming sentiment and the inspirational presence of Bellwether Farms, an innovator in sheep cheesemaking.

No less interesting are the statistical outliers: with over a million dollars in revenue from sheep, Mono County is credited as a major producer. That number owes much to the continued pattern of transhumance, with sheep herd owners (who are often of Basque origin) trucking herds to the eastern Sierra, where the large bands (typically totaling over 1,000 animals) are herded by a shepherd (who these days is Chilean or Peruvian). Over a hundred years ago, that south-to-north movement was a common pattern in seasonal grazing, with bands of sheep from the southern San Joaquin Valley, or the Mojave Desert, moved up the east side of the Sierra, crossing the range by any of a variety of mountain passes between Yosemite and Lake Tahoe, and returning southward along the western Sierra Nevada foothills until the point of origin was again reached. Strikingly, in those early twentieth-century days, it was the shepherds who were more likely to be Basque. But with time and thrift, the group that once characteristically constituted the day-to-day workers are now the ranch and livestock owners.

A second outlier is Sutter County, with $1.7 million in sheep-based revenue (2008). While sheep graze into the Sutter Buttes (a notable volcanic formation that erupts from the Sacramento Valley), the concentration of sheep in Sutter County owes much to the equally pronounced localization of northern Indian stone fruit farmers in the county. Generally Sikh, these farmers are a marked presence in peach and French plum cultivation and historically have been especially notable consumers of lamb and mutton. Sometimes, an agricultural crop can say much about ethnicity and race and demand in a specific area.

As ruminants, sheep are efficient eaters and lack traits (obsessive climbing, rampant curiosity) that can make goats less than perfect companions in a suburban or semirural setting. Although wild predators are of concern to the sheep rancher, often the greatest risk and damage comes from pet dogs whose owners unthinkingly allow them to escape or run free. Dogs in packs can be frenzied and efficient killers, and sheep, which react poorly to stress, will flock together when pressed but then quickly panic and bolt. Dealing with predators is a difficult issue—the cause of death for fully a third of sheep deaths in 2004 is listed as "predation." Government regulations prohibit a number of experimental antipredator measures attempted in the 1960s and 1970s, so ranchers have taken to inserting burros or llamas with sheep, since they share the sheep's aversion to dogs, wolves, and coyotes and defend their "flock" if predators threaten. Other ranchers use guard dogs as defenders, including the Great Pyrenees or the mastiff. Fencing, penning sheep at night, and indoor lambing are used, too. No solution is perfect, for predators adapt; the consequence is that sheep numbers decline. And few things in agriculture are so dramatic as the huge decline in sheep in California.

Plate 44. Unshorn sheep in March seem somewhat startling, but this pair near San Luis Obispo suggest the potential value of wool—too long languishing at depressed prices—in a market that has not favored raising sheep and lamb for several decades.

Sheep and lambs are in a difficult state, nationwide, and California shares in that dilemma. Farm-gate revenues for sheep and lambs in California came in at $57 million, sixtieth among agricultural products—lower than dates, and just above vegetable seeds. Although picturesque to a fault, sheep and their offspring are not doing particularly well in the nation's agricultural economy. A graph of the countrywide count of breeding ewes and lambs manifests a marked dropoff. The wool clip is way down—other countries can produce wool far more cheaply—and total wool production has essentially been on a nationwide decline since 1931. Although sheep are still clipped for comfort and sanitary reasons, many sheep ranchers simply leave the wool where it falls, or give it to the shearer. California produced some 10 percent of the wool in the United States, about half its production of 15 years ago.

California shows an interesting regionalization in sheep and lamb production. The inventory is up 6 percent from 2008 to 2009, with 660,000 sheep and lambs. The big-revenue counties for sales of lambs and mature sheep for mutton are Fresno, Imperial, Kern, Merced, and Solano—in the Delta. What shows up there are sales of slaughter animals, in effect. But undifferentiated "revenue" appears in the consolidated report of the county agricultural commissioners. There, Solano reappears, but so do Sonoma, Marin, and even Riverside—and those figures represent revenue from sheep dairying, perhaps from some wool, and income from the sales of liquid milk. The production and sales of sheep cheese is yet another set of numbers that is difficult to extract, since it appears in another set of ledgers and is barely reported. Suffice it to say that a check on artisan cheese production in California shows at least a dozen sheep cheese operations, with more adding their names to the roster month-to-month.

TURKEYS *Meleagris gallopavo*

FAMILY: MELEAGRIDINAE.

RANK: U.S. #8; CALIFORNIA SHARE 6%.

The turkey suffers from an unfortunate reputation. The wild turkey, a North American native, is the progenitor of the domesticated turkey, and although sighting a rafter (the proper term, not a flock or a gobble) of wild turkeys is cause for delight, the same can actually be said of the domesticated version: the turkey is a great bird, these days generally white, with a deliberation to its movements reminiscent of an unbalanced small-town mayor. Turkeys are fed up to slaughter weight in long thin barns with great fans on the ends that provide fresh air and expel contaminants. Unlike chickens, which often never leave their cages, turkeys may make sojourns outside, and a yard of white turkeys strutting for one another is something to behold.

Domesticated turkeys need not be solely commercial ventures; they may be raised in backyard settings, or in small confined A-frame structures that look like the pens of fighting cocks. The largest concentration of turkeys in California is in the San Joaquin Valley, in Fresno, Merced, and Stanislaus counties, each of which boasted over $60 million in turkey sales in 2008. Kings,

Tulare, and San Joaquin counties are each additional substantial contributors for an industry that totals more than $250 million in farm-gate value in 2008.

That total has turkeys as the thirty-second commodity in agricultural production value in California. Turkey production is conspicuously balanced, more or less equal, through the top 10 states that produce turkeys, and is certainly a far cry from the sort of dense concentration of hog production within the borders of Iowa. In 2008, California produced 16 million turkeys, barely half the total from the early 1990s, although improvements in diet and technique mean that even though the number of birds is down, the total poundage of meat produced has decreased only a little. The concentration of large-scale turkey growing is pronounced—there are just three major firms in-state—and production is decidedly integrated, with breeding, hatching, milling of feed, processing, and growing done by a network of company-owned subsidiaries. Not much is bought from outside the localized corporate network, in other words.

The domesticated turkey is unable to fly far, because of its weight, and a tom (male) turkey can live for up to 10 years if spared the hatchet. Hen turkeys are significant egg producers; turkey eggs were a favored food (like the birds themselves) in Aztec Mexico. Discovered by the Spanish in Mexico, turkeys were brought to Spain and elsewhere in Europe, and distinct breeds developed in time. There are accounts, in Midwestern narratives, of turkey drives being mounted to move the birds from place to place—another image worth conjuring up. Turkeys are fast growers, weighing a quarter-pound at hatching and growing to as much as 37 pounds at 22 weeks. They convert feed to flesh well and muscle up to the animals Americans are familiar with at Thanksgiving.

There is a pronounced "alternative" movement emphasizing the advantages of raising heirloom or heritage turkeys. That effort features free-range birds from breeds such as the Bourbon Red, the Jersey Buff, the Blue Slate, the American Bronze, and the Narragansett that are at a considerable visual departure from the Broad-Breasted White that dominates nearly all industrial turkey production. Heritage turkeys are advertised as being raised in conditions more likely to meet with public approval. This can double or triple the per-pound cost of a turkey, but in an urban market, the growers are finding an increasing number of markets.

Tree Crops

A giant in bearing acreage, year in and year out, tree crops are a 500-pound gorilla of California agriculture, able to throw their weight wherever they want. Not only do tree crops cover a sizable acreage in the state, they produce an enormous amount of agricultural wealth. The roster is impressive: within the top 30 agricultural products, judged by farm-gate revenue, are 11 tree crops—almonds, oranges, walnuts, pistachios, peaches, lemons, cherries, nectarines, avocados, and plums and prunes (now officially dried plums), with a 2007 value surmounting $7.4 billion (USDA). The total of all fruit

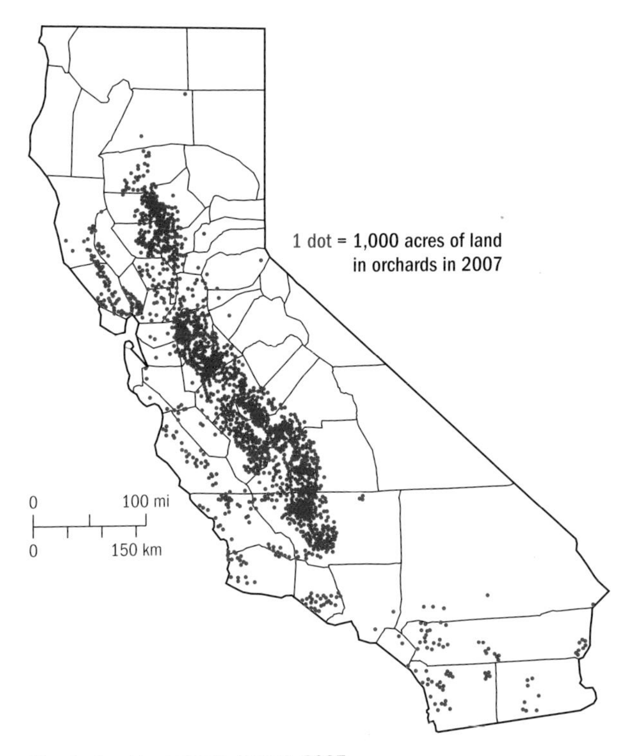

Map 5. Land in orchards (acres), 2007.

and nut crops is nearly $11 billion, including grapes and kiwifruit. And it is worth noting that nursery products (third on the list) are valued at an added $3.2 billion, although only part of that is for landscaping in trees and palms. Money isn't everything, however. Tree crops share another virtue, should that be a consideration: they are a form of what the economic geographer J. Russell Smith, following Greek and Roman precedent, called "a permanent agriculture," which means they are enduringly rooted in the soil.

Although the establishment of an orchard or a grove can be a messy process, once planted, trees require relatively little manipulation, and certainly nothing on the scale of the ripping, plowing, discing, harrowing, and land-leveling so often involved with field crop agriculture. Not all is heroic. An almond orchard planted to one variety may be ripped out because that cultivar is no longer popular; all the wood—all trunks, branches, and foliage—are soon uprooted and mounded to dry in the open field. But, like as not, another variety will arrive in its place. Permanence is relative; consider the European world, where olive trees dating 2,000 years old are still in place and producing. Olives are in California, too—another successful introduction. No California olive is more than a couple of hundred years old, or is ever likely to be, in view of the swiftness with which new varieties and cultivation techniques arrive. The super-high density plantings will have a useful life of perhaps 30 years; they take a beating in mechanical harvest.

The soundness of not disturbing soil too much has a certain peace and respectfulness to it. A benefit for owners is that once it is in place, a successful orchard keeps producing, year after year, without requiring an enormous amount of work. Southern California was in part built on that imagery and mythology, with the orange grove as the symbol of "a better Mediterranean." That magnetism and attraction has not entirely faded.

ALMONDS *Prunus dulcis*

Pls. 5, 45, Fig. 1, Map 6

FAMILY: ROSACEAE.

RANK: U.S. #1; CALIFORNIA SHARE 99%.

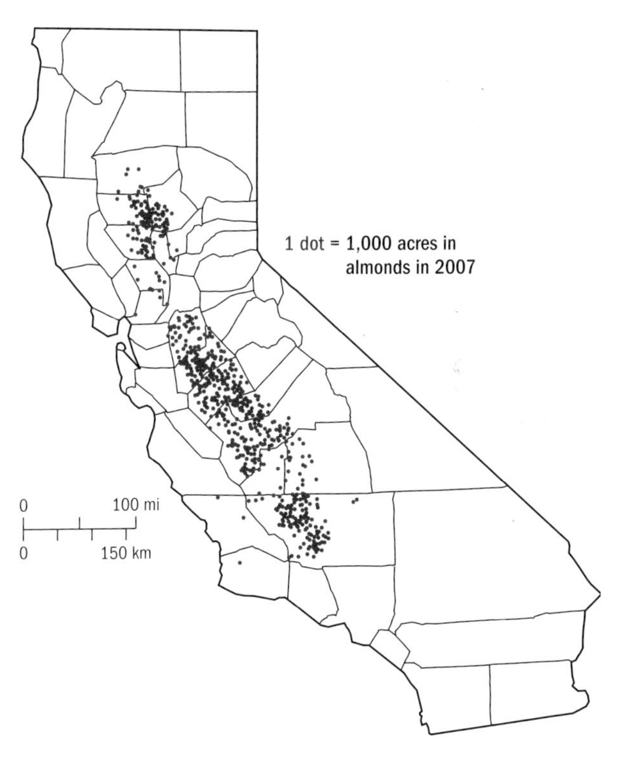

Map 6. Almond acreage, 2007.

The production of almonds, *Prunus dulcis*, is a $2.4 billion industry (2008) and the behemoth of California tree-nut crops, with two-thirds of a million bearing acres (680,000) planted in almonds, and competition only from walnuts (218,000 acres) and pistachios (114,000 acres and rising). From Red Bluff in the upper Sacramento Valley extending south through the San Joaquin Valley as far as Kern County, almonds are a frequent and robust new arrival (since the 1960s) on the agricultural landscape. The rosary-like chain of missions near the California coast offered a less than ideal climate for almonds, which prefer a warmer growing ambiance with reduced moisture. Raw heat makes the Sacramento and San Joaquin valleys a superlative almond-growing venue, planted now to huge acreages. Because tree crops are generally less destructive of soils and can be efficient water users, there are ecological advantages to almonds over field crops. A member of the Rosaceae family, almonds, among tree crops, are most closely related to peaches. Technically, almonds are stone fruits, though they are eaten as nuts, and almost 100 percent of almond production in the United States issues from California, with Kern, Fresno, Stanislaus, Merced, and Madera Counties atop the county-by-county leader board.

Dominance does not stop there; beyond topping U.S. production, California markets provide about 90 percent of the world's commercial production of almonds, and an inspired marketing campaign and deep penetration into foreign markets makes this the most successful of California's agricultural exports. Nonpareils are the largest almond cultivar in California, accounting for two of every five acres, but the Carmel, Mission, Merced,

Plate 45. No part of the almond production cycle is quite so startling as the emergence of the drupes (pods that hold the almond shells and, inside that, meats), which eventually split to reveal drying nuts emerging inside.

Price, Peerless, Thompson, Butte, Ne-plus, Fritz, Padre, and Monterey varieties are significant, too. The differences are relatively subtle, and largely for the expert agronomist to sort out. Although the rise in pistachio acreage in California has been rapid, nothing compares to the planting boom in almonds. A half-billion pounds were produced in 1996; a billion pounds produced in 2002; 2008 production is estimated at 1.5 billion pounds, and the number of trees planted has risen commensurately. A price dive in the 1980s marred the rise, and marginal acreage and less favored varieties in the upper Sacramento Valley were torn out; since that half-decade hiccup, another 200,000 acres of almonds have been added, largely in the San Joaquin Valley.

Few aspects of California agriculture are quite so dramatic as almonds at the end of winter or, even later, in the spring–summer harvest season. The earliest-blooming of fruit and nut trees, almonds in February generate ample white blossoms, and although in the United States the trees are valued for their nuts, in Japan the budded branch of the almond (like a branch of cherry or peach) is a prized addition to ornamental arrangements. By mid-March, the trees are leafed out, and the first nuts are visible. In a certain irony, the favored rootstock for California almonds, which are generally grafted onto seedling rootstocks, is from peach seedlings. Hulls split in July, and at that point, the tree leaves are chaotic and almonds are not a particularly sightly item, especially in groves that can go on for miles, as along the I-5 corridor. As an almond drupe matures on the tree, the hull dries and splits before harvest, signaling near-maturity.

Production starts at 3 years and reaches a maximum in 6 to 10 years, but the tree continues to produce at heavy levels for decades. Once the kernels start to dry on the tree, the almonds are ready to be gathered. Harvest of almonds is now heavily mechanized. Mechanical tree shakers are used on all but the youngest trees, and padded jaws on a reciprocating arm shake nuts from the branches. Younger trees are harvested by hand-knocking. Nuts can either be left on the ground for a few weeks until they dry or harvested into a span of tarpaulins that is extended below the shaking arms, with the nuts collected at harvest and moved to dryers where the curing and cracking process is completed. Nuts may also be bleached for color improvement, and then treated by salting, roasting, or flavoring.

The process of almond harvesting is dusty, dirty, and chaotic. In the field, mechanical shakers are clamped onto the trunks of trees and work them hard. This, on trees that have not seen rain for several months, produces an unsightly dust cloud. If almonds are allowed to fall to the ground (which is sometimes, although ever less commonly, treated with herbicides to denude vegetation from the tree understory) to dry, they are reclaimed by a mechanical vacuum machine that generates a trail of dust like a vacuum cleaner with its hoses hooked in backward to drive dirt out instead of sucking it in. With the high value of the California almond crop, harvest machinery and techniques are fast-evolving.

The tree is small to medium in size, with a spreading, open, and somewhat chaotic canopy, usually pruned (sometimes mechanically) to a height of 10 to 15 feet in commercial orchards. Almond flowers are similar to those of peaches and other *Prunus* flowers, with five petals. A distinctive feature of almonds is pollination, or the lack thereof: they are self-incompatible, and so require cross-pollination. Pollinators, usually honey bees, are essential, especially considering the cold, wet weather in which almonds generally set flower. This has produced a particular dilemma for California growers since around 2000, when an only partly explained die-off of entire bee colonies threatened the setting of entire almond crops. The varroa mite is thought to be the guilty party, but in 2005, bee colony numbers were at 25 percent or less of the historical total; the rule of thumb is that two colonies are needed for each acre of almonds. This shortage in pollinators led to year upon year of bee hive transport from one part of the United States to another to sate demand for almond pollination. Bee colonies appear, in 2009, to have rebounded, and current thinking is that the colonies at risk were victims of diet—literally, poor nutrition brought on by insufficient variety in bee diet, and most of all a proliferation of low-protein pollens. Improving the variety of feed that hives are exposed to may be the cure, although an expensive one. Bee hives are rented to almond growers, and it is a common sight to see dozens, if not hundreds, of hives together in a clearing in the middle of a grove of almonds. Honey from almonds is considered unpalatable, and the honey is either turned to industrial use (baking, for example), or left in the hives for the bees.

The marketing of almonds by Blue Diamond, by the more recently arrived Emerald Nuts, and by other producers has made California almonds an international commodity. By 2007, not only did California produce a sizable part (more than 80 percent) of the world's almond supply, but over two-thirds of that almond production was exported. Helping assure that this all goes smoothly is the Almond Board of California, a USDA marketing arm created in 1950. The Almond Board is arguably the most successful of California's marketing organizations, with rivals only in such venerable predecessors as Sunkist. For years, the television Blue Diamond advertising mantra of "a can a week is all we ask" was hugely successful, and domestic consumption remains strong, although nowadays 50 percent of exports go to Europe, and 30 percent to Asia.

Almonds are a native of the Middle East, from Turkey to Pakistan, with a history as an Islamic cultivar that the prefix "al-" itself strongly suggests. Almonds were likely domesticated about 3000 BC, although they appear in the lower strata of Greek archeological sites as far back as 10,000 years ago. Almonds arrived in California with early Spanish expeditions and first appear in the chronological record of Mission Santa Barbara, in the late 1700s. The United States is the leading world producer, followed by Spain (which grows the Marcona and other varieties less common than those of California), and then Syria, Italy, and Iran. In short, this is a classic Mediterranean-type climate tree, which thrives with careful attention. In 2008, Spain imported 14 percent of California's exports; the delectable pan-Mediterranean treat marzipan, which is a sugar–almond mixture, clearly remains a favored Spanish creation, sufficient to introduce a blip in world consumption. References to almond–sugar mixtures appear as far back as the *The Book of One Thousand Nights and One Night*, and variants on marzipan are considered delectable, especially in the Christmas season, in Italy, Hungary, Germany, Persia, Mexico, and parts of China. Romans showered newlyweds with almonds as a fertility charm; in Sweden, cinnamon-flavored pudding with an almond inside as a prize is a Christmas custom. At times made from almond milk, horchata is a favored sweet drink, utterly distinctive and with superior thirst-quenching qualities, that is much favored in Spain and Latin America. Almonds, in short, are recognizably part of a thoroughly globalized phenomenon.

Being frost-sensitive, almonds in California grow mainly in the Sacramento and San Joaquin valleys and in bearing acreage are among the fastest-growing crops in the state. Three items, perhaps tellingly in terms of the state's agricultural health, threaten the further spread of almonds within California. First is overproduction, a perennial possibility given the sorts of grower attention that almonds have received. Regulation is a second, with a massive debate afield over a USDA ruling in 2007 that requires almond growers to pasteurize almonds. Health food stores and raw-nut enthusiasts often prefer raw almonds, and will pay up to a 40 percent premium for the non-pasteurized nuts, to be processed into almond butter or eaten plain. If nuts are not pasteurized, there is a small chance of exposure to salmonella; the debate is ongoing about the regulations, which in 2008 affected 1.6 billion pounds (yes, billion) of almonds, worth more than $2.4 billion. The trend is distinctive; an almond tree can produce for 50 or more years, so the production trend points distinctly upward, except for one final glitch—water supply, which poses the third threat. Much of the recent acreage in almonds is planted west of the I-5 corridor, in the San Joaquin Valley, an area chronically at risk of federal and state irrigation water reductions. Since 2007, some almond growers have of necessity massively pruned back their trees, hoping for wetter years in the Sierra and northern California to increase flows to water delivery systems.

Visually, almonds are a somewhat unkempt crop, their branches splaying and racing skyward, and almond groves lack the formidable understory darkness of an established walnut orchard, with that characteristic walnut graft line, nor do they look like the up-and-coming pistachio, which is distinguished by the light-colored bark of the tree and a certain lithe elegance. The closest tree in appearance to the almond is the peach, which is unsurprising, considering their relatively close genetic relationship. But the peach is a late producer, whereas the almond gets on the job early. If you are unsure what you are looking at, slow down and take a good look: the almond pods, or drupes, are distinctive, and the peaches no less so. That makes telling them apart a little bit less of a guessing game. But whereas the numbers of peaches are settling or heading downward, almond tree numbers are definitely on the uptick.

APPLES *Malus domestica*

Pls. 16, 27, 46, Fig. 1

FAMILY: ROSACEAE.

RANK: U.S. #5; CALIFORNIA SHARE 4%.

Apples are a quintessential American crop, but no less characteristically are they Californian. Bringing in $123 million in 2008, California ranks fifth in United States apple production, which has undergone drastic shifts in the last several decades as varieties long considered ironclad favorites were replaced, on the one hand, by new commercial varieties such as the Pink Lady, Gala, and Fuji, and, on the other, by heritage apples originally developed a hundred years ago or more that have been rediscovered in Missouri, Arkansas, and the northeast and restored to production. This shift comes with considerable expense, since tearing out old varieties and planting to different apples is expensive, both in labor and materials—not least because it halts production until the replanted orchard matures (in five to seven years) and begins yielding again at cost-effective levels. For all that, the reddish-hued bark of the apple tree, the characteristic leaf, and the distinctive fruit—growing red, green, or yellow, as it matures—is a sign of an area given to productive tree crops. Ever since Johnny Appleseed became a folkloric figure in the United States in the 1860s with his Swedenborgian gospel and his dissemination of wild apple varieties through the Midwestern landscape, apples have loomed large in American agricultural life and legend.

For all the hubbub surrounding the symbolism of its production, apples are a fruit much valued by their producers; in California, arguably only oranges, avocados, and grapes can match the mighty apple. There is plenty of argument about whether the Edenic fruit that led to Adam and Eve's eviction was an apple—UC Berkeley Sinologist Edward Schafer claimed that notoriety actually belonged to the peach, and Big Sur novelist Henry Miller argued that the orange should be blamed, based on Hieronymus Bosch's famous triptych, *The Garden of Earthly Delights*. In other cultures, the "blame" falls to pomegranates or tamarinds. That said, in American lore, the "Tree of Knowledge" supposedly yielded apples, which is no small part of the reason that teachers are traditionally awarded apples by their students; vernacular phrases such as "the apple of my eye" only serve to reinforce the favoritism.

An added benefit of apples is their durability. With careful cold-chilling, apples can be stored for months, and the smaller whole fruits find ready markets in school lunches and among dietetically inclined adults given to calorie-watching. Apples for pies and baking can be either special varieties, or well-kept eating cultivars. The varied appearances of apples, especially heritage or heirloom varieties, make for a visual feast, and strolling in fall through the apple section of a well-supplied grocer is a journey into the heart of genetic plasticity and variation. Apples are a diverse fruit: some growers have shifted to dwarf varieties specifically so they can host U-pick operations (for which the clients do the picking) at reduced risk of falling from ladders; visitors are charged by the pound, or by the bucket, and provide the labor. Specialization in exotic, multicolored, or heritage varieties is common in such operations, and because apples require cross-pollination, sporting varied apples within an orchard is not a problem.

Juice from apples is a product with diverse value streams: hard apple cider is favored by some; applejack (with a higher alcohol value) meets the fancy of others. Although apple brandy (in the spirit of the French calvados) is a rarity in the United States, there are a handful of California producers. And that includes only the fermented options. The Gravenstein apple, strongly tied in its production to an area west of Sebastapol in Sonoma County, was first popular not long after the Gold Rush began in 1850 and produced apples eaten raw or cooked that are also suitable for juicing. In a bold but highly

Plate 46. An Arkansas black apple at the Chileno Valley Ranch in Marin County suggests qualities that make it a prized eating apple when the grudging trees produce, yielding a fruit showing color shading from lively, assertive red to a hue of cardinal nearing black.

successful marketing move, S. Martinelli and Company, based since 1868 in the Pajaro Valley in Monterey County, began apple cider production. Switching to non-alcoholic cider during Prohibition, their sparkling cider is a signature beverage for alcohol-disinclined adults and a modern-day mainstay for children's parties.

"Pomes" are a member of the Rose family (which also includes crabapples, pears, pomegranates, Asian pears, and quince) and when sliced open are identifiable because of a papery central cavity that surrounds the seeds. Some parts of California are too warm for reliable apple cropping. To produce well, the apple orchard requires prolonged winter chilling, with 1,200 to 1,500 hours under 45 degrees Fahrenheit. Thinning of apples is necessary and desirable to improve the quality of remaining fruit and protect branches from overload, which can make limbs susceptible to breakage in unexpected summer or fall storms when heavy with fruit.

Cooling requirements scatter the production of apples throughout California, wherever there are isolated valleys with pronounced cold air drainage. Somewhat oddly, this is one crop that truly needs extremes in climate that would irk many Californians: without the chilling, apples will set an inadequate blossom and fruit supply. Various rootstocks are available, and maturation of fruit varies from 70 to 180 days; at the long end, some fruit will not mature before fall frost. Pockets of apple orchards appear in Mendocino and Sonoma, and, prominently, in the Sierra Nevada foothills east of Sacramento. (The Apple Hill Growers Association, centered in Camino, now has 50 member ranches.) Mountain regions of Southern California, including Tehachapi, Yucaipa (in Riverside's San Jacinto Mountains), and Julian in San Diego County, were once apple strongholds, but are less so today. The Sacramento Valley and the San Joaquin Valley counties of Kern, Fresno, San Joaquin, and Madera are currently California's main apple producers—Sonoma County is sixth, but has dropped steadily. Along the Gravenstein Highway in Sonoma, it is common to drive by apple orchards in fall where windfall fruit has dropped from the trees to form a deep and redolent understory of fermenting apples—a sad sight, and testimony to the plummeting popularity of some once-prized apple varieties, and the ascent of others.

Whereas older orchards on standard rootstocks might have 50 to 100 trees per acre, newer plantings often use dwarf rootstock (which also reduces labor costs for harvest), with 180 to 500 trees per acre. The trend is definitely toward denser plantings of smaller trees. The California Apple Commission, based in Fresno and established in 1994, controls apple marketing orders in California, replacing several older associations chartered in the 1930s. Production of apples in California is a telling story. Production began during the Gold Rush, with varieties brought to California from New England and the Midwest, which accounts for production in Sonoma, and in the Sierra foothills. Bearing acreage peaked in 1926, with 52,000 acres. But strikingly, the actual production of apples peaked 70 years later, in 1994, with 525,000 tons produced on just 34,900 acres. Production in 2008 was a little more than half that, at 19,500 acres planted, yielding 165,000 tons. Every year since 1997, apple acreage has dropped, with acreage converted to suburban homes, or into higher–dollar value crops such as grapes.

APRICOTS *Prunus armeniaca L.*

FAMILY: ROSACEAE.

RANK: U.S. #1; CALIFORNIA SHARE 95%.

For connoisseurs, nothing approaches the first bite of a fresh, ripe, sun-warmed apricot, and that delectable eating experience—plus an early ripening—keeps apricots a much-sought-after staple of the California U-pick industry. Not so oddly, once their shape and growing habits are taken into account, apricots are a member of prune family, a stone fruit (or drupe), and they grow readily in California; a decline in acreage is due to reduced interest, not lack of vigor. In form and pruning style, the small- to medium-stature tree is relatively unremarkable, with heart-shaped leaves two or three inches wide to five inches long, and identification is easiest with a close look at the white or pink flowers, which come at the start of spring. The apricot fruit is distinctive but green for a good part of its early growth cycle, until its hue changes to yellow, and then assumes the familiar rich dusky shade of orange recognizable to those whose childhood treasures included a set of crayons kept in good order, with "apricot" the choicest of colors.

The apricot in California has a long history that begins with its introduction in the Spanish era; records exist of apricot harvests at Mission San Jose in 1792. Dried, canned, juiced, or eaten fresh, the apricot is a relatively delicate fruit and is readily bruised, which makes it difficult to ship, and sensitive to rough handling on display. California produces more than 90 percent (2008) of the U.S. crop, although domestic production is ninth in the world, with Italy, Spain, France, Pakistan, and Morocco the first five. California acreage has headed steadily downward since 1928, when there were 83,000 bearing acres of apricots in California. In 1935, 3,000 small growers in the Santa Clara Valley, who tended 18,631 acres of apricot orchards, accounted for a quarter of the state's crop. Soon afterward, though, urbanization pushed growers eastward into the San Joaquin Valley, in Stanislaus, San Joaquin, and Merced counties, as Jan Broek has written. By the late 1960s, Silicon Valley had replaced the orchards, and fruit trees were remembered only in the names given to malls and highway arterials, such as Blossom Hill Road. In 1980, total state bearing acreage was down to 24,100 acres; by 2008, less than half that, 11,100 acres, came from 300 growers (50 of them major). Grapes, alfalfa, tomatoes, almonds, and pistachios are the main crops replacing apricots—when subdivisions are not the active usurper of orchard acreage.

The tree's name comes from the Spanish *albaricoque*, which is in turn taken from the Arabic *al-burquk*, although curiously, in Argentina and Chile, the apricot is referred to as the "damasco," which recollects its historic prevalence in Damascus, Syria. From origins in northeastern China, near the Russian border, apricots were brought westward through Asia Minor, and from Armenia into Europe and North Africa. Although the tree is sometimes considered delicate, in fact it is more cold-hearty than the peach; the apricot is, after all, native to areas with cold, dry winters. An initial risk is a hard spring frost, which can kill the early blossoms. Heat and sunburn are more a problem, and the trunks of young apricots are often painted with watered-down white latex paint to reduce borer infestations and sun damage. Cuttings from a strong-bearing apricot can be grafted onto peach or plum rootstock, which provides the tree improved growth characteristics. Best-known California varieties include Poppy, Earlicot, Lorna, Robada, Katy, Tri Gem, and Helena, but the most common are Pattersons, Blenheims, Tiltons, and Castlebrites. Harvest takes place from May to mid-July, and the trees have a productive life of 20 years, bearing from age five onward. With a relatively short production peak in tree life, growers contemplating planting replacement trees or converting to other crops face market pressures, and more than a few are changing to other tree crops or grapes, or selling to developers. The apricot production for 2009 is forecast to be down 14 percent from 2008, when it brought in $35 million.

Apricots are a fruit of many a use. Home gardeners and backpackers relish dried apricots, and drying techniques are now improved, reducing or eliminating a use of sulfur that was once common in dried fruit. In eighteenth-century Britain, apricot oil was used medicinally, and laetrile comes from apricots seeds—itself a controversial subject. Apricots can be made into wine and brandy, something more popular abroad than in California; there are those who attribute aphrodisiac qualities to the warm apricot. Despite acreage and production decreasing, the apricot remains a valued product in California, and one of the state's distinctive tree crops. But the total value of the yield is a third that of apples (another less-than-prominent crop now), a tenth the value of avocado production, and just 6 percent of the value of orange production in California. Only a quarter of the crop is sold fresh—the rest goes to canning, drying, or concentrate. But sighting an apricot orchard in Westley (Stanislaus), or in the few groves remaining in the suburbanizing outskirts of Patterson, where many Santa Clara Valley growers moved in the 1950s, remains a thrill.

AVOCADOS

Mexican — ***Persea americana*** **var.** ***drymifolia*** **blake**

Guatemalan — ***P. nubigena*** **var.** ***guatemalensis*** **L.**

West Indian — ***P. americana*** **mill. var.** ***americana***

Pls. 17, 47, 120, Map 7

FAMILY: LAURACEAE.

NOTES: HYBRID FORMS EXIST BETWEEN ALL THREE TYPES OF AVOCADO.

RANK: U.S. #1; CALIFORNIA SHARE 92%.

There's no denying that one charm of avocados is their texture: an oily, yet firm smoothness that when eaten is utterly distinctive and, to many, profoundly addictive. A tall tree with a distinctive dark green shade and leaf, avocados grow from San Luis Obispo County on the Central Coast to the Mexican border (and beyond). San Diego County represents nearly two-thirds

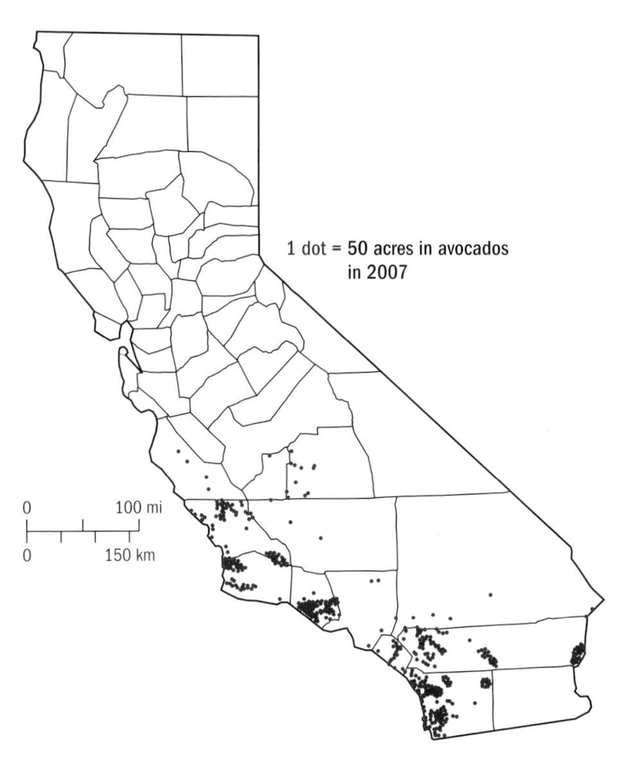

Map 7. Avocado acreage, 2007.

of California production, although a hard freeze there in 2007 destroyed 10 percent of the state's orchards. Avocados trees are sometimes spoken of as coarse and needy: orchards require full sun, demand protection from high winds, want shelter from downslope cold air flows, prefer well-drained loose soil or sandy loam, and are often grown on hillsides. Grafted plants fruit much more quickly than unimproved varieties. A single tree can produce 500 pieces of fruit; a more typical yield is 150, but each of those can bring $1 or more at retail, depending on the competing volume of fruit shipments issuing from other growers.

The remarkable transformation in California avocado growing came with marketing. Crafting a market for tree fruit that was

Plate 47. The handsome form of pendulous Hass avocados in a North County San Diego grove is familiar, even though—like many cultivars—avocados come in varied shapes and sizes, even within California.

native only to the Western Hemisphere, and therefore quite unfamiliar to the Europeans who arrived with Columbus and after, was no simple matter. Long a food staple in Mexico, the avocado was all but unknown north of the border until the late nineteenth century. Finding a U.S. market for the avocado was something begun in California, where specimens of the exotic fruit trickling in from Mexico, Guatemala, and Brazil caught on and spread. By 2009, nearly half of United States households bought avocados. Creating such demand from a standing stop is a departure from convention, but the note of success was an anthem that California avocado producers were more than willing to embrace. The avocado advertising effort brought other commodity groups, such as kiwifruit producers, to watch the avocado campaign with envy.

Marketing avocados is not for the faint of heart. The California Avocado Commission recently claimed that two-thirds of yearly avocados sales in the United States—40 million pounds—came in the week leading up to the Super Bowl football game, presumably foreshadowing their reappearance in short order as guacamole. Snopes.com starchily debunks that figure, dismissing it as hyperbole, suggesting that 5 percent (eight million pounds) is more likely than 67 percent. But the myth stays alive. Actual peak avocado use comes at Cinco de Mayo, with 14 million pounds consumed—still an impressive total. But the point is a lively one: avocados attract a fanatical following that seeks fruit at its peak of ripeness, essentially unconcerned about variety or source. Avocado partisans are a rabid crowd, and they want their fix. Although avocado production undulates on a two-year cycle, since 1997 the average sales are about $300 million a year, a rate that puts avocado consistently among the top 20 dollar-earning California agricultural products. California generates 85 to 90 percent of the U.S. production (the rest is from Florida, with a trace added from Hawaii). Sales incursions from south of the border make California growers nervous, since the state sits at the northern edge of optimal avocado habitat—especially with 73 percent of imports hailing from Mexico and the overall total of imported fruit rising yearly.

Avocados are a member of the laurel family, originating in Mexico, Central America, and South America. The varieties show considerable genetic and regional variation, and hybridize readily. The original date of first avocado use is lost in distant time, thought to be between 7000 and 5000 BC; domestication probably came about 500 BC. The name in English is taken from the Spanish *aguacate*, which borrows from the Aztec word *ahuacatl*. After the avocado was established in California, residents took to the new crop with élan, thanks in part to grower enthusiasm. In one significant way the avocado orchard resembled the orange grove and the grape vineyard—the avocado grower could own a small acreage (10 acres is even today a typical holding) and yet be part of an elite agriculturalist society producing fruit that was much sought after and cherished—and of reliably high value.

The avocado tree is striking in its dark green color and size—an unkempt evergreen that can reach 80 feet. Leaves of the Guatemalan variant have medicinal uses; Mexican types can exude a scent of anise when stepped on or crushed; West Indian varieties are scentless. The leaves, heavy and waxy, are slow to compost and can form a dense understory cover. Brazilian avocados can grow to such a large size that travelers on foot have to be wary of falling fruit, which can weight more than two pounds each—enough to bring on a concussion if the descent has a clear path from tree to unprotected head. The variety of avocado-eating experiences in the world reflects the diversity of fruit and different local custom. Avocados are eaten in varied ways; monounsaturated fat in avocados can be regarded enthusiastically by dieters. Brazilians will sometimes eat avocados with ice cream; Filipinos puree avocados with sugar and milk (Indonesians layer in chocolate syrup) for a dessert drink; the fruit is generally eaten when past ripe—when the meat of the avocado is starting to turn is usually when a sharp knife halves the fruit, the large pit is popped out, and the two halves are peeled whole, or a spoon is used to scoop the flesh from each half.

There is controversy about where the avocado came into the United States; Florida bids for the honor (in 1833), but that claim is disputed, and the first confirmed presence and harvest was in Santa Barbara, California, in 1871. The Fuerte variety grew popular soon afterward. In the California of the 1950s, two dozen varieties of avocados were packed and shipped, but Fuertes then accounted for more than two-thirds of production. The main avocado cultivars today are Hass, Fuerte, Zutaro, Bacon, Gwen, Lamb Hass, Pinkerton, Puebla, Duke, Creamhart, Lula, and Reed, a roster that includes many now-recognized hybrids. The Hass variety, quite popular because it preserves well, ships, and doesn't bruise readily, is hardly the most succulent and memorable, but it slowly became the leading variety produced in California by the late 1970s and now dominates production, with better than a 90 percent share. Growers in Florida typically favor less oily varieties. West Indian types produce glossy, round, smooth fruits that can weigh up to two pounds. Guatemalan varieties are pear-shaped or ovoid, with hard-pebbled skins that turn blackish-green when ready to eat. Mexican avocados are small, with fine skin that turns bright green or black when the fruit is ripe. Trees grafted onto strong rootstock can produce in one or two years, compared to a decade delay in production from seedlings. Fruit stays on the tree for extended periods of time. Mexican-type trees ripen in six to eight months, whereas Guatemalan varieties may take up to 18 months to ripen. Fruit can continue enlarging on the tree even after maturity. That leads many growers to leave fruit on the trees until harvest is convenient, but raises the risk of frost damage in the advent of a cold snap.

An avocado grove can be a formidable sight. With store prices for individual avocados often in excess of a dollar per fruit, growers are deservedly nervous about the purloining of their fruit, whether by casual visitor or an organized crew that can strip a corner of an orchard in just a few hours. "No Trespassing" signs are to be heeded. That said, avocado harvest is no easy business. Although some trees are pruned low so that fruit can be picked from near ground-level, fruit from taller trees is gleaned by pickers who wield special shears, called "clippers," and ladders up to 30 feet long. Pickers place fruit in large bags fastened around their shoulders, which hold 30 to 50 pounds per bag, and fruits are then placed in larger bins that hold 600 to 800 pounds each. Quick cooling and a cleaning follows, and fruit is packed into cartons known as lugs. Avocado presentation and sales is itself an art form; Calavo Growers, based in Santa Paula, California, is a sophisticated marketing organization with extensive research into new ways to present their product.

Bearing acreage since 1993 has fluctuated 10 percent on either side of 67,000 acres, and the value per ton likewise edges above and below $1,750—not far from a dollar a pound. In terms of long-term trend lines, avocado production is on the upswing: acreage in 1969–1970 was just 18,000 acres; in 2007–2008, it topped 65,840. San Diego, Ventura, Santa Barbara, Riverside, and San Luis Obispo are the five largest producing counties, but even Monterey County has 221 bearing acres, and Los Angeles has a meager, but still active, 81 acres.

CHERRIES *Prunus avium*

Pls. 35, 39, 48, 123

FAMILY: ROSACEAE.

RANK: U.S. #3; CALIFORNIA SHARE: 14%.

To think of California as a major cherry producer might seem to defy logic, but that happens to be true. Although Washington State is regularly the largest value sweet cherry producer, Michigan, Oregon, and California joust for second and third place, and the bearing acreage suggests that California may someday eclipse Washington in sweet cherry acreage, if not in overall production (a humid climate having its advantages). California's sweet cherry acreage is on the upswing, with 30,000 acres of bearing trees in 2007, generating $155 million, although predictions for the 2009 crop indicate a decline in acreage and in production.

Few crops compare to sweet cherries in visual beauty and popular appeal. Tart (or pie) cherries (*Prunus cerasus*) are another matter, with California a nonentity in national production. But sweet cherries are a favorite in U-pick operations in the heartland of California cherry growing, which is the Delta region at the western confluence of the Sacramento–San Joaquin river systems. Commercial orchards exist too, and it is these that loom largest in the official statistics. The production data from U-pick farms are less detailed than for commercial growers, so it is safe to assume that California cherries total more than is counted in county agricultural commissioner reports. A tree that is routinely planted near town, the better to be accessible and usable, the sweet cherry offers its own bittersweet story: sizable plantings persist along the Hwy. 101 corridor near Morgan Hill and San Jose, but risk desiccation and disappearance as hunger for buildable real estate replaces interest in cherry fruit. But from May well into June, cherry season is on.

The origin of sweet cherries scrolls far back in human history, perhaps to prehistoric times in the area between the Black and Caspian seas of Asia Minor. Although a member of the genus *Prunus* (in the Rose family), cherries are distinct from their other stone fruit relatives: plums, apricots, peaches, and almonds. Favored by both Greek and Roman societies, dried cherries were an essential part of the legionnaire's diet, and the trees were favored in roadside plantings not just for their fruit, but also for their wood. Sweet cherries traveled to New England with Puritan settlers, and to California with Spanish

Plate 48. Cherries are a tantalizing, prolific crop, if short-seasoned. Rainier cherries, here U-pick specimens, can also be found in markets.

missionaries. Cultivars brought to Washington, Oregon, and California—typically as slips to be grafted onto other rootstock—remain significant. Bing, Napoleon, Rainier, Lambert, and Van are the most important cultivars, but other varieties are included because cherries are self-incompatible and need high success in fruit set to develop a commercial crop. Pollinating trees of exotic varieties are interplanted as every ninth tree or so, with honey bees the main pollinator. Generally, cherries are grafted onto rootstock, and new rootstock varieties imported from abroad promise higher yields than have traditionally been possible. Pruning can take various forms, and is done to increase yields and to make the trees easily entered by ladder for harvest, which is done by hand. The harvested fruit has a short shelf life, and requires gentle handling to prevent bruising.

Cherries are not easy trees to raise and keep thriving, requiring well-drained soils to avoid suffering from "wet feet." Sweet cherries are susceptible to various rots and viruses and, like peaches, plums, apricots, and nectarines, require a stiff chilling spell to set flowers and produce consistently. They will sunburn, and applications of dilute white latex paint will give a particularly well-maintained look to the lower trunks, and make for a signature appearance (and one shared with apricots). If the spring season is dry, irrigation may be needed, along with netting, to protect fruit from birds once the cherries start forming. Research on production in California is supported by grower fees paid through the California Cherry Advisory Board.

Although cherry trees in the wild can rise to 50 feet, they are pruned to a moderate height in cultivation, reaching 15 feet or so. Planting is not dense, at 100 trees per acre, and the distinctive reddish hue of cherry trunks (if unpainted) is a signature feature. Dwarf and semidwarf rootstock is used to limit height, which makes harvest easier. Cherries are not just eaten fresh, they are also juiced, dried and sold at retail, and frozen. But the unmodified fruit is increasingly touted, like blueberries, as an antioxidant, with assorted and newly discovered health benefits, including production of melatonin, known to affect natural sleep patterns. Cherries have other and curious uses, including the ever-popular Maraschino, which can be either a sweet or a sour cherry, decolorized and steeped in Marasca, a liqueur made from the fermented juice of wild cherries. Curiously, in some circumstances, cherry blossoms are held in greater regard than the fruit. The white flowers and their branches are prized in Japanese and other flower arrangements; Little Tokyo, in Los Angeles, has an annual Cherry Blossom Festival. The harvest season of this showcase tree and fruit is brief but well noted by fruit fans in California and within the export region.

CHESTNUTS

North American sweet chestnut — ***Castanea dentata***

European or Spanish chestnut — ***Castanea sativa***

FAMILY: FAGACEAE.

UNRANKED.

The nut of the chestnut tree was a staple food for humans for millennia, providing high-quality calories and protein for early humans and animals, and today is still considered by some writers to be the third most important nut crop in the world's temperate zones, after peanuts and the coconut. Less than 1 percent of the world's production comes from the American chestnut, but the European variety is found in various parts of the United States, including California and along the Northwest Coast. California is not a large chestnut producer, but interest is on the increase, with more than 600 acres in production, and demand is so strong during the fall months from Thanksgiving to New Year's that imports from abroad are a constant, with chestnuts sold in bulk in ethnic markets throughout the state. This is a crop on the upswing, and were that not reason enough to include chestnuts in a field guide, the dramatic double products of a chestnut tree—valued nuts and superlative timber—add to the crop's significance.

A member of the Fagaceae family, which includes oaks and beeches, the chestnut, like so many crops, came from Asia Minor to Europe, and onward to America. Paintings from the fifteenth century show pigs grazing chestnut mast in the forest understory, and peasants might eat two to five pounds of chestnuts a day. In parts of Italy today, polenta is still made from chestnut meal, and in European cities chestnuts are a favored street food, roasted in kettles into the winter months. The European chestnut (*C. sativa*) has a large nut with acceptable taste, but the American variant (*C. dentata*), more commonly known as the sweet chestnut, produces small nuts with superlative flavor. Both trees can grow to great height, although examples being raised for nut production are generally held to 35 to 40 feet; trees raised for timber can grow to 100 feet. The American chestnut provided foodstuffs in copious quantity to Native Americans, and a quality of wood and lumber unequaled by any other timber source; its loss is strongly felt, and furniture crafted from American chestnut wood commands a premium in antique auctions. Groves of American chestnut were once ubiquitous, but in the 1930s, a chestnut blight fungus thought to have originated in East Asia destroyed some four billion American chestnut trees, and only in the last dozen years have efforts shown some success at recreating blight-resistant chestnuts from the native strains.

Crossing between species of chestnuts is not only successful, but going to new cultivars is essential to produce effective pollination of chestnuts. The Colossal, a European variant, is grown with Silverleaf, Nevada, Eurobella, or other trees as pollinizers. A single variety of chestnuts will lead to low production; having different cultivars at hand is crucial. Examples of *C. dentata* crosses are grown in the Sierra Nevada foothills, along the North Coast, and in parts of the Sacramento Valley and the Delta, generally in areas of well-drained soils avoiding wet conditions. County ag commissioners' reports indicate just 214 acres harvested, producing 174 tons of chestnuts worth $621,000 in 2006, but a sizable acreage did not report. Trees must be trained and pruned to develop a desirable canopy structure, but they can be grown in plantations, 14 to 40 feet apart, although mature trees may have to be cut out if the grove is planted at the tighter spacing. One pollinizer is needed for every eight to 10 trees of the main cultivar, though the distinction between varieties is a subtle matter for the nonspecialist to discern. Distinctive leaves, a red-brown color, and the distinctive burr—a spiny capsule that holds one to seven nuts—are signature features of chestnuts, and in California we may expect to see more of these formidable trees in the future.

CITRON

Fingered or Buddha's hand citron — ***Citrus medica***

Mediterranean citron, and the etrog of the Hebrews — ***C. medica* var. *ethrog***

Pl. 49

FAMILY: RUTACEAE.

UNRANKED—CALIFORNIA IS MAJORITY PRODUCER.

Plate 49. The Buddha's hand fingered citron is as dramatic as its presence, and as distinctive, and is often used to perfume rooms—or vodka.

Citron is a rare, but utterly distinctive, California crop, generally used for the manufacture of candied peel. Although citron is considered the progenitor of all citrus crops, the variety best known in California and grown in the southern states is Fingered citron, better known as Buddha's hand citron. The startling shape belies a no less dramatically pungent aroma, and although citron has scant commercial production, it can be candied and served as a dessert treat, made into preserves, or pickled; citron zest adds fragrance to alcohol (especially to otherwise neutral vodka), and the interior can be processed to generate soluble fiber, found in its thick white interstitial flesh, known as albedo.

There are religious uses for all varieties of citron, extending to origins in India or Yemen. The oldest documented use in ritual is in the Jewish Feast of Tabernacles, which calls the fruit etrog. Aside from mention in the Torah (Lev. 23:40), Jewish scholarship suggests that the citron came from Egypt, where evidence of domestication does exist, during the Exodus. Citron is documented in Hebrew artwork and archeological excavations, although the Near Eastern variants (the etrog) tend to lack the ribs and digits of the fingered citron. The etrog is traditionally wrapped in soft flax fibers and stored in a special silver box.

Genetic evidence suggests that all other cultivated citrus cultivars derive from four ancestral types: the citron, the pummelo (or pomelo), the mandarin, and the papedas. The thick, watery pulp is an adaptation to the dry component of the monsoon cycle. Known in antiquity as the "Persian apple," the citron moved west with the Persian leader Cyrus the Great, and the first Roman greenhouse-like structures (with a sort of protective "glass" formed from sheets of mica) shielded the citron from winters in northern Italy.

Small and thorny, highly fertile and a successful colonizer, citron is a curious cultivar: a fascinating fruit with distinctive features and sparse commercial use, and therefore with a small likelihood of rapid spread. Nonetheless, citron is grown in coastal San Diego, and in added areas of the Southland offering an equable climate, since the tree is frost-sensitive. Once blossoms are set, fruits 6 to 12 inches in length grow and begin to split at the end opposite the stem and curl, so that carpels appear, seeming very much like human fingers. As the fruit grows, the rind exudes a remarkable resinous smell, until maturity is reached in late fall or early winter.

The rind can be grated with a microplane or peeler, as is done with lemons, and the peel added while cooking in the same way as lemon peel. Within California's Chinese community, the Buddha's hand citron is popular (as in China) for ritual presentation on altars at temple or at home. A bonsai form may be grown in pots; Japanese households will sometimes purchase "bushukan" at New Year's for good luck, setting the citron atop mochi cakes or placing it on display instead of flowers. Considered a sign of prosperity, Buddha's hand citron is often seen near cash registers.

DATES ***Phoenix dactylifera***

Pls. 7, 50, 121

FAMILY: ARECACEAE.

NAMES: DATE PALM.

RANK: U.S. #1; CALIFORNIA SHARE 82%.

Date palms are the agent for producing a wonderfully eccentric California crop, and the date palm in California embodies an ancient adage that the palms thrive with "their feet in water and heads in the sun." Dates are the fruit of the date palm, a marvel of sweetness and varying texture, and to bite into a date is to journey far back in time, to human origins and our curious, if not obsessive, hunger for the sweet. Add to that the places where date palms grow in nature—namely, oases, canyons, and washes where a water supply exists—and a certain fascination is complete. Wherever grown, there had better be water, because date palms are profligate drinkers. Tall and straight-trunked palms, dates can reach up to 60 feet in the wild, but cultivated palms are kept to around 30. They grow, appropriately enough, in the California desert, with other isolated oases in San Diego County and a scattering of date palms extending north into the Southland. An oddity of the commercialization of the date palm are the ladders that are frequently left hanging down from the crowns. They are attached to each tree and left there to reduce setup time for the countless trips up and down the palm that are necessary to complete the required steps from cultivation through pollination and harvest. Dates are anything but a labor-minimal crop.

The center for commercial production is in the Coachella Valley of eastern Riverside and in Imperial County. Of course, towns with names such as "Mecca" and "Thermal" are likely to have dates. But so do Palm Springs and Palm Desert, and even today they remain a date palm paradise, even as more and more people come to the desert for winter warm and summer hot. The 220,000 date palms estimated to have been in the Coachella Valley in 1980 are reduced in number, replaced by golf courses and suburban housing projects. But Shields Date Gardens, begun in 1924, still exists in Indio with its vaunted "Mother Tree," a beacon for the curious and for date fanatics.

An important Near Eastern crop, and grown across North Africa to Morocco and into southern Spain, date palms have an ancient history and religious significance in several world religions. Dates, with yogurt or milk, are often consumed as a first meal after dark during the fasting of Ramadan, something noted by Captain Sir Richard Francis Burton in his travels to Mecca. From their Mediterranean-edge presence, it was entirely reasonable that date palms would make the move to California and Arizona, which are the two large U.S. producers of dates. In farm-gate value, dates are fifty-ninth, worth $53 million in income, with commercial production in 2008 of 28 thousand tons. Only 6,000 acres are planted to dates in California, on the upswing, but an appetite for dates remains an on-again, off-again matter; fanatics love them, but others are less enthused. Small

Farm Initiatives, a federal effort, has encouraged some growers to lease shares of their trees to individuals who will receive their portion after harvest in October. Although Riverside and Imperial counties are the sole commercial producers, any traveler in east-central San Diego County, on the way to Joshua Tree, has likely stopped for a date shake at the Hadley Date Gardens, a famed roadside stop near Cabazon, whose beguiling logo is "Sweetness Is Our Nature."

The three distinct varieties of date cultivars are broadly classified as soft (Barhee, Halawy, Khadrawy, Medjool), semi-dry (Dayri, Deglet Noor, Zahidi), and dry (Thoory), and they are determined by the glucose, fructose, and sucrose content. Although only a dozen or so date varieties are common in California, more than 4,000 are known worldwide. Most commercial plantations rely on cuttings to reproduce date palms,

Plate 50. With butcher paper covers to shield fruit from sunburn, insects, and birds, commercial dates are a delicate, yet highly valued crop in the Coachella Valley, where date gardens draw those who are curious about California's production of an otherwise alien crop.

since reproduction by seed produces as many males as females, and telling them apart when young is not easy. Although dates in the wild are wind-pollinated, in plantations it is usual to assist in the process and guarantee far higher pollination success. Dates can take up to seven years after planting to reach production, and are harvestable a year or two later. When mature, date palms can yield 150 to 300 pounds of dates per harvest, ripening slowly through time, for a total of some 9,000 pounds per acre. The ripening fruit is thinned, and the remaining dates are bagged to protect them from birds and from inclement weather, for rain can spoil dates.

Dates were introduced to California in 1890, although the dates' establishment in Mexico was well before that. Finding the right environment took trial and error, and trees are now planted about 25 feet apart—as instructed in ancient Islamic texts. A major drop-off in date production a decade ago was resolved with discovery that soils were too compacted, and cover crops that fixed nitrogen were planted to help with nutrition of the palms. When the dates come, they hang in fronds, and several hundred to a thousand dates dangle from each branch. The stages of date development bear Arabic names—kimri (unripe), khalal (full-size), rutab (soft and ripe), and tamr (sun-dried to ripeness). Little wonder that among followers of Islam, many regard the date palm as the Tree of Life spoken of in the Bible's Book of Genesis.

EXOTICS

Mangos, guavas, passionfruit, lychee, longan, and many others

In an agricultural field guide of a reasonable length, some crops have to be left out for reasons of length. But their exclusion from any detailed discussion is not to say they aren't interesting crops, or fascinating in their own right. Some tree crops are exotics, meaning that they hail from distant origins and are not established in commercial numbers. In fact, nearly all tree crops raised now in California were once exotics themselves; technically, anything not here in 1492 is an exotic. True, the English Walnut is grafted onto rootstock from the native Black Walnut, and most of California's grapes are grafted onto rootstock that was originally from the eastern United States, a last resort when the phylloxera mite began to damage European grape rootstock in the late nineteenth century. But those and a few other examples aside, exotics attempted as introductions, and in small numbers, offer an interesting look at what California can do—and perhaps at crops that may be all the more important in the future, subject to climatic warming and potential shortages in water supply.

Because many of the most interesting exotics being attempted are from tropical or subtropical climes, San Diego County and the Southland in general constitute a hotbed of agronomic experimentation. As is common, there are two fronts: one is with the professional nurseries, where exotic fruit trees in particular are often brought as ornamentals. Those are the work of experts—and the only wish is that there are no untoward introductions such as the mulberry, which when introduced to Phoenix in the 1930s turned a hypoallergenic paradise into a land full of the most noxious pollen and allergens known to humanity. But a second source is backyard experimenters, who bring in slips or colms or grafts and expertly go to work with fruit crops—sometimes with, but perhaps more often without USDA approval.

The results of experimentation with exotics are easily seen through two superlative sources for information on exotic fruits in California. First is both a standby and a standout: Cooperative Extension specialists form an intricate and well-engaged network of interested experimenters who are always on the lookout for new crops, some being tested by the USDA satellite facilities located in various states. They are particularly well plugged in to the crucial National Plant Germplasm System, based at UC Davis with stations in several parts of California. The Extension Specialists never cease to amaze. But perhaps the best source is the second: *Fruit Gardener*, the publication of the California Rare Fruit Growers, published bimonthly to a subscription-based fan club. What emerges as possibilities, from a serious group of fruit fanciers, never ceases to amaze.

Recent fruits discussed in the pages of *Fruit Gardener* include a typical mix. Some of the discussions are of fruits established as viable in California, usually of varieties not yet produced commercially. Variations on the pomegranate (*Pafiankas*) are discussed in the context of threats to the Germplasm program at the

Wolfskill Experimental Farm along Putah Creek, west of Davis. But other examples of "grown but not at commercial levels" include cherimoya ($1.7 million, 2008), guava ($993,600, 2008), mangos, rare peaches, longan, passionfruit, jujube, lychee, and bananas. Bananas have a special piquancy in California, because travelers down Hwy. 1, near Santa Barbara, might stop in the early 1990s at the coastside town of La Conchita, where Doug Richardson used to sell bananas grown just up the hill at his charmingly named "Seaside Banana Gardens and Palapa Fruit Stand." A buffering microclimate kept La Conchita warm, securing it under a high cloud cover, and there the bananas thrived—until two successive landslides wiped out not only the hillside plantation, but several houses.

A second body of fruit discussed are plants grown elsewhere that are considered possible candidates for arrival in California: the wolfberry or goji (*Lycium barbarum*), an Asian staple, is one; starfruit, or carambola (*Averrhoa carambola*) is another, as are longan (*Dimocarpus longan*) and sapodilla (*Manilkara zapota*) (both are grown in less than commercial quantity). The engkala (*Litsea garciae*), a fruit native to Borneo, is examined, as are additional varieties of dates (like pomegranates, discussed in this guide). Cherimoya (*Annona cherimola*) and soursop (*Annona muricata*) are grown experimentally in California, but the markets are not ready for much more than home consumption. The color photographs in *Fruit Gardener* tend toward the vivid—or luscious, depending on your hunger level as you turn the pages looking at the images.

Given moderate temperatures and extremely rare freezes in Southern California, a great deal can be grown—and that is especially true, for some reason, of National City, a plateau just south of the body of downtown San Diego where rare fruit growers abound. Innovation will come from such sources, and fruit commercialization will allow the hobbyists who began it all to nod their heads, even if acknowledgment is slight, knowing that they did the work that made a new crop possible.

FIGS *Ficus carica*

Pl. 51

FAMILY: MORACEAE.

RANK: U.S. #1; CALIFORNIA SHARE 99%.

A grove of fig trees is a somewhat startling sight, each a low-growth form of tree, perhaps closer to a large deciduous shrub, but with characteristic medium-green and deeply veined leaves that form almost a shroud around the tree. Fig acreage in California has decreased in the last five years, but the crop is still distinctive, with large leaves and flexible branches. Fruit of the two darker fig varieties are on trees from May to December; the trees bearing lighter-colored fruit have a shorter season. Trees are grown to 20 to 30 feet and set fruit readily on their own with the help of wasps; some varieties yield two crops per year, one maturing in mid-summer, the second in late summer or fall. While considered a fruit, the "fruit" of the fig is a flower that is inverted into itself, and the fruit—of course, also called "a fig"—is the only fruit to fully ripen and then partially dry on a tree.

Properly processed, figs are the essential ingredient to a favored American food, the Fig Newton, but documentation of fig growing dates to Classical times, with archeological evidence of the eating of figs earlier still. Desiccated ancient fig remnants were discovered in the Jordan Valley, in an early Neolithic village dating from 9400 to 9200 BC, and the fig may have been among the very first instances of agriculture. In Greek and Roman society, the fig was considered an antidote to sundry ailments, and Roman tradition has it that Bacchus introduced the fig to humankind, which made the tree sacred. Because of that, Roman statues traditionally were crowned with fig leaves. Less happily, the asp that killed Cleopatra is said to have been delivered in a basket of figs. It is left for the California Fig Board to note that although the fig was not the source of the forbidden fruit in the Garden of Eden, it is the most talked-about fruit in the Bible. After their rebellion, Adam and Eve sewed together fig leaves, "and made themselves aprons." There is enough discussion in Classical literature for experts to recognize what fig variety is described, one more indication of the favoritism bestowed on figs in times past. Although traditionally harvested when mostly dry on the tree (or after falling from it), the fig can be picked earlier, and the split Mission or Calimyrna fig floated in cream is a dish dating back to antiquity.

The varieties of figs grown in California are diverse, if less than adventurous. Commercial fig varieties include the Black Mission, a dark-colored fig; the Calimyrna (the Turkish Smyrna variety, renamed when it came to California); the Kadota; and the

Plate 51. With a crunchy texture (thanks to their seeds) but a smooth and easy taste, figs are a favorite California tree crop, with a formidably ancient history of cultivation worldwide.

Adriatic (each of which is an amber or blond fig) and the Brown Turkey, each with a varying season. These amount to only a few of the hundreds of known fig cultivars grown in California. As the name would suggest, the Mission fig came with the padres, and there are accounts of figs being sold in Spanish California in the early 1800s. Figs are marketed in various ways: as the filling of cookies and pastries, sliced or compressed, diced, as a juice or concentrate, made into jam, presented as dried fruit, or purchased fresh for eating out of hand—none of these is quite as exotic as the uses in Greek, Roman, Egyptian, or Biblical times, but they are worth noting. A resolutely Mediterranean product, figs fit the strictures for those following the so-called Mediterranean diet. Their fan base remains strong.

The bearing acreage, never particularly large, is decreasing—to 9,400 acres in 2008—with production averaging 39,000 tons per year—still a lot of figs. The main problem with fig production is spoilage: the ripe fruit is not easily transported and once picked does not keep well. Madera and Merced, in the San Joaquin Valley, are the main producing counties, but figs can be found almost anywhere in the lower two-thirds of California. Northern California is too cold for a thriving crop. Figs are seen particularly often as small groves of a few dozen trees; locating any particular "center" for fig production is difficult. The value in 2007 was $25 million, about the same worth as kiwifruit, but just half the value of apricots. Clearly a valued crop in times past, the fig may yet rebound. It certainly has the cachet—and the history.

FRENCH PLUMS

SEE: PLUMS.

GRAPEFRUIT *Citrus x. paradisi*

Pl. 22

FAMILY: RUTACEAE.

NOTES: A CROSS OF *C. MAXIMA* AND *C. SINENSIS*.

RANK: U.S. #3; CALIFORNIA SHARE 14%.

Grapefruit is a curious obsession among its devotees: a subtropical citrus tree that produces a strongly bitter fruit. The tree has no special look, is of medium height (45 feet), and is simply citrus-like in terms of foliage or structure. What is distinctive about the grapefruit is its fruit, an oblate sphere of yellow flesh that looks like an orange or a pummelo (or pomelo) run amok. There is a reason for this: the grapefruit (like so many citrus varieties now prized and praised) is a hybrid of the pummelo (*Citrus maxima*) and the sweet orange (*see also* Oranges). Discovered in Barbados in the mid-eighteenth century, it arrived in Florida in the 1820s. The grapefruit is particularly important as a progenitor, since its descendants include the tangelo and the minneola.

Sensitivity to cold offers the limiting factor for grapefruit spread—aside from the acidity of the fruit. Florida and Texas both outproduce California, and Arizona is a fourth producing state, though at lower levels than California. California production has been downward, to between 11,200 (2008–09) and 9,400 (2009–10) cartons (the unit of measure for citrus). The notable story in California grapefruit is location, location,

location. By far the majority of grapefruit is produced in eastern Riverside County, taking advantage of the high temperatures that make Palm Desert such a delectable destination. San Diego and Imperial counties are also producers, but the surprise is in the migration northward of grapefruit from the desert to the San Joaquin Valley. Taking advantage of microclimates that afford protection from frost, both Kern and Tulare counties are respectable producers of grapefruit. Total value of California production was $67 million in 2007, but production has been declining in California as also in Florida, because of disease.

The major pleasures in sitting down to a grapefruit are fourfold. Some address the grapefruit as they might an orange, peeling and then dividing it into segments. Fie on that—connoisseurs understand that eating grapefruit is a diversified process, each with its delight. First, there is the color on cutting into the fruit: will the flesh be pink, red, or white? The second pleasure is in preparation. Although the runcible spoon is averred by some to be a nonsense term invented by Edward Lear for his best-known work, *The Owl and the Pussycat,* that poem does feature a runcible spoon, a term that was applied a couple of decades after Lear's death to a spoon with three or more teeth at the end. These nasty teeth, it turns out, are perfectly suited for the careful dissection of a choice grapefruit; they make it possible to separate the meat of the fruit from the membranes that segment it, and the dentition allows for cutting the flesh from the pith and rind. The penultimate charm is in dealing with the sourness—is honey, sugar, or brown sugar to be added—or will the tartness be taken straight? Finally, those with hands of a certain size will take the grapefruit, suitably emptied, fold it in one hand, and squeeze as hard as possible, sometimes assisting with a second hand, over a bowl or wide-mouthed glass, removing the (now) sweetened juice from what remains adhered to the peel. The art of grapefruit deconstruction is best taught to children at an early age, and leads to much hilarity at the breakfast table as small hands fall short. The process converts breakfast into an almost aerobic exercise.

While grapefruit have a local market, much of the production nationwide is exported; the United States is by far the world's largest grower, followed by China and South Africa. California cedes to Texas the "red" market in grapefruit, with patented

Ruby Red and Rio Red varieties that have not spread in any significant degree to California. The Marsh, Star Ruby, and Oroblanco hybrid are popular cultivars. An interesting side effect of grapefruit is its utter healthfulness: the "grapefruit diet" was popular in the 1980s, and may yet return because of the fruit's low glycemic index, which helps the body burn fat. A less helpful effect is drug interactions, which put those taking beta blockers or a variety of other compounds at some risk. But grapefruit is not believed to block the effect of the drugs; quite to the contrary, it accelerates their effect, increasing bioavailability, which leads doctors to warn some patients with hypertension to avoid grapefruit. Unfortunate, but something to watch for, subject to superior medical advice. Grapefruit are always dramatic fruit to spot in the field. Like many citrus, they can be "stored" on the tree for extended periods; until picked, they do not start to decline.

KUMQUATS ***Fortunella,* var. species**

Pl. 52, Table of contents photo

FAMILY: RUTACEAE.

NOTES: GENUS IS DISPUTED; SOME FAVOR A RETURN TO THE GENUS *CITRUS.*

UNRANKED—BUT CALIFORNIA DOMINATES.

Kumquats (also spelled cumquats) are invariably a surprise when found in the field, where they are trees of modest height (under 15 feet). Bright yellow-orange, with oblong fruit that looks like a small, stretched, orange, kumquats stand out from the evergreen and glossy-green leaves of the trees they grow on. Originally from China, kumquats were an ancient and appreciated fruit in Chinese literature. They arrived in the United States in the mid-1850s, and today are cultivated primarily in Florida and California. Within California, they grow as far north as the San Francisco Bay Area, resisting frost better than any other citrus variety, although they do best in warmer regions.

The kumquat fruit is generally eaten raw. The pleasure of eating a kumquat is very much a tribute to the cultural mix of its ancestors. It is at once a sour and a salty fruit. The rind is particularly distinctive in its mouth-puckering qualities (it shares this with the loquat, with which it is sometimes confused). Culinary

Plate 52. Considered to be among the oldest antecedents to modern-day citrus crops, the kumquat, with its modest market, is often used to provide desirable attributes in readily hybridizing citrus fruits.

uses are various, once the fruit is found: it is made into jams and preserves, jelly or syrups, and—claims one source—a slice is used as an alternative to the olive in a martini. Because of the color, the kumquat is very much prized in holiday seasons in Asia, where the color orange is considered both festive and propitious.

Production in California in commercial quantities is limited to San Diego County, where kumquats brought in $873,000 in 2008. Aside from a remarkable taste, the kumquat's most interesting trait is the modest battle being fought over its genus, which is *Fortunella.* That name is a tribute to Robert Fortune, a Scotsman who was an envoy of the British Horticultural Society dispatched to China to collect "curiosities." Proficient in Chinese, he disguised himself as a peasant so he could travel to forbidden sites. Fortune made four trips to China and one more to Japan and brought out tea cultivars and growing techniques for tea, helping to break the Chinese tea monopoly. He is credited with introducing bonsai to the larger world, and he transported the kumquat to England in 1846. Kumquats have been called "the little gems of the citrus family," which belies their movement from the genus *Citrus* to *Fortunella* in 1915. But perhaps the genus should be permitted to stay as it is.

LEMONS *Citrus limon*

Pls. 22, 53, Map 8

FAMILY: RUTACEAE.

RANK: U.S. #1; CALIFORNIA SHARE 90%.

Lemons, in song and sourness, are a highly successful crop in which California dominates U.S. production. Unlike much of California citrus, which made a move from Southern California to the San Joaquin Valley with notable success in the 1940s, the lemon has a favorite spot in California, half a state away along the coast in Ventura County, nestled between Santa Barbara and Los Angeles in what fruit growers and climatologists might describe as the coastal–intermediate region. The region includes

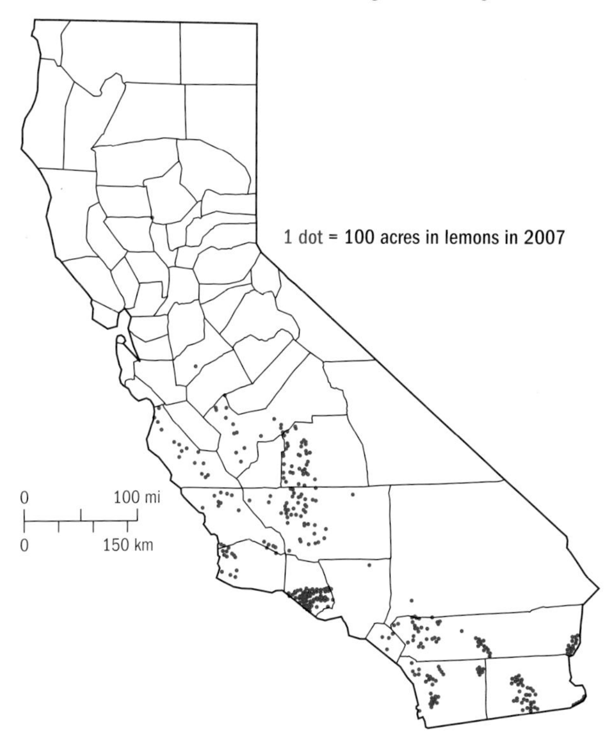

Map 8. Lemon acreage, 2007.

the Heritage Valley, with towns of Piru, Santa Paula, and Fillmore near the Santa Clara River, which are still significant producers of lemon and other citrus. (Piru is where novelist Helen Hunt Jackson set the home of Ramona, the protagonist of her most famous novel of the same name.) By itself, Ventura produces one-third of farm-gate value U.S. lemon output; the country earns 54 percent of lemon revenue in California. In turn, California has 80 percent of the acreage in the United States, with Arizona in a very distant second spot. Although lemons do not break the top 10 list of California agricultural products, they are raised on 47,000 acres (2009–10) and in 2008, lemons accounted for $428 million, which places lemons twentieth in gross value of agricultural farm-gate production.

Few other backyard fruit trees equal the charm of the lemon. Fruit from one year can winter over and still be usable a second year as the blossoms of a following year's fruit go into bloom. For reasons that boggle the mind, livestock and wildlife are fascinated by lemons, and watching a goat or deer take a lemon and eat it with juice dripping from its cheeks is something for a YouTube-closeup. Technically, the fruit of any citrus plant is a hesperidium, a fleshy and bright berry that ranges widely

Plate 53. Markedly concentrated in coastal zones of California, lemons are frost-sensitive, yet prolific producers when kept to an appropriate climate; seen here in San Diego.

in size, shape, and juice quality. The waxy leaves of the lemon have a distinctive citrus odor, but that isn't half the story. As the Peter, Paul, and Mary song has it, the scent of the lemon flower is almost impossibly sweet, in utter contrast to the fruit itself. Like many citrus varieties, lemons come in diverse cultivars. The Lisbon and Eureka, the Ponderosa, and the Meyer are commercial favorites. The Meyer lemon is a cross of the lemon with an orange, or possibly a mandarin, that is named for its discoverer, Frank Meyer, who identified the hybrid in 1908. Tough to ship, the Meyer lemon is rarely seen in commercial production but is a backyard gardener's favorite, and is widely used in the production of marmalade, pies, and tarts: cooking mutes the sourness yet boosts the sweetness of sugar used in cooking.

Although lemons are not Ventura County's largest agricultural crop (strawberries take that honor), they do account for a sizable share of that wealthy agricultural county's income. But lemons are also produced in Tulare, Riverside, Kern, and Imperial counties, and, to a lesser degree, in Santa Barbara, just to the north of Ventura. The lemon was hardly a native crop; Columbus is credited with bringing lemon seeds on his second voyage (1493), and lemons spread with the Spanish colonial presence in the New World. Miners during the Gold Rush era were prone to scurvy (like Britain's sailors in the same era), and as a preventative, lemons (and limes) sold in mining towns for $1 each or more. Lemons are a worldwide crop, and thrive in temperate Mediterranean-type climates from Israel to Spain and from Greece to Algeria and Morocco. Although yields falls below the production levels preferred by commercial growers after trees are 25 years old, there are examples of lemon trees still producing after 150 years.

An enduring wonder of lemon is its uses and the allusions drawn from its sour taste. A full catalogue of uses would be too long to spell out fully. Lemonade is a start—in its genuine form, not from a powdered mix—and the symbolism associated in the United States with the sidewalk lemonade stand as a training ground for capitalism is too rich to knock; lemonade offers a first contact with the rules of supply and demand; the next stop is presumably Ayn Rand's *Atlas Shrugged*. Lemons are an ornament for drinks—a slice of lemon is habitually added to a water glass in restaurants of certain aspirations. The use of lemon to prevent scurvy was a discovery of 1747 that benefited both the British Navy and Napoleon's troops. The use of citrus as an antiscorbutic was revisited by Kevin Costner in the otherwise less than convincing 1995 film *Waterworld*. Lemons are used in cleaning, to arrest discoloration when cooking, and to deodorize hands or counters when working with garlic or onion. Lemons can bleach pretty much anything, including human hair; some tellers will moisten fingers with lemons when counting out money, and lemons are favored for aromatherapy. Nor would any paella be served in an authentic restaurant in Spain without a wedge of lemon as garnish and supplement. That does not quite amount to ubiquity, but the lemon is close to an essential item in gentle living.

LIMES

Mexican lime — ***Citrus aurantifolia***
Persian, Bearss, or Tahitian lime — ***C. x latifolia***
Sweet lime — ***C. limetta***

FAMILY: RUTACEAE.

UNRANKED—ALMOST EXCLUSIVELY CALIFORNIA.

The very word "lime" is a spur to controversy: there are about 15 different "limes," each of a different species, and outside the United States many have distinct and significant uses. In general, limes in California refer to the Mexican lime, *Citrus aurantifolia*, which is also known as the Key lime, the West Indian lime, or the Bartender's lime. Although an essential part of California cuisine (in particular, some raw fish dishes, and the culture of tequila drinking), the low-statured (6 to 13 feet tall) but well-spined lime is less than a huge factor in the state's hierarchy of citrus, accounting for just $1 million, all that attributed to San Diego County by the ag commissioners' data in 2008. Limes do not feature among the top 70 crops in revenue earned.

That said, the fruit of the lime is a wonder. The tree has handsome, spreading branches, and a very tough set of axillary spines that do damage to the unwary. It was lemons and sweet limes (*C. limetta*) that "solved" the scurvy epidemic that afflicted the British Navy in its years of supreme authority over the global seas (which explains the generic, and in the day presumably less than slanderous description of Britons as "limeys"). The lime, as a citrus, has a widespread distribution, with India, Mexico, and Argentina leading world production, followed by Brazil, Spain, and China, with the United States in seventh place. Cultivation of the lime is thought to have originated in the Indo-Malayan region, but the fruit was likely carried west to North Africa; it was unknown in Europe until Crusaders took it north and west in the twelfth century. Limes were in cultivation in Haiti by 1520, and the fruit became naturalized in the West Indies and Mexico. From the U.S. point of view, the crucial step was its migration to the Florida Keys in the early 1800s. When Florida pineapples were destroyed in a 1906 hurricane, limes were the substitute crop, and Key limes became a signature crop, pickled in saltwater and sent to Boston. The acidity of lime (as with the lemon) makes it a superlative player in food dishes in which cutting down a cloying graininess is helpful; both Thai and Andalusian (southern Spanish) cooking benefit from the effect of limes on the tastiness and texture of foods.

Lime trees are smooth-wooded, with bumps and wattles, and the leaves are smooth and dark green, with a distinctive scent. Possessed of a stronger and sourer taste than lemons, they are used as a contrast in taste, and when acidity is essential. Limes are, for example, the preferred citrus for making ceviche, in which the high acidity in lime juice literally cooks or pickles raw fish. The recipe for *poisson cru*, or Tahitian tuna, would be meaningless (and perhaps less than safe) without limes. Lime zest is used in recipes in which its distinctive tang is required, and enhances the flavor of any dish that benefits from acidity: barbecue sauce, marinades, fish stews, and (of course) Key lime pie.

The lime's known limitation as a commercial crop should not be taken as an absolute; backyard lime trees are routinely in evidence through much of California. There are two varieties of lime that are common in the state: the large-fruited Tahitian type is the Bearss lime (*Citrus* x *latifolia*; in Florida, it is the Persian lime); the smaller lime is the Mexican, or Key, lime, and is the more common and the more acidic. Either is a suitable addition to a repast.

LOQUATS *Eriobotrya japonica*

Japanese medlar, nispero

FAMILY: ROSACEAE.

UNRANKED—ALMOST EXCLUSIVELY CALIFORNIA.

The loquat (the term refers to both the tree and its fruit) is a remarkable presence in California, a handsome and dark-hued evergreen with a rounded crown, a relatively short trunk, and branches that are whorled at the tips. The tree is not particularly tall, growing up to 30 feet, but can be kept shorter, and although it is prized internationally for its fruit, the tree is quite ornamental, and owes its origins in California and Florida to landscaping. Its profusion of white flowers, quite fragrant, is dense and in season can nearly cover the tree.

The loquat originated in China but was naturalized more than 1,000 years ago in Japan. Loquats are now grown in the Middle East, Europe, Africa, Brazil, and Hawaii, and throughout California. The loquat is considered a subtropical fruit but survives in a wide climatic range; cool weather is actually helpful to fruiting. Growing loquats close to the coast maintains an equable climate that suits the tree. Commercial loquat production plants the trees 12 to 24 feet apart, and some winter cover crops are used in the understory. Bearing loquat trees are heavy nitrogen users, so manure gleanings from stables often go to loquat plantations.

The variety introduced first to California was the Giant, brought as an ornamental in the 1870s because of its tropical look. But public fascination with the loquat spread. There are allusions to promotion of loquat-eating by a Southern California spiritualist commune in Placentia, the Societas Fraterna, that 100 years ago was held to be "notorious for its unusual sexual practices." The commune's leader, who called himself Thales, named a variety of loquat the Gold Nugget; another variety is the Thales. As befits a nineteenth-century arrival to California, there is a great deal of horticultural improvement since, much of it from local fanciers. A rather more accomplished loquat breeder was Charles P. Taft, who lived near the town of Orange, and produced many of the better-received varieties. Orange-fleshed cultivars include the Golden Red, Gold Nugget, Early Red, Premier, Tanaka, and Wolfe; white-fleshed loquats include the Advance, Champagne, Herd's Mammoth, Chatsworth Victory, and Vista White. Each has distinct attributes and impassioned local constituencies.

The key with the loquat is its taste, which is delightfully exotic. In Southern California, the fruit ripens in April, in May farther to the north. Loquat production at a commercial level is small—the fruits do not transport well—but the flavor is a cross between a passion fruit and a guava, sweet and sour at once; on biting into one, up to four quite large, inedible seeds are unearthed. Loquats may be served in fruit salads; some make wine from it; still others put the fruit to work in baking.

A curiosity of loquats in the United States is their land-use and "ownership" status in the Southern California yard. The horticultural editor of the *Los Angeles Times* suggested in 1899 that every yard should have a loquat, which led to obedient plantings throughout the Southland. The fervor faded in a generation, but trees did not. Often unrecognized now in California, the loquat fruit—which can be quite prolific, when ripe—is in many cases simply allowed to drop to the ground; livestock and wildlife such as deer make quick use of it. Back in the 1930s, the fruit was better known, and there were elaborate rules for taking loquats, constituting an informal but customary tree tenure, as recorded by geographer Homer Aschmann. Children could not go into backyards to gather loquats (leastways, not without permission), but trees in the front yard were fair game. As a consequence, almost no loquat fruit ever reached its bright yellow-orange maturity; instead, children would pick underripe fruit to prevent others from getting it first. The conventional result was stomachache, but presumably also a good lesson.

MANDARINS AND MANDARIN HYBRIDS

Clementine — ***Citrus clementina***
Mandarin — ***Citrus reticulata***
Satsuma — ***Citrus unshiu***
Tangelo — ***Citrus x tangelo***
Tangerine — ***Citrus x tangerina***

Pls. 22, 54, Table of contents photo

FAMILY: RUTACEAE.

RANK: U.S. #1; CALIFORNIA SHARE 48%.

Spotting the mandarin or tangerine tree in the field can produced a startled reaction; after mile upon mile of sweet oranges, the mandarin appears dwarfed by the larger orange. The trees are not dissimilar; the mandarin can be as tall as a sweet orange, depending on the variety. Like other citrus, the bark is relatively smooth.

For such an attractive and easily eaten fruit, the mandarin and its close relations (clementines, satsumas, tangerines, pummelos, tangelos, Murcott, and mandarinquats) sow a great deal of taxonomic chaos. The genetics and interbreeding of the class of oranges with a thin, loose peel is complicated, not least because "tangerine," which was for generations the standard term, is less favored by growers now, who side instead with the name "mandarin." To be accurate, the name "tangerine" was used for a specific cultivar, the Dancy mandarin, which was imported from Tangiers, and subsequently called the "tangerine." But mandarin is the currently agreed-upon name for all fruits in the category. Also in the group is the W. Murcott (derived from the Afourer variety of Morocco, and released to growers in 1993), the tangelo, and various crosses, including the "mandarinquat," an outcross with the kumquat. Because a significant part of California's citrus production is exported, it bears noting that the mandarin (and kin) are much more popular in Asia than in the European markets.

The so-called zipper skin of the mandarin makes it highly popular for lunches and among children, and there is some opinion within the citrus industry that loose-skinned fruit may mark the future path of the citrus industry. The United States dominates North American production but is dwarfed by Brazil, Spain, and Japan. All countries grow less than China, which yields a third of global production.

Plate 54. The loose skin characteristic of the mandarin is a feature often passed on in hybrids, as here in an Indio mandarinquat, seen in the University of California's Lindcove Citrus Research Station orchard.

The mandarin orange is more cold-tolerant than the sweet orange, and the fruit is relatively cold-hardy. But like all citrus, the mandarin family is a true subtropical crop, growing in a band of 30 to 40 degrees latitude on either side of the equator. The cultivars include the mandarin, the tangerine, and the satsuma. Added varieties exist, many of them hybrids, including several recognized in California only in the last decade: the King, Murcott, Clementine, and Temple. Hybrids of the mandarin with grapefruit are called tangelos; California produces few. The mandarinquat is a hybrid of the mandarin with the kumquat, and can go in either category. Determining which citrus is which, within this entry, is not easy. In an palpably candid entry, a Purdue University horticulture professor notes that "citrus classification [is] based on morphology of mature fruit, and is considered confused at present."

California produced 13,400 cartons of tangerines and mandarins in the 2008–2009 season, more that 60 percent of the U.S. commercial crop, with Florida the second-rank producer and Arizona a distant third. For years, Florida had the largest share, but no longer. Production in California is split between the San Joaquin Valley (Kern is the largest producer, Tulare second) and secondary production in the desert regions of Riverside and Imperial counties, with coastal groves in San Diego and Ventura. Harvested acreage is 30,000 acres for 2009–2010 is on a marked upswing, and worth $286 million (2008).

NECTARINES

SEE: PEACHES AND NECTARINES.

NURSERY CROPS (TREES)

SEE: NURSERY AND GREENHOUSE CROPS, FLOWERS AND FOLIAGE.

OLIVES *Olea europaea*

Pls. 28, 55, 125

FAMILY: OLEACEAE.

RANK: U.S. #1; CALIFORNIA SHARE 99%.

A modest tree with a gnarled trunk cloaked in dusty-gray foliage: that's the olive in its usual manifestation. The olive is among the most ancient tree crops on earth; perhaps it is the oldest of all cultivated trees, with more than 2,000 cultivars, or variants, each named and respected within its place of origin (Italy alone has more than 300 cultivars). California has no native olives, but with hot and dry summers several dozen varieties were adopted, beginning in Spanish–Mexican times, most quite successfully. There are 750 million olive trees cultivated in the world, 95 percent within the Mediterranean region. California has the mouse's share—under 5 percent of that quotient—but essentially all the commercially cultivated olives in the United States are Californian. In the last decade, California has turned from canning edible ripe olives, a curious form of the fruit, to generating some of the best olive oil and eating olives produced in the world—fully a match for the European, African, and Near Eastern equivalents. Considering that olives are the common ingredient in table olives, ripe olives, and olive oil, a surprising degree of mutual disdain exists among the producers of each sort of olive-derived crop; specialization may take a toll.

Olives are a most curious item. Utterly ancient, in terms of cultivation, olives are complicated to deal with and to process. As anyone who has bitten into a raw olive knows, they require significant and persistent processing to become edible; the glucoside contained in olives, exceptionally bitter, has to be broken down for olives to be consumed, which requires either persistent rinsing in fresh water (over months of time) or treatment with wood ash or lye or other substances (including salt packing) that break down phenolic compounds and oleuropein, a bitter carbohydrate that makes the raw olive inedible. Ripe (black) olives, other eating olives, and olive oil are all processed differently.

Black ripe California olives, usually canned, are a distinctive crop with a longstanding fan base. Californians have long promoted eating the black olive, which is allowed to ripen on the tree to a red or dark-red shade before being picked. Fruit destined for the ripe olive market, which can be from several varieties, are placed in a lye solution when picked. The olives are then softened and processed. The traditional European treatment employs a lye made from wood ash, ideally from oaks. That extremely basic solution purges the glucosides from the olives, in effect cooking out the inedible compounds using an ancient chemistry.

What turns the California olives black is air bubbled through the olives after "the bitter principal" is removed; the addition of introduced oxygen completes the change, and helps hold the pitch-black color. There are two centers for California ripe olive production that also happen to be home to the two ripe olive processing plants in California. One ripe olive hearth is in the Sacramento Valley (Sacramento, Glenn, Tehama, and Butte counties), and the other is in Tulare County (centered around Lindsay), in the central San Joaquin Valley. Production quotas and quality for ripe olives are under the watch of the California Olive Committee. Until recently, about 85 percent of California olives went to curing and canning. The remaining olives went to table eating, or a small residue to olive oil. Spain, for example, is the opposite—90+ percent of the national olive harvest goes to oil. But the proportion directed toward olive oil in California will increase.

Production of olives in California varies by purpose. First, California has essentially all the U.S. production of olives. The olives come, however, in three forms: ripe and green (for eating by hand) and crushed (to make olive oil). The table olive, which can be green, or mottled, or a black (ripe) olive, is a deposit on greatness. But to get there, it must be modified to capitalize on its value. To become edible, green olives are either processed by repeated rinsing, or by addition of wood ash or lye, which breaks down the unpalatable elements in raw fresh olives. These techniques, all post-harvest, take time. Black olives are scattered on salads or pizzas or eaten in hand. Green olives are brined or vinegar-cured, and are purchased in jars, or selected from urns or vats in stores, where they are kept in marinade. There are also salt-cured olives, unusual but delectable, that bring elevated prices and high demand. All these have undergone processing, and processed olives have a peculiar genius for taking on the flavor of whatever they are cured in—the addition of lemon, pepper flakes, *herbes de Provence*, thyme, rosemary, or vinegar to a brine solution can make all the difference, adding a piquancy unique to each maker's mix.

Although canned (or jarred) olives still outnumber olive oil production by a ratio of 4:1, that will likely change. With care, olives picked green from the tree can be crushed and processed and yield top-flight olive oil. The oil is not only a simple and perfect product, it is a staple of cultures where the olive is naturalized, and when produced with care is worth a mint. Although there has been artisan-produced olive oil in California for centuries, the commercialization of olive oil production is quite a recent phenomenon. Between 1996 and 2004, olive oil production in

California increased 168 percent, and it continues growing. In 2008, the California Olive Oil Council estimated that olive oil production from 400+ growers topped the benchmark level of 500,000 gallons. On the other hand, the United States does import 65 million gallons of olive oil each year, so a half-million gallon production in California will not significantly dent world demand—California is able to produce only 0.6 percent of the olive oil consumed in the United States. There is both interest and demand for California extra virgin olive oil. Scandals linked to the adulteration of Italian oil have not helped the cause of European Union olive oil—part of the problem is a $150 per ton grower subsidy on top of what processors pay for a crop.

As a culinary practice, making olive oil easily dates back to 2000 BC, as found in the export records of Egypt and Phoenica. Some credit the birth of Athens to an olive tree that sprang from a rock cleft in a battle between Athena and Poseidon. Pliny catalogues 15 olive varieties cultivated for oil in his era. An olive tree in Crete is claimed to be 2,000 years old. Two live trees in Galilee are over 3,000 years old—and still produce. There are literally thousands of cultivars of the olive; the Internet offers a seemingly unending list—over 1,100 names. California-grown varieties include Manzanillo, Mission, Arbequina, Koroneiki, Picholine, Picual, Sevillano, Ascolano, and Barouni, but there is no lack of choice even after these.

In the nineteenth century, California produced olive oil in fair quantity, but the market was considered marginal, especially when use of other vegetable oils around the country blossomed with federal support. Olive oil production slowed to a minimum until the 1990s, when up-to-the-minute European techniques for making olive oil were refined in California with some technical additions and a cottage industry in premium olive oil making went large-scale. When olives are processed green for oil, native phenolic compounds add acidity and make for a distinctive green or leafy taste; if allowed to stay on the tree longer (especially in the case of varieties that add flesh with age), the oil will be trend toward the more buttery and mild.

Spain is the largest global producer of olive oil (36 percent), with Italy following at 25 percent. In 2008, legislation signed into law in California conformed olive oil labeling and grades to standards of the International Olive Oil Council. With that, California joined the international convention for grading its olive oil.

Plate 55. Olives came to California with the missions, but until recently were used largely for the production of ripe (black) olives. However, olive oil production and green olives are beginning to regain a prominence they had 100 years ago.

California producers are now bound by the nomenclature and standards of other olive oil–producing countries, which establish acidity and taste rules for extra virgin, virgin, pure olive oil, olive oil, and olive–pomace oil. Packaging will change to match.

Olive trees, by nature, bear fruit in alternate heavy and light years, producing a large crop one year and a short crop the next. The farm-gate value of olives averaged from 2005–2009 was $60 million, which would put olives about sixtieth on the "value" list of California ag products. Olive harvests run from September into November, and the harvest traditionally has involved labor-intensive hand work, although there are interesting experiments afield. Machine harvesting of olives is a developing technology, and in California, agriculture and technology are never far apart. The Oroville-headquartered California Olive Ranch in 2006 expanded onto an additional several thousand acres outside Artois, in the northern Sacramento Valley, and planted acreage using a superhigh-density (SHD) planting system, with 550–670 olive trees per acre, neatly hedge-rowed and trellised, looking more like grapes than olives. Using Arbequina, Arbosana, and Koroneiki varieties, the olives are pruned to about a two-meter height, and a mechanized harvester passes down the rows picking one tree every three seconds. With techniques imported from Spain, the operation anticipates doubling extra virgin olive oil production on-site to an average of 263,000 gallons per year, milling the olives for oil in facilities in Artois and Oroville. In 2009, they are already producing more than 50 percent of the olive oil in California. Little wonder that Glenn and Butte counties show a sudden spike in harvested acreage. Ending on a perhaps telling note, the California Olive Ranch is owned by a dozen investors from Spain.

ORANGES *Citrus sinensis*

Pls. 6, 17, 22, 29, Acknowledgments photo

FAMILY: RUTACEAE.

VARIETIES: NAVEL AND VALENCIA.

RANK: U.S. #2; CALIFORNIA SHARE 30%.

The orange tree is never difficult to recognize, with its extraordinary long dark leaves and moderate height (sometimes top-pruned to keep fruits accessible to pickers), growing in neatly maintained groves. Orange-growing through the last 200 years was the ultimate California prestige crop; among devoted grove owners, it remains that. If perhaps vineyards may have come to surpass oranges, that probably has more to do with the present-day locations of premium and superpremium vineyards within an hour's drive of the coast, whereas sweet oranges now are overwhelmingly grown in the San Joaquin Valley (Tulare, Kern, Fresno, Madera, with more than $1 billion in navel and valencia production in 2008), exiled from Southern California by the water challenges and the pressures of urbanization and suburban expansion in the 1940s and 1950s. Even today, the clean regularity of an orange grove, seen from the air or on Google Earth, shows a crop that when maintained has a clean and precise order to which few other crops can aspire. Avocados, for instance, which are often grown in upland areas near citrus, are a slightly chaotic, scruffy crop, and unapologetically so. The contrast could not be more stark.

In the day—and for oranges, that meant from the 1880s into the 1940s—there was nothing so prestigious as coming to own a small orange grove—10, 20, 40 acres—and living the good life along one of the dozens of lines where interurban trolleys delivered residents from Southland towns such as Pasadena,

San Dimas, Highgrove, Riverside, and Monrovia to downtown worksites in a snappy 45 minutes or less. In Southern California, and especially Los Angeles, Orange, and Riverside counties, oranges hit a peak of 165,000 bearing acres in 1945. That year marked the largest statewide acreage planted to oranges, and acreage has dropped ever since. In 2008, just 77 acres in Orange County were harvested, and Los Angeles reported nary an acre.

Despite the dramatic change of venue to the San Joaquin Valley in the last 50 years, orange-growing does big business among California tree crops. Although the 184,000 acres of all oranges (2009–2010) is dwarfed by 710,000 acres of almonds, income from oranges is just three notches down on the "value" list, at eighth in farm-gate value, with $1.1 billion in revenue from navel and valencia oranges in calendar year 2008, according to ag commissioner reports. Firm citrus numbers are difficult to come by, because some statistics are kept from July of one year through June of the next. Production also varies considerably from year to year, depending on fruit-set and weather, and nationwide orange acreage is down 200,000 acres from recent historic highs in 1997–1998.

The sweet-orange production in California is split between two major varieties, navel and valencia oranges, which are divided about 3:1 in value. Florida offers the competition, with slightly more dollar-value total production, but Florida produces more valencia (juice) oranges than navels. The difference in the varieties is significant: the navel orange was a mutation (a mutant blood sport) without seeds, and with a distinctive dimple (actually a conjoined twin) that looks like a "navel." Discovered in 1820 at a monastery in Bahia, Brazil (some texts claim Salvador, Brazil), cuttings of the navel were sent by a missionary to the National Agricultural Garden, then in Washington, D.C. After years of cultivation, cuttings were shipped in 1870 for grafting to an inquiring client in Riverside, California, and the navel orange business was begun. With this checkered history, the navel is variously known as the Bahia, the Washington, or the Riverside navel orange—a tribute to its stops along the way. That 1870s arrival spawned an outsize interest in the easily peeled navel.

The comparative advantage of the valencia, by contrast, is late-season production, which means the grower who has both can have a longer growing and harvest season. For all that, oranges are often left on trees after they are apparently ripe, because they will go unspoiled for several months after peak sweetness is reached. This is done because prices will fluctuate with deliveries to the packing plant, so holding onto a crop can produce better income. Then again, a hard freeze will ruin fruit left on an orange tree, and pickers must be contracted with some time in advance of harvest, so the waiting game is always a gamble. Ventura and San Diego counties ($46 million in oranges in 2008) are again the beneficiary of a coastal-modified climate that makes freezes less likely—but because near-coastal growers assume that they will not have hard freezes, they are somewhat less prepared when deep chills come than are orange growers in the San Joaquin, who anticipate the occasional cold snap. Frost–freeze warnings from public sources and weather reports provided by private meteorologists are crucial to citrus growers, who can see an entire crop wiped out in a deep freeze. And the resources and countermeasures available are not as varied now as 40 years ago; smudge pots burning old crankcase oil are forbidden for air-quality reasons, but growers get by.

It's likely that the definitive influence on orange-crop value in California, and to an extent for all citrus, was the creation of Sunkist Growers, the orange cooperative formed in 1893 to unite local associations into a single statewide not-for-profit cooperative. Before that, distribution was intermittent, quality control suspect, and revenue weak. Chartered originally as the Southern California Fruit Exchange and later as the Fruit Growers Exchange, the Exchange began using Sunkist in 1908 for marketing, and officially shifted the name of the Exchange to Sunkist in 1952. Advertising began in 1907, and orange sales increased by 50 percent; with time, other citrus crops were added to Sunkist's efforts. Orange crate labels were both a successful merchandising tool and a dramatic expression of grower identity, and an estimated two billion labels went out on the end panels of wooden orange crates. Districts were formed, and within each were local associations. Packing houses were created in each district to guarantee oversight and a central organizing point, and among growers, orders were prorated to keep fruit moving. Quality control and marketing were major emphases, and Sunkist was incorporated in 1895. The reach extended into the San Joaquin Valley in 1905, and after the 1906 San Francisco earthquake, when wood supplies dried up, the Fruit Growers Supply Company was created to handle timber ownership, logging, company towns, and sawmills, all an artful vertical integration to create boxes or crates for fruit. In 2009, Sunkist still owns 360,000 acres of California forests. For growers whose families have been Sunkist members for several generations, it is almost impossible to convey the degree of loyalty and appreciation lavished on the brand name; they recognize how a trademark has changed lives. If nonmembers are less sanguine, estimates are that 6,000 of California's citrus producers are signed up under the Sunkist umbrella, with total revenues averaging $1 billion from 2004–2008, 45 percent of that from exports abroad.

Sweet oranges include the Valencia (first found in the Azores, and perhaps of old Portuguese origin), the Washington navel, the Cara Cara Pink navel (found in Venezuela in the 1970s), and the always-dramatic Moro, Tarocco, and Sanguinelli blood oranges, known for dark-red streaks within and for an especially distinctive juice. Oranges, like all citrus, have segments known as carpels that can be pulled apart as the fruit is peeled. The bitter orange, *Citrus aurantium*, not common in California, is another species that is mainly cultivated for canning, to provide rootstock for less vigorous species, and—as with the Seville orange, or *bigarade*—that is raised for marmalade, compotes, and liqueurs. Bergamot oranges (*C. aurantium,* subsp. *bergamia*) is yet another Italian cultivar that is used to produce bergamot oil, key to the aroma of Earl Grey tea and some perfumes.

Valencia and especially navel oranges are the dominants in California agriculture, and with aggressive testing of new varieties these are likely to remain on the scene, earning the gratitude (and sometimes frustration, when weather turns inclement) of their grove owners.

PEACHES AND NECTARINES

Peach ***Prunus persica***

Pl. 56

FAMILY: ROSACEAE.

RANK: U.S. #1, CLINGSTONE AND FREESTONE PEACHES; CALIFORNIA SHARE 100% AND 54%—OVERALL, 70%.

Nectarine ***Prunus persica*** **var.** ***nectarina***

FAMILY: ROSACEAE.

RANK: U.S. #1; CALIFORNIA SHARE 98%.

Peach and nectarine orchards are ungainly, disorderly things, even when carefully tended. Like their relative, the almond (another drupe), the trees throw out branches and lack the neat look of walnuts or cherries, or even the dusky order of olives. But peaches are a significant producer, worth some $498 million in 2008—nineteenth in value. Nectarines clock in at twenty-sixth, worth another $284 million. Both are members of the "Family of Flowering Plants," Rosaceae, which includes apples, pears, strawberries, roses, and apricots. The family tends toward the spectacular during flowering season, and peach blossoms in particular are a visual feast. Peach and nectarine flowers attract bees, and are quickly pollinated by wind or honey bee. If there is a downside, it is the relatively short lifespan of peach and nectarine trees in commercial production: 15 to 20 years.

Peaches and nectarines are closely related, with the nectarine a bud variation that yielded the "fuzzless peach." This makes the nectarine a smooth-skinned cultivar of the peach, distinct by only one recessive gene. In fact, the nectarine occasionally produces peaches, and the peach on occasion grows nectarines. Peaches in California, to make things slightly more complicated, come in two related forms. The clingstone peach, which holds tight to the pit, is used generally for canning, juicing, or drying, or goes to baby food. By contrast, the freestone peach readily breaks away from the pit and is generally eaten in hand. California produces a majority of both varieties of peaches, but in nectarines and clingstones, all but 1 to 2 percent of the U.S. production hails from California. All three varieties—clingstone, freestone, and nectarine—do best in environments with well-drained soils that have abundant nitrogen and water, and require a fair amount of effort thinning and keeping pests under control.

Producer of a succulent fruit, the peach tree was adopted early in American life. Peaches appear in the inventory of crops at George Washington's Mount Vernon. Peaches came to California with the arrival of the Spanish and took hold. The varieties raised in California vary according to the harvest season of each: freestone varieties include Babcock, Fay Elberta, Forty-niner, Loring, O'Henry, Rio Oso Gem, and Veteran. They are shipped in standard two-layer, tray-packed "panta-pak" cardboard boxes. The clingstones are more concentrated into fewer cultivars because of processor preference, but include Fairtime, Indian Blood, La Feliciana, Nectar, and Suncrest. Although the producers are varied and scattered through the Sacramento and San Joaquin valleys, there are half a dozen processors of clingstones, and peaches canned in various densities of syrup are the result. They tend to be concentrated in the middle Sacramento Valley, where clingstones predominate. An annual Peach Festival in Marysville is attended by 50,000 area residents in a good year.

Peaches come from two different areas of California, the San Joaquin Valley (Fresno, Tulare, Stanislaus, Kings), and from the Sacramento Valley (Sutter, Yuba). Although some freestone peaches are grown north of Sacramento, they are more common to the south; clingstones are particularly effective crops in the central-north Sacramento Valley, where the soils are especially deep. There is equally a cultural divide in their keepers. Many of the largest clingstone (and French plum) producers in Yuba and Sutter counties are South Asian Indians (many of them Sikh), and there are Sikh temples in the vicinity as a result. Specialization in agricultural production is not just a matter of crops; on the human side, a sort of chain migration into the same business is common.

Home gardeners will gravitate to individual peach trees, or to small orchards planted to a few dozen trees. Because the fruit is delicate, it takes a brave orchardist to mount a U-pick operation with peaches. Other, more robust fruit generally fares better. An advantage of both peaches and nectarines is the availability of semidwarf or dwarf rootstock, which limits height of the mature trees. Fruit size is unaffected; a flowering-only (nonfruiting) variety of peaches also exists, for fans of flower arrangements and ornamental display rather than fruit yield.

The nectarine is not a recent arrival on the scene; it dates back 2,000 years and was known in Antiquity as the nut of Persia. The name is owed to the nectarine's stop in Greece, where nectarine juice was considered a drink of the gods. Experts forthwith attached "nectar" to the fruit's name, and it stuck. Nectarines were not common in early California, arriving around Gold Rush times. Varieties were so delicate that they bruised with any travel, so the California nectarine industry began at a commercial scale only after a sturdier fruit was developed in the 1950s; today, it is concentrated in Fresno, Tulare, and Kings counties. Current varieties come, like peaches, in white and yellow shades, and their names are evocative: Arctic Star, Arctic Glo, June Pearl,

Plate 56. A clingstone peach is a distinctive crop in California, grown predominantly for processing (canning) in the Sacramento Valley, ceding other regions to the freestone (fresh market) peach.

Arctic Queen, and Fire Pearl (for white nectarines) and Summer Bright, August Red, and September Red (for yellow nectarines). Nectarines tend to be denser and somewhat sweeter than peaches, and few go to processing. The dollar value of peach production in California is sizable, with freestones cashing in at $318 million (2008), and clingstones $159 million. With nectarines, collective worth of the three was $782 million.

PEARS *Pyrus* (more than 30 species)

Pls. 18, 57

FAMILY: ROSACEAE.

RANK: U.S.: #2; CALIFORNIA SHARE 29%.

The adage of a friend whose master's thesis addressed leakage through the levees of the Delta endures: pears like wet feet. Indeed, San Joaquin and Sacramento county remain two of the major producing areas for pears in California, along with an upland pear-growing district in the North Bay's Mendocino and Lake counties. Near the edges of Delta levees in eastern Contra Costa, Solano, and San Joaquin counties, pears are seen right by the lower edge of levees that were built at the behest of a remarkable labor commitment in the nineteenth century to hold back waters of the there-conjoined Sacramento and San Joaquin rivers. Where spring waters run high with water seeping from gopher holes, undermined soil strata, and weaknesses in the engineered firmament, leaks trickle from the Delta waters into the adjacent levee bottoms that lie as much as a couple of dozen feet below sea level—the pressure of a 20-foot drop from river level to levee bottom ensures that any weakness will produce a slow (or quickening) leak, puddling at the levee edge. There, with amazing consistency, grow pears, content to exploit the fragility of human engineering.

In value, there is considerable variation in pear production from year to year. In 2008, 14,000 acres of pears brought in $106 million. In 2008, however, Fresno County was the biggest single producer of Asian pears (raised in China for 3,000 years)

Plate 57. Although pears retain value and interest as tree crops, the market is less than kind to them, as with apples.

on a paltry 900 acres across that county, yet the price gleaned in 2006 for the robust Asian pear (*Pyrus serotina*) in Fresno was an exemplary $23 million: almost $2,800 a ton. This brings to mind another of the modest economic lessons of California agriculture: when demand exceeds supply, poised growers can mint money. But by contrast, Bartlett pears in 2006 produced one of the grimmer sights in California agriculture: harvested fruit dumped alongside the road, with 10,000 acres of pears left unharvested because of market limitations and labor shortages, with prices in August and September averaging $287 a ton—barely a third of the July 2006 price of $759/ton, and just one-tenth the price of the Asian pears in Fresno. The sense of waste, and the aroma of discarded and fermenting pears, were everywhere. Considering the spectacularly attractive scene that a branch of pear blossoms offers in flower arrangements or a well-ordered field, the decay of the pear industry seems an abuse of faith and aesthetics. Pears produce for an average of 50–75 years, although some are still active after a century. The main problem is spacing and efficiency; Bartlett orchards planted long ago on a 20- × 20-foot block are woefully inefficient by current standards, and replanting would substantially increase the trees-per-acre tare. But at what cost?

Closely related to apples in physical structure and growth patterns, pears come in assorted species, more than 30 altogether, with further cultivars at the subspecies level. This reflects in part the difficulties of negotiating plant taxonomy among "lumpers" (who combine similar plants) and "splitters" (who are all for creating new species based on morphological differences). The Roman Pliny, in his *Natural History*, catalogued more than three dozen pear varieties. In California, three basic types of pears are grown: European or French pears, including Bartletts, Bosc, Seckel, Comice, D'Anjou, and red pears; Asian pears, sometimes known as the "apple-pear" because of their drier texture and ability to take handling; and hybrids with varying attributes. The United States is a net exporter, and California is a substantial pear producer. In Gold Rush days, it used to be said that miners would come by boat to the Sacramento Delta towns of Locke, Groveland, Isleton, and would start bidding for pears before the paddle-wheels stopped turning.

Pears may be consumed in various ways: dried, as juice, fresh, and canned. There are significant ancillary uses of the trees themselves, with pear wood a favored material for woodworkers. The trees are large and require a generous spacing (and although they do fine in saturated soils, they prefer well-drained lighter soils). Bartlett pears (*Pyrus communis*) constitute about 60 percent of California production and are picked at a more advanced stage of development in order to increase their salability. Bartletts have several advantages: they are self-pollinating, so they do not require honey bees for pollination, and the fruit does not ripen acceptably on the tree, so they actually have to be harvested when green, and then allowed to mature off-branch.

In their day, pears were a prestige crop: they never quite reached the summit height of oranges, but they were close contenders. Collectors of crate labels revel almost as much in the labels for pear crates as for orange crate labels, and some display a great deal of ingenuity, if not quite the artistic genius of the Schmidt Lithograph Company, based at Second and Bryant in San Francisco, at the turn of the nineteenth century. The firm's artwork was inspired.

Pears are tall, even stately trees, but they have a tendency to grow, and to grow *up*. As a consequence, every few years, pears are topped with a rolling buzz saw brought in to chop off the top branches and limit a pear orchard's height to something manageable—perhaps 30 feet, instead of the potential 50-foot height. To see those four- to six-foot-diameter blades whirring is to observe a teenage boy's horror movie script in practical form.

On 14,000 acres (2008), California produced 250,000 tons of pears in 2008, valued at $106 million—about 30 percent of the U.S. total. California is surpassed in production by Washington State at $171 million; Oregon is also a significant producer. In California, about equal quantities of pears are dried and processed or eaten fresh. Pears will keep for several months after picking if they are cooled promptly and kept in refrigerated storage.

PECANS *Carya illinoinensis*

FAMILY: JUGLANDACEAE.

RANK: U.S. #8; CALIFORNIA SHARE 2%.

Pecans in California are a crop regarded with affection, although they register only as a minor blip in the agricultural value spectrum. In 2008, some 3,500 pounds of pecans were produced in California, worth a respectable $4 million in 2008 but only a tiny fraction of the national pecan production of $269 million. That said, there is pronounced demand in California for pecans, and the large deciduous tree, which can grow up to 130 feet in height, is attractive. Almost half of California's acreage in pecans is maintained in Tulare County, with the remaining acreage scattered around the San Joaquin Valley. The main problem with the spread of further plantings is climatic: humidity and high summer heat are needed, and the trees are very large, larger than walnuts or chestnuts.

The most distinctive feature of the pecan, aside from its nuts, is its status as a native tree of south-central North America, ranging from Mexico south to Jalisco and Veracruz and north to the lower Midwest of the United States. The trees were introduced to Europe in the seventeenth century, where the wood was valued perhaps more than the buttery nuts. The long-lived pecan tree grows best in deep, well-drained soils, and a hot climate is essential to maturing the nuts, which take at least 180 days to mature. "Pecan," according to taxonomic guides, derives from an Algonquian word that means a nut best cracked with a stone, a tribute to the hardness of the shell.

PERSIMMONS

American persimmon — ***Diospyros viginiana***

Japanese persimmon (kaki) — ***D. kaki***

Pl. 58

FAMILY: EBENACEAE.

RANK: U.S. #1; CALIFORNIA SHARE 99%.

The persimmon is an odd fruit, shining bright orange to almost red as it matures, with a firm flesh that can seem forbidding. Its internal texture is almost like pudding, high in tannin in some cases, but nonastringent and so delicious that in the *Odyssey*, those who ate the persimmon forgot about returning home and stayed to eat the fabled "lotus," which most Classical scholar-agronomists conclude was the persimmon. Some suggest that it was the "fruit of the gods." The persimmon is actually classified in the ebony wood family (Ebenaceae).

Persimmons can be eaten fresh or dried, raw or cooked. Eaten fresh, they are quartered and are eaten like an apple. There is

Plate 58. The dusty persimmon stays on the tree here in Tulare County—an import from China that is grown largely in the San Joaquin Valley as a specialty crop.

nothing quite like the yielding, popping, texture of the fresh persimmon; most who eat it like it, but the barrier to consumption is inexperience, rather than intolerance. Once picked, persimmons can be held at room temperature and will continue ripening. There is a strong tradition for cooking persimmons in the United States, especially in desserts such as persimmon pudding.

The persimmon is anything but insignificant in California agriculture—it brought in $27 million in 2008, on 2,312 acres, strongly concentrated in the Sacramento Valley (Fresno and Tulare) but with added acreage in San Diego, Sutter, and Riverside for a wide distribution through the state. The trick with persimmons is to allow them to mature sufficiently to overcome the tannins in the fruit, and to give them time to soften. Because the persimmon is a relatively new crop in the United States, cultivation techniques are still evolving.

PISTACHIOS *Pistacia vera*

Pl. 59

FAMILY: AVACARDIACEAE.

RANK: U.S. #1; CALIFORNIA SHARE 96%.

The pistachio was introduced to California in 1904 as a hobby crop, although an experimental planting at the Chico Plant Introduction Station was established in 1917. Federal and state records of production were not kept until 1977. Pistachios took time to catch on. From 1,700 bearing acres in 1976 (some 26,000 acres were not yet bearing), pistachio acreage has exploded to 118,000 acres, with another 30,000 acres planted but not producing commercially in 2008. The graph bears imagining. The yield is 3,000 pounds per acre (averaged over four years, since the pistachio bears heavily in alternate years). At $561 million in farm-gate value (2008), the pistachio is seventeenth on the list of California crops, growing best in the driest parts of the San Joaquin Valley (Kern, Madern, Tulare, Fresno, and Kings counties). The pistachio's rise in crop value is meteoric, placing it atop longtime contenders such as peaches, lemons, avocados, nectarines, plums, and cherries, all of them among the

Plate 59. Few crops have the innate magnetism of pistachios, which all but ask for an audience as they ripen from green, go to yellow with a pink blush, and see their outer skins split open, exposing the nut.

top 30 dollar-value crops. Among tree crops, only almonds, oranges, and walnuts rank above pistachios. In 2008, California was the world's second largest producer of pistachios, with nearly half the production sold abroad—133,000 tons, in 2007–2008.

The pistachio tree has a slow-growing, spreading form, and both tree and nuts are things of quite extraordinary beauty, with the nuts (like almonds and peaches, the pistachio is a drupe) growing in a long dangling cluster that through time shifts hues from green to yellow to pink, growing more opaque as it matures. Pistachios grow best in arid environments, tolerating saline soils; the main enemies of the tree are elevated humidity and saturated soils. Pistachios are dioecious, with male and female trees, and they must be planted near one another.

Pistachios are impressive producers; a tree will produce 50,000 seeds in a two-year cycle. There are two main problems with pistachios. First is slow growth, with trees requiring up to five years of training to assume the preferred shape and another several years to produce a full crop. The second issue is a pistachio's sizable water consumption. Fifty gallons of water may be needed per tree per day from late July to late August for good-quality kernels with split nuts to be produced. Much of that water is administered by drip irrigation to pistachios, since plantings are relatively new and the dry-season irrigation required is substantially more than almonds, and nearly twice the water needed by citrus trees. Fertilized by wind, pistachios require both a female and a male tree, of distinct varieties (the Kerman [female] and Peters [male] are used), and are raised on one of four distinct rootstocks, each of a different tree species.

Pistachio nuts grow as a single seed inside a thin shell, surrounded by a thin hull. As nuts ripen in fall, the hull separates from the shell and reveals the kernel inside. The nuts mature to ripeness with prolonged hot summers, which makes coastal California a poor area for pistachios. Nuts are harvested by hand from young trees—up to 10 years old—by knocking off the fruit with a pole. As trees grow older, they are harvested by reciprocating shakers of the sort used to harvest almonds, although the pistachio trunk cannot be worked quite so vigorously. The one crucial detail is quick processing after harvest, to prevent bruising of the fruit—producers used to dye pistachios red to disguise mechanical damage done to nuts in the harvest process. The degree of shell splitting and the color of the nut inside contribute to the price given for a crop—the deeper the green of the kernel, the higher the payment. Nuts in 2008 brought $1.94 a pound, so a high-quality harvest is of substantial value.

For a crop native to west-central Asia (Iran, Turkmenistan, and Afghanistan), it is easy to believe that political unrest from the 1970s onward influenced California growers to embrace dramatic experiments with commercial-scale plantings. With vigor, they launched a run at the world market. Early growers joined forces in the California Pistachio Association and taxed themselves in support of research and marketing. In part, the scientific inquiries of the California Pistachio Research Board were a result.

While growers fund studies and agronomic research, a separate lobbying wing is maintained, with grower support. A successful and aggressive marketing organization can take odd tacks. Lobbyists for the Western Pistachio Association, in the WPA's Spring 2009 newsletter, refer five times by full name to the "Islamic Republic of Iran," adding another prod by chastising Israel for importing pistachios from Iran by way of Turkey. The WPA has sought to have Iranian exports cut out of the world supply and has insisted that the country of origin be required information on packaging—and in the late 1980s, a duty of up to 318 percent was imposed on non-American pistachios. Presumably, an argument can be made that all is fair in commerce. But comments that might be considered anti-Islamic are not heard from other producers in California of crops for which competing imports come from the Middle East or Central Asia: dates, for example, or apricots, or even almonds. International economics is a field clearly far from immune to the playing out of food politics.

PLUMS AND DRIED PLUMS — *Prunus domestica*

Pl. 116, Map 9

FAMILY: PRUNUS.

NOTES: PLUMS ARE FRESH MARKET; FRENCH, OR EUROPEAN, OR SUGAR PLUMS ARE FOR DRYING; UNTIL 2000, FRENCH PLUMS WERE REFERRED TO AS PRUNES. *P. DOMESTICA* INCLUDES MOST PLUMS AND PRUNES, BUT AN EXTREMELY WIDE VARIETY OF SPECIES IS EDIBLE.

RANK: U.S. #1, FRESH (CALIFORNIA SHARE 95%); U.S. #1, FRENCH (CALIFORNIA SHARE 99%).

Seen from an adjoining road, plums are wonderfully pendulous fruits, and both plums and French plums (née prunes) remain

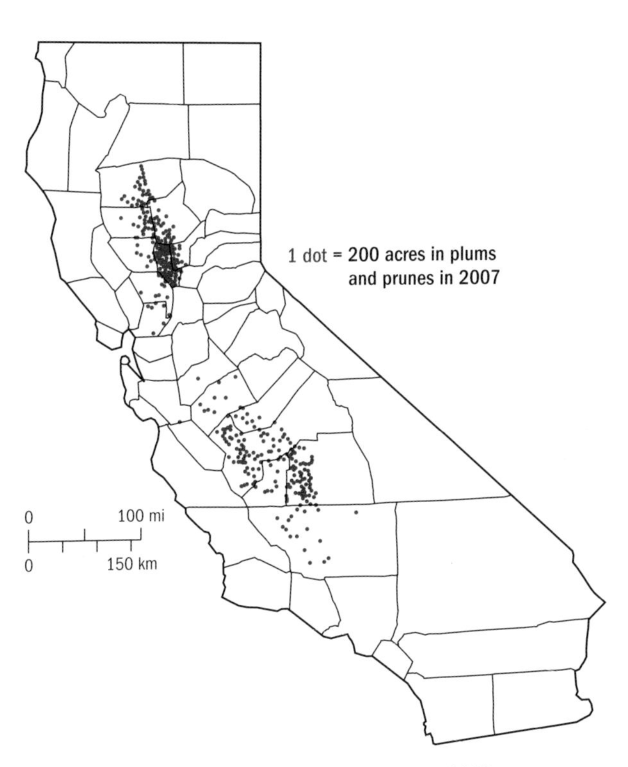

Map 9. Acreage in plums and prunes (dried plums), 2007.

green until relatively soon before they are harvestable. Genetically intimately related, plums and French plums are two grandiose fruits, each delicious. They each brought in nearly a quarter-billion dollars in 2008, but the distribution of their production could hardly be more divergent—the French, or dried, plum is grown, overwhelmingly, in the middle to upper reaches of the Sacramento Valley, while fresh-market plums are just as prevalent in the San Joaquin Valley.

The plum (fresh market) has undergone change as the favored varieties altered with time, although both are stone fruits of the same genus and species; the essential difference is a matter of semantics and image-packaging. The fresh-market plum is a cling fruit (difficult to separate from the pit), whereas the French plum, which is usually dried, is a freestone, which comes easily from the pit. But the French plum, it deserves noting, is a prune by any other name, having undergone an official (federally sanctioned) name change in 2000, when marketing groups worried that "prunes" sounded too gerontological. The California Prune Board changed its name shortly thereafter to the California Dried Plum Board. In this case, the devil is not in the details—the name change is preferred by the buying public. Both fruits, and their parent trees, are sizable moneymakers in California. Then again, they have long been moneymakers.

The uses of plums, of whichever stripe, are diverse. The cultivars tend to be organized by the color of the plum when ripe, which can variously be red, purple, green, or yellow-orange. Plum juice is readily fermented into plum wine, or can be distilled (and is, in the United States) into what is known as plum brandy in this country, but abroad is slivovitz (from the Damson plum), rakia, tzuica, or palinka. And, in a matter that gained considerable national news with President Ronald Reagan's medical diagnosis of colon cancer, fiber-rich plums (the fiber is largely in the skin) are, indeed, respected for their laxative and health-giving effect.

Plum acerage is concentrated, and strongly so, in the San Joaquin Valley counties of Fresno, Tulare, Kings, and Kern, which dominate all production. Equally, but separately, dried plums come from the Sacramento Valley, where (as with peaches) they are overwhelming raised by South Asian farmers, mainly Sikhs, with Sutter, Yuba, Butte, Tehama, and Glenn counties occupying the first five spots. In 2006, plums were worth $265 million, and dried plums $260 million: a push, as they say in the gambling trade. If one were to combine their value, then the unified category of "plums" would be eighteenth in California farm-gate value; as it is, they are twenty-eighth and twenty-ninth, respectively. By 2008, the value was much reduced.

Fresh market plums offer a different delight than prunes. Their tang and acidity make them a pleasure for a snack, although plums are notoriously difficult to transport. Predictably, there are many variants. Over a dozen Japanese varieties of plums account for some 70 percent of the U.S. production of fresh plums; they include the Red Beauty and Black Beauty and the Santa Rosa, Queen Rosa, Black Amber, Casselman, Angeleno,

Simka, and Laroda. Other varieties are favored more for canning than for fresh eating. But let it be said: there is nothing quite like getting hold of a Santa Rosa plum recently plucked from a tree and biting in hard but carefully, to avoid the pit. The pop of skin giving way and the gentle slide of teeth into the plum is an experience that validates human sensuality.

The acreage of both French plums (2008: 64,000 acres) and fresh market plums (2008: 29,500 acres) have stayed relatively unchanged since 2005, although the value has varied substantially: for French plums, $262 million in 2006 and $117 million in 2008; for plums, $108 million in 2006 and $54 million in 2008. The 2009 California prune crop is forecast at 170 thousand tons, up 32 percent from the 2008 crop, overwhelmingly on the Improved French Prune cultivar, although other varieties include the Sutter, Tulare Giant, Moyer, Imperial, Italian, and Greengage. The largest and best-known dried plum (prune, French plum) producer is Sunsweet Growers, with its headquarters on Hwy. 99 in Yuba City, which controls more than two-thirds of the dried plum (prune) market worldwide. Within the industry, market research may have suggested that "prunes" had an old-fart sound to it, but in fact, some of the most chic cuisine in the world relies on prunes, which are vital to both sweet and savory dishes, and are an essential ingredient of North African tagines. Who, after all, can forget the 1990 episode of *Star Trek: Next Generation,* when Guinan (Whoopie Goldberg) introduces Klingon Starfleet officer Warf to prune juice: "A warrior's drink," he announces, with delight. And he is right.

POMEGRANATES *Punica granatum*

Pl. 60

FAMILY: LYTHRACEAE; ALSO PLACED IN PUNICACEAE.

RANK: U.S. #1; ESTIMATED CALIFORNIA SHARE 80%.

Let it be said: it is the view of the authors of this book that for its flowers, its unkempt growth form, and its enchanting fruit with arils (the red flesh that surrounds small white seeds) that are both decorative and delicious, the pomegranate offers an almost unlimited charm. Fortunately, a large literature now attributes a variety of health-enhancing features to the pomegranate, whose day may have come. A tree that does well in varied soils with relatively sparse irrigation, and that starts easily from cuttings and in just two or three years yields fruit capable of giving pleasure to an audience from the very old to the utterly young, the pomegranate is a crop to be respected. We want you to know that we were there first, however.

The pomegranate, from the French *pomme garnete*, or "seeded apple," is a small tree or shrub, growing up to 25 feet tall (but with dwarf varieties extant). The tree has shiny foliage and a long flowering season, with flowers in a shade of such vivid orange-red that they are almost entirely hummingbird-pollinated; a pomegranate can bloom two or three times in spring. Native to Iran and the Himalaya Range in northern India, the pomegranate was first cultivated in ancient times and spread widely through Asia, Africa, and Europe, with fruit used much as it is today. Granada, in southern Spain, was named after the fruit, so much prized in the Islamic society that controlled Granada until 1492. The fruit does best in well-drained loamy soils but can tolerate heavy earth. The fruits crack open with the first autumn rains, but should be harvested earlier; fruits are ripe when they make a metallic sound when tapped.

Plantings in California are generally in the central reaches of the San Joaquin Valley, mostly in Tulare, Fresno, and Kern counties, with small plantings in Imperial and Riverside. The pomegranate was introduced to California in 1769 by Spanish missionaries, and naturalized readily. The description of a garden at Mission San Buenaventura (Ventura County) in 1792 speaks to the presence of pomegranates, apples, figs, oranges, grapes, and peaches. But in a striking display of adaptability, the pomegranate grows as far north in the United States as Washington County, in central Utah. "Wonderful" is California's most common cultivar, Granada another. Foothill Early, Ruby Red, and Spanish Sweet (or Papershell) round out the varieties seen with regularity, the last three more commonly witnessed in gardens than commercially. The fruit is picked by hand, usually through two or three rounds, or in sequential U-pick visits to the orchard.

Production of 250 growers on 8,630 acres in 2008 totaled $41 million, which makes pomegranate one of the larger small crops (another 5,000 acres are up, but not yet bearing). Pomegranates are sold as fruit, to be eaten by hand, but also as

Plate 60. Every stage of pomegranate life, from flowers describable only as "pomegranate red" to the mature fruit, offers a spectacle. Pomegranates have boomed in California with the successful marketing of their juice.

garnish, decorations, jelly, and juice—which is taken to have especially advantageous antioxidant properties. In fact, when POM Wonderful, a San Joaquin-based cultivator of Wonderful variety pomegranates, began marketing the juice, consumption rose 10-fold, thanks in considerable measure to a brilliant and concerted marketing campaign reminiscent of the early days of Sunkist (*see* Oranges). Pomegranate syrup is sold commercially as grenadine, and pomegranates are, in some circles, made into wine. The fruit stores well, as do apples and squash. The traditional means of eating pomegranates is to score the fruit several times, vertically, and lift out from the rind the clusters of juice sacs, which are then eaten. The tree branches, bark, and fruit rind contain up to 26 percent tannin—the roots more—and were once important in the production of Moroccan leather. In short, a wonderful crop.

PRUNES

SEE: PLUMS AND FRENCH PLUMS.

WALNUTS *Juglans*, var. species

Pl. 61, Map 10

FAMILY: JUGLANDACEAE.

NOTES: MOST COMMON CULTIVAR IS *J. REGIA* (ENGLISH OR COMMON WALNUT), WHICH IN CALIFORNIA IS OFTEN GRAFTED ONTO CALIFORNIA BLACK WALNUT ROOTSTOCK, *J. NIGRA*.

RANK: U.S. #1; CALIFORNIA SHARE 99%.

More than half of the counties in California (34 of them) report producing walnuts in 2008, and, it merits remembering, that is in commercial quantity. The acreage is scattered through all of the Sacramento and the San Joaquin valleys, somewhat more densely so in the Sacramento Valley. That is no problem: walnuts have all the virtues of tree crops and almost no discernible flaws. Commercial walnuts are further found in Sonoma, Lake, Monterey, and San Luis Obispo counties—only the desert and extreme-north counties lack walnut acreage. It's a tree crop with stature and remarkable endurance. The trees are tall and green, and thinned to 30 to 50 trees to an acre, they form an almost complete canopy over the ground and create a remarkable darkness underneath that is cooled by the irrigation waters that are sluiced over the orchards by strong drip irrigation or, in many cases (depending on the age of the planting), by flood irrigation. The utterly distinctive graft line where the English walnut slip was grafted onto a native black walnut rootstock (*Juglans hindsii*, or a cross with *J. regia* called Paradox) shows 6 to 24 inches above the ground: an instantaneous sign that this is a walnut, and not some other nut tree or fruit tree crop. The cicatrice is signature.

Among California tree crops, walnuts are the grand old man. They come in at twelfth in the farm-gate value chart; the central San Joaquin Valley has the edge, but the Sacramento Valley is on its trail, and were it not for the disruption and a furloughing of agriculture required by the I-80 corridor, the Sacramento Valley might occupy the dominant spot in terms of walnut production in California. That physical corridor displaced a good bit of historic production, but who was to gainsay the federal government in its insistence on establishing a national defense highway

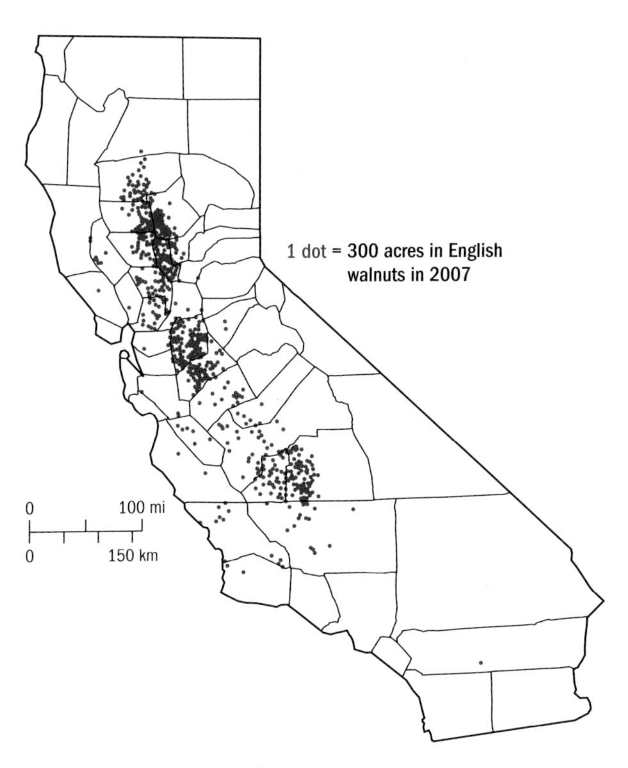

Map 10. Walnut acreage, 2007.

through the middle of California's richest agricultural soils. In 2007, 218,000 acres of walnuts in California produced a yield of $751 million, a clear three-quarters of a billion dollars. There are over 4,900 walnut growers, and about 60 processors of walnuts. A fifth of all walnuts sold were in-shell; the rest were shelled. The one- or two-pound bags of walnut meats sold at Costco were All-California.

The close friend of the California walnut crop is Diamond Walnut of California, established in 1912, and in its success an early equivalent of citrus champion Sunkist. With the Walnut Marketing Board ("California Walnuts"), formed in 1933, Diamond provided a support mechanism for walnut growers. The crop, even today, succeeds because growers have confidence that there will be a market, with research ongoing. A steady diffusing resolves any oversupply through marketing organizations that support the walnut industry as they have for nearly a century. The aesthetics of walnut-growing are equal to the profit. The tree is instantly recognizable, and walnuts are the only tree crops in California (except for the rare pecans and chestnuts) that will completely dominate a landscape, in a tradition of excellence.

To call *Juglans regia* the English walnut amounts to a bit of a purloining; it was long before that the Persian walnut, or the regal (= *regia*) walnut, and is likely of Near Eastern origin. The largest walnut grove in the world is Shahmirzad Orchard in Iran, at nearly 1,800 acres, according to the FAO. In the Renaissance, the walnut was thought to be the perfect cure for mental ailments, following on the so-called "Doctrine of Signatures," which held

Plate 61. Nothing says "walnut" quite so loudly as a distinctive base where native black walnut rootstock gives way to grafted (added) English walnut–bearing wood. One tree is notable; a grove is signature.

that plants were effective as herbal medicines in direct proportion to their resemblance to human body parts—the walnut salved the human brain and the emotions. The English walnut is so called because in the nineteenth century, English merchants had a near-monopoly in world walnut commerce. Two varieties are the most common in California, now: the Chandler and Hartley walnuts, which together account for nearly 60 percent of production. More than 30 additional varieties are planted, but none at the same scale. *J. regia* is useful for both timber and for its nuts, a dual use that has benefited humanity for perhaps 4,000 years.

The walnut is an orderly crop, but it is also insistent. Once walnuts begin to split their green hulls in August and September, the orchard floor is rolled and cleaned, nuts are shaken free of the trees using a mechanical shaker, and the walnuts are blown into a row so that mechanical harvesters can pick them up for cleaning and hulling. The most beautiful walnuts go to the in-shell market—holiday cracking of walnuts is a tradition feted in musical performance and folklore. The less comely walnuts are cracked to produce shelled walnuts, with just the meat of the walnut available. That is then sorted by quality, intactness, and size.

There are two centers for walnut production in California, but plenty of outlier counties as well. The first and largest center is in the San Joaquin Valley. The first three counties in production are San Joaquin, Tulare, and Stanislaus. A tier of the next four, in farm-gate value, are Butte, Tehama, and Glenn counties; Kings County (eighth) then intrudes, but the following three counties are again in the Sacramento Valley: Yuba, Yolo, and Sutter. The Sacramento Delta, however, is a third producing region. The Sierra Nevada foothills are substantial walnut producers, and so are Napa and Santa Clara counties. A statewide crop, indeed. The value of California walnut production in 2008 was $754 million, from 218,000 acres. Revenue was up by almost $200 million from 2006. Problems can follow plenty; so many walnuts were produced in 2008 that the USDA announced that they would buy $30 million of walnuts to help avoid a glut.

Vine, Bush, and Trellis Crops

The vine, bush, and trellis crops are intermediate in height, with bush and trellis crops three to eight feet tall, and vines at ground level. They occupy a middle terrain in the roadside agriscape. Some crops are readily recognized—artichokes, grapes, and nopal come to mind. Others are more difficult: beans can be either pole beans or bush beans, and they grow quite differently; melons, squash, or pumpkins can be trained upward or allowed to sprawl. Marijuana is not often seen, and if spotted should be treated with caution, but is not particularly difficult to recognize for anyone born in the last 50 years.

There are subtleties, though. When spotted, are grapes wine grapes, table grapes, or raisin grapes? Recognizing the right trellis system can make that identification easier, but sometimes location has to be factored in (although a site such as Lodi, in San Joaquin County, produces sizable quantities of table grapes yet bills itself as the "Zinfandel Capital of the World," and indeed happens, among other things, to be the birthplace of wine pioneer Robert Mondavi). "What is grown where?" is a legitimate question.

The value of the crops in this section will not necessarily match that of tree crops or animal products, but it is substantial nonetheless. All grapes are worth nearly $4 billion (2008). Tomatoes, which also can be trained upright or allowed to spread laterally as vines, depending on the variety and whether they are for processing or fresh market, are worth about $1.3 billion. And marijuana, if that is considered a part of California's agricultural production (and it generally is, though there are no believable statistics kept) is worth somewhere between $20 billion and more than twice that (these numbers are explored in the "crop" discussion). That would make cannabis the single largest value crop in California (by far), and one possibly worth more than all the other agricultural products combined.

California dominates U.S. production of various artichokes (99 percent), grapes (all types; 91 percent); kiwifruit (97 percent); processing tomatoes (95 percent); mixed melons (72 percent); raspberries (61 percent)—even marijuana production is certainly more than 30 percent of the U.S. total. Many of these can be considered signature crops for California—when people think of California, they think of these. They also represent a sizable export market, and income from outside coming back into California.

ARTICHOKES ***Cynara cardunculus* or *C. scolymus***
Globe artichoke

Pls. 34, 62, Title page photo

FAMILY: ASTERACEAE.

NOTES: SINCE 1993 *CYNARA CARDUNCULUS* IS THE PREFERRED NAME; ACCORDING TO GERMPLASM RESOURCES INFORMATION NETWORK, *C. SCOLYMUS* REFERS BACK TO *C. CARDUNCULUS*.

RANK: U.S. #1; CALIFORNIA SHARE 99%.

Perhaps nothing so captures the initial exotic charm of the artichoke as a pregnant fact: the first official California Artichoke Queen was Marilyn Monroe, in 1947. For a thistle, a member of the sunflower family, to find its cause boosted by an aspiring film heartthrob suggests something of the beauty and the beast. To see artichokes in the field isn't a complicated business, since most are planted as perennials. Along the Central Coast, you need to drive just a bit inland from the Pacific Ocean along the approaches to Castroville, Watsonville, Prunedale, or Salinas: there will be artichokes. Few vegetables are so quintessentially Californian as an artichoke, which thrives on—indeed, requires—frost-free coastal areas with cool foggy summers. A bout at high heat turns artichokes woody, lessening their charm. Although artichokes can stand cool temperatures, down to the upper twenties Fahrenheit, they do best with an equable climate. The marketable portion of the artichoke is what most people are familiar with—an immature flower that is picked, sorted, quick-cooled, and packed. How it is eaten becomes a matter of taste, training, and artistry.

Although the first commercial artichokes were planted near Half Moon Bay, down the coast from San Francisco, the modern industry began in the 1920s in Monterey County's Castroville. Since 1959, Castroville's Artichoke Festival has kicked off each harvest season in that Pajaro Valley town, where a physical gap in the Coast Ranges draws cool ocean breezes and fog toward the inland communities of San Juan Bautista and Hollister. Little wonder that Castroville's welcoming arch spanning main street touts it as the "Artichoke Center of the World." If that title is a bit hyperbolic—Spain, France, Italy, and several additional

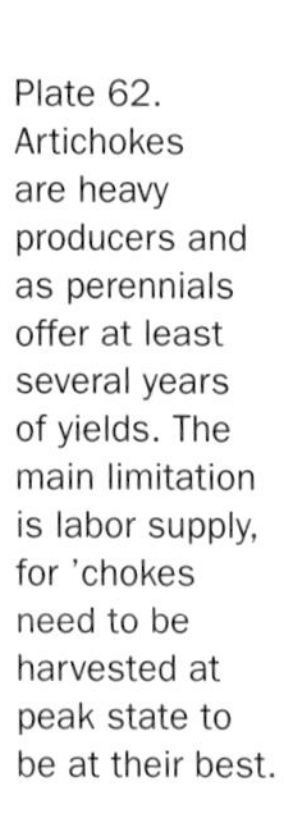

Plate 62. Artichokes are heavy producers and as perennials offer at least several years of yields. The main limitation is labor supply, for 'chokes need to be harvested at peak state to be at their best.

countries export more artichokes than California—California does control 99 percent of the U.S. production, and 86 percent of that came from Monterey County in 2007, where 8,200 acres of artichokes brought in $50 million.

Artichokes are among the oldest foods known, cultivated in the Mediterranean Basin for thousands of years. Theophrastus, a Greek philosopher and naturalist from the third century BC, wrote of artichokes being grown in Sicily and on the Italian mainland. After Classical times, they fell from discussion until artichokes saw a resurgence of respect and interest in sixteenth-century Italy. Arriving in North America with French immigrants to Louisiana, artichokes later traveled with Spanish clerics and quartermasters to California.

In the field the artichoke is a curious thing. An intriguing shade of bluish-green, the plant grows to a height of six feet, and can be three to four feet wide. It packs defenses—a variety of prickles and sharp points—that leave no doubt about its origins as a thistle. Within California, the Green Globe cultivar takes about 90 percent of the parent stock, and that variety is deep-rooted, a perennial capable of producing for five to 10 years, and does require some irrigation. While artichokes will produce year 'round, the largest part of the California harvest is between March and mid-May, since yields drop as summer heat comes on—which it does, as far inland as Salinas. Generally, a field is harvested once or twice a week, depending on the weather; if raised as perennials, a field of artichokes may be harvested 30 times a season.

The harvest of artichokes is labor-intensive and and is done almost exclusively by Mexican immigrant labor. Crews canvass the field, walking up the rows and wielding sharp, curved knives that slice off an artichoke plus three inches of stem, which the field workers immediately toss into the *canastas*, metal-framed packs carried on their backs. The yield can be up to 550 cartons per acre—not a bad income, with artichokes costing around a dollar apiece. Typically, the sorting and packing is done along the edge of the field. Packs weighing up to 80 pounds are emptied into wooden bins, where the artichokes are eased onto a conveyor belt for a trained crew to grade by size, quality, and appearance. As cartons are loaded, workers sort the 'chokes by size, from jumbos (18 per carton) to petites (72 per carton). From the field, the cartons are placed on pallets and trucked to a cooling plant, where they are hydrocooled—moved into a vacuum chamber and cleaned with frigid water, which plummets temperatures. From there, the crop—packed, sorted, and cooled—heads out to market.

Part of the charm of artichokes is their very oddity. Although the novice may be stumped by how to prepare, cook, and approach eating an artichoke, the veteran artichoke connoisseur knows exactly what to do, whether artichokes are to be consumed as hearts or quarters, canned or conserved (a quarter of the crop), or eaten fresh—and 75 percent of artichokes are harvested for fresh eating. For an artichoke recently off the field, it's a matter of adding humid heat, softening the flesh, and digging in with whatever condiment appeals: butter, horseradish, mayonnaise or aioli, Hollandaise sauce, or more exotic fare. The reality is, they're all good. The experienced hand must teach the novice how to cope with the inner leaves, the choke, and the delectable heart. Passing on such skills has entertainment value, but even more intriguing is sharing a first experience, looking at a live artichoke field.

BEANS ***Phaseolus* var. species**

Fresh market, and dry or field beans

Pls. 63, 111, 112

FAMILY: FABACEAE, ALSO LEGUMINOSAE.

RANK: U.S. #3; CALIFORNIA SHARE 9%.

Beans are not the most dramatic of crops, but they come in an almost unending variety, which makes identifying beans more interesting than might initially seem to be the case. A nitrogen-fixing legume, beans are widely regarded as a beneficial crop to plant when soil needs refreshing, or as a cover crop. They are a staple food in many cultures, and are well integrated into California cuisine, drawing on cultural traditions ranging from the European (*cassoulet*, *fabada Asturiana*, Portuguese *feijoada*) to Central and South American—is it even feasible to create a burrito sans beans? Commercial bean production is widespread, though less in evidence in California now than in the past. Beans come in two variants: dry and fresh. The dry beans start out fresh, with beans inside growing pods, but are kept on the vine until the pods dry and the beans inside can be winnowed: the USDA category "beans, all dry edible" includes much—black, blackeye, cranberry, garbanzo, dark red and light red kidney, baba lima and large lima, pink, and the always popular "beans, other, dry."

Superlative sources of protein, dry beans were domesticated in Central and South America some 7,000 years ago and moved northward through Mexico, into what is now the American Southwest. Native Americans who were practitioners of agriculture, such as the Hopi, evolved a sophisticated form of cultivating the seeds of beans, corn, and sometimes squash in arroyos or stream channels. Indigenous farmers would make use of subsurface water, and the crops would benefit from the intermittent summer rainfall that was usual in the so-called Arizona Monsoon rains. Beans, corn, and squash were sometimes referred to as the "Three Sisters," and provided a high-quality diet of foods that could be dried and stored for seasons when fresh food was less readily available. Dry beans in California have a checkered history, with a recent decline. Peak production was reached in 1955; dry bean production of all varieties in 2008, totaling 960,000 hundredweight of beans, was less than a quarter of that record high. In fact, the 59,000 acres harvested in dry beans in 2008 is under half the acreage harvested in 1999.

Some dry beans are tall and sprawling, like garden bean varieties, but typically the parent plant of a dry bean crop will grow into an erect, yet bushy plant. Although some are vine-like, in general, dry beans are relatively short-statured, growing up to two to three feet high. Pods grow along the length of the stem, and each contains a few seeds, which are collected for harvest, often by threshing. What is marketed are the ripe seeds of the crop.

Fresh beans come in diverse forms (over 130 varieties are known). The main varieties of fresh bean in California are

Plate 63. Long beans, in a distinctive black hue, are a form of snap bean that is harvested in immature form and sold as fresh market beans—one of an almost unending variety of fresh beans.

variants on the green bean. Those fall into three main categories: the snap bean, with a flat pod; the stringless (or French) bean, and the string bean, or runner bean. In the field, these are either bush beans, or pole beans (sometimes called climbers). Bush beans are 15 to 24 inches tall, and they set pods, which can be harvested. Pole beans are instead trained on trellises or on stakes, grow tall, and set pods over a long season. The pods are harvested when immature; if allowed to grow to full size and length, they will yield beans. It is the pods that are purchased, cut up, cooked, and eaten. Varieties include wax beans, green beans, French beans, the yard-long bean, the hyacinth bean, the winged bean, and the string bean, which goes by "snap bean" in the California vernacular.

California is a significant producer of fresh-market snap beans. There is no genetic difference between dry beans and the snap beans (green beans) grown in a garden, or harvested commercially as fresh beans; they are different cultivars, or varieties, and do well in different circumstances. Fresh market snap beans brought in $46 million in revenue in 2008, and on acreage decreased from two years earlier. California produces about 7 percent of U.S. snap beans, and, as with dry beans, is a significant consumer. The coastal counties (Ventura, Monterey, Santa Barbara, San Diego, and even San Mateo and Santa Clara, in the San Francisco Bay Area) are the main producers of fresh market snap beans.

Although the San Joaquin Valley produces the largest proportion of dry beans in California, the Sacramento Valley and the Delta (Solano and San Joaquin counties) each have a share, and there is some production from Riverside and San Bernardino counties. Dry bean production is spread around California—not evenly, since the San Joaquin dominates, but there is no monopoly, as with some crops. Total value produced of all dry beans was $63 million in 2008.

BERRIES

Blueberries — ***Vaccinium*, var. species**

Boysenberries — ***Rubus ursinus x idaeus***

Red raspberries — ***Rubus idaeus***

Blackberries — ***Rubus fruticosis***

Pl. 64

FAMILY: ERICACEAE (BLUEBERRIES); ROSACEAE (BOYSENBERRIES, RASPBERRIES, AND BLACKBERRIES).

RANK: U.S. #5, BLUEBERRIES (CALIFORNIA SHARE 6%); U.S. #2, BOYSENBERRIES (CALIFORNIA SHARE 3%); U.S. #1, RASPBERRIES (CALIFORNIA SHARE 61%).

SEE ALSO: STRAWBERRIES.

Berries are a fine sight from roadside: a dense congregation of canes rising up, more or less trained into respectable vine-like rows. The fields tend to be relatively small areas of a few acres, in deference to the labor requirements involved; although mechanized berry harvesting machines exist, their use is neither perfected nor widespread. Some berry fields are covered with an odd netting-like canopy whose purpose isn't entirely clear—until you think about the natural enemy of berries: birds. It takes a walk through a field to get a full sense of just what is being grown, and with a first glimpse of the fruit, all becomes clear. Berries are harvested frequently—as often as daily—and the perishable crop is sorted right into the pressboard cartons or clamshell cases seen in stores, many labeled with the Driscoll name—a single producer with a long history and a sizable share of the California market. The fruit departs the field on the backs of flatbed trucks and goes to a cooling plant, and the goods are then distributed to market. For berries, time is of the essence.

A catch-all category, "berries" excludes strawberries, the seventh most valuable crop in California (it's not at all the same berry) but includes blueberries, boysenberries, blackberries, and raspberries, all of which are significant contributors to the California agricultural economy. Other berries that are not discussed here figure less prominently in the California agriculture pantheon: gooseberries, bumbleberries, currants, olallie blackberries (olallieberry), marionberries, and loganberries. In California, berries are significant producers: blueberries brought in $30 million in 2007, and the delectable boysenberry $2.5 million. Raspberries are the comparative heavyweight, at $285 million. What distinguishes these crops is demand: in 2007 boysenberries brought in $0.72 a pound; blueberries, $1.83/lb; raspberries, $2.88/lb. In each case, that is a good bit more than fresh market strawberries.

Berries issue from perennial canes, produced from an established rootstock that must be carefully pruned for maximum yield, as with grapes. Once berry bushes are established, they spread with a vigor ranging from tentative (usually because of climatic issues)

Plate 64. High-yield blackberries make California a highly significant producer, although not quite equal in farm-gate value to Oregon (except when it comes to raspberries, for which California is first).

to relentlessly voracious. Canes grow one year (primocanes) and the next year produce blossoms and fruit on the floricanes; once they have fruited, the depleted canes can be cut back in fall—and if the orchardist fails to do that, the result can be a thicket that is intimidating to work through. Many berries are modified through cultivation and selection to have thorn-free versions—generally a good thing, because a well-established blackberry can sport thorns seemingly capable of piercing Frodo's mithril chain mail singlet.

A crucial factor in berry production is the number of chill-hours required to set a crop. Blueberries were once thought to need a prolonged cold period before they would set fruit. Low-chill varieties are known now and are planted in various parts of California, from Monterey to Ventura counties. The cultivars are various, but the success is beyond argument. Blueberries are grown with undaunted success in the San Joaquin Valley, with Tulare, San Joaquin, Fresno, and Stanislaus counties putting out $57 million in blueberries in 2008. They grow all the way to the coast, however, and mixed orchards of diverse berries are common: the presence of blueberries, gooseberries, blackberries, and raspberries in one field makes for a pronounced marketing advantage.

Boysenberries (cultivated from nursery stock) or the more robust and feral blackberries are a triumph in California—in fact, blackberries sometimes seem to dominate every road verge and property line. Both do exceptionally well in cooler coastal regions, including the North and Central Coast. There are two forms in California—the erect and the trailing blackberry, depending on how strongly self-supporting their canes are. Like many berry bushes, blackberries can live for many years, and they produce often copious amounts of fruit on short lateral shoots from the floricanes. California production lags behind Oregon, which has 80 percent of the U.S. yield. In a sidenote, the Oregon crop goes overwhelmingly to processing, and California actually dominates the market in fresh market blackberries, according to USDA statistics, which do not distinguish with any precision between varieties.

Unmistakably, the big-ticket item in California's berry world is raspberries. A relatively delicate plant, with precise climatic preferences, the raspberry does not travel especially well, nor does it keep long—but the raspberry's friends are many. Raspberries do especially well in Santa Cruz County, growing on southern exposure in the Santa Cruz Mountains, with just the right amount of winter chill. Acreage sizes for raspberries tend to be small; many are raised organically on plots under 20 acres. High in antioxidants, including vitamin C and phytonutrients, raspberries bring a premium, and often produce two crops a year, although careful selection of varieties makes production for eight months a year possible. Almost all of California's raspberry crop goes straight to the fresh market. Income from two coastal counties, Santa Cruz and Ventura, was close, in 2008, at $106 and $85 million; Monterey County produced another $24 million in raspberries for a total (including a small production elsewhere) of $214 million.

CANNABIS

SEE: MARIJUANA.

CUCUMBER

SEE IN: FIELD, ROOT, AND ROW CROPS.

GOOSEBERRY, CHINESE

SEE: KIWIFRUIT.

GRAPES *Vitis vinifera*

Raisin, table, wine

Pls. 9, 21, 65, 66, 106, Fig. 4, Map 11

FAMILY: VITACEAE.

RANK: U.S. #1, ALL; CALIFORNIA SHARE: 91%.

To say that grapes have an ancient charm is understatement at best. Grapes satisfy human interest and food needs in ways so diverse that only a few other ancient foods can match. After all, grapes were long eaten in-hand as table grapes, dried as raisins to be carried on journeys as nutritious food, and fermented variously into wine or into fortified port or sherry, or distilled into high-alcohol spirits, starting

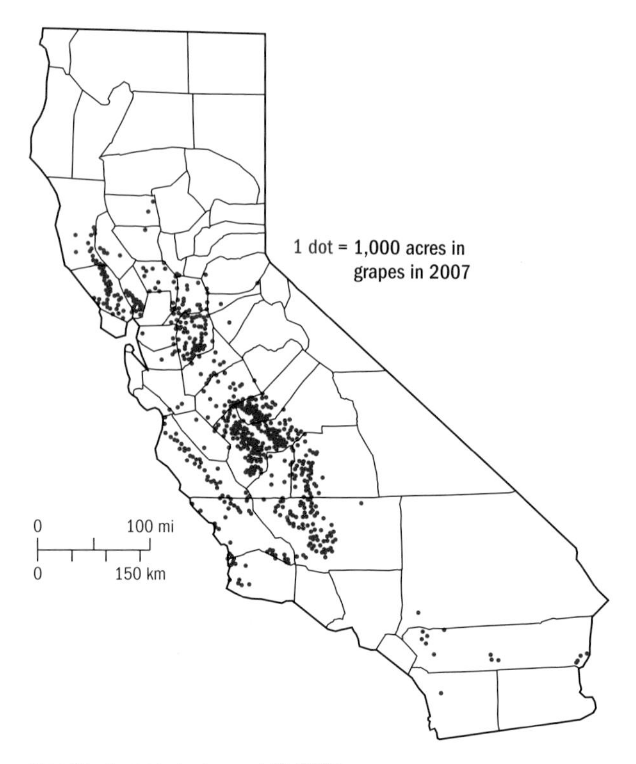

Map 11. Acreage in grapes (all), 2007.

with cognac or brandy but extending to dozens of additional forms. All grapes in 2008 were worth a tidy $4 billion—the second-ranked crop in dollar value in California agriculture. Grapes are a crop that moves: add together the exported wine, table grapes, and raisins, and at hand is an export value worth $1.6 billion, which places it second on the exported crop list, after the ever-formidable almond. A vineyard is a special place—part cool, part hot, part shaded, part absorbing the energy of a relentless summer sun—the sort of landscape that instructs you to think about where you are. And honesty compels an admission that there is something toothsome about the shape of grape. First, there is the beguiling form of their clusters; then there is the grape skin, which provides such protection and containment, and which definitely pleases the eye. In a guide, grapes can be treated in different ways—here, the various forms are combined, since the essential details of grapes are shared, but each major form of grape consumption has its own subsection.

Distinguishing traits of the grape, whatever its use, are the cordons, cane, and spurs, those shoots that grow from a woody trunk or stem, which can be pruned in various ways according to the needs of production. The purpose of pruning and training is twofold: to get keep the grapes themselves off the ground and out of harm's way, and to maximize exposure of the plant's leaves to the sun, which provides the energy for fruit production. The variation in pruning, training, and trellising styles is in truth almost as great as the varieties of grapes themselves, and there is no simple best practice for handling grape production, only old ways and new ideas. Raisin grapes and table grapes have their own pruning and trellising systems, passed on through experience; and in time, experts will produce additional techniques. A head-trained, spur-pruned wine grape, typical of old grape plantings in Europe, may look during winter like a dark branch sticking up from the ground; a T-pruned or cordon-trained trunk may be cut back to a three- to four-foot-high trunk with two gnarled woody laterals stretching in opposite directions from the trunk along wires or other supports. Or there is the Four-Armed Kniffen system, or the Geneva Double Curtain—look carefully, and take your pick (at some point, trellises start to sound like competing martial arts schools). The tools are no less specific than the training and trellising, including the hand-held pruning shears called a secateur. But the results, in the form of a trellised plant, are recognizable—even in an aerial photograph or satellite image. For anyone who is more interested in the opposite scale of study, *Nature* published the full genome sequence of *Vitis vinifera* in 2007.

Grapes are billed as a healthy food, typically meaning that they contain various phytonutrients. While flavonoids and the carotenoid lycopene are saluted, the rising star in grapes (and not just wine) is resveratrol, an antioxidant compound found chiefly in grape skins, although also in some other plants. Its effect on aging, arthritis, hearing loss, Alzheimer's disease, and various additional maladies will no doubt be studied further, but if there were not enough other reasons to consider consuming grapes in various fashions, those may be added. Botanically, grapes are a true berry that grows on the perennial deciduous vine, *Vitis vinifera*. The Concord grape is altogether another species, *Vitis labrusca*, and is rarely seen in California.

The Mission grape came to early California in 1769 and stayed. A lovely fresco painted during the Mission period on a wall at Mission San Fernando shows California Native Americans harvesting grapes from a massive vineyard; regrettably, it crumbled in the 1971 earthquake and exists only in photographs. Native grapes grew in eighteenth-century California, and some theorists argue that the Mission grape was actually a hybrid of the Spanish *Vitis vinifera* with the native California grape, *V. girdiana* or *V. californica*. At the moment, the theory is just that, but undoubtedly will be explored. What is not argued is the depth of time involved in grape culture: it dates back to some 8,000 years ago, to 6000 BC. Grapes seeds dated 4000 BC exist in

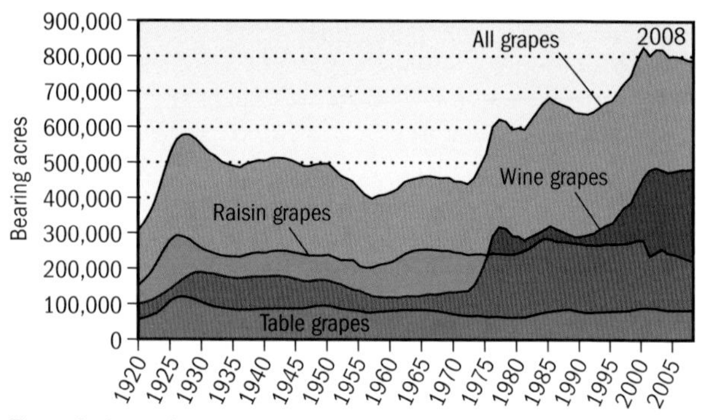

Figure 4. Acres devoted to grape production in California, 1920–2008.

human settlements in Italy and Switzerland, and traders brought grapes to Egypt in 3000 BC. Viticulture was established in Greece by 2000 BC, and it was the Romans who developed the first trellis system to raise grape vines off the ground and protect the fruit.

WINE GRAPES: A vineyard at work is something to behold. Symmetry abounds, presenting complementary and delicate, yet forceful lines of wire and vine, accented by a seasonal feast of colors: spring light-green turning to emerald, in the dead of summer, shade-shifting to red and brown as leaves fade in fall, bowing to the buff gray wood of pruned vines in winter. Elegies to the grape in literature date back 5,000 years to the Epic of Gilgamesh, evidence of the uninterrupted acclaim given to the vineyard (and certainly wine); finding ingenious means to wrest interesting beverages from grapes has its own profound history.

Relatively easy to grow, grapes are an attractive crop; but growing them well, and making great wine from them, is altogether another matter. An elaborate culture surrounds vineyards and the wine trade. If orange growing was long the doyen of California cropping—a permanent agriculture tied to a tree crop that permitted owners of small holdings to live on their acreage and commute to work, yet participate in the Virgilian idylls on agricultural life—then owning a California vineyard and producing premium or superpremium wines represents, in prestige and presence, an up-to-the-minute rival to the culture of the orange. It sometimes seems that everyone who is (or was) anyone—CEOs, CFOs, CIOs, directors, presidents, board chairs, Oscar and Emmy award-winners, MVPs, wannabes—has a winery in Napa or Sonoma or Monterey, or wishes they did.

Few California vineyards (although more and more all the time) have the mansions and estate homes and castles of Bordeaux or Tuscany. Many sites are simply where grapes are grown; once harvest is complete, the fruit is carted off, and the drama settles. The excitement of production, from crush to bottling, often takes place elsewhere. What is recognizable are the grapes, green-leafed and up to six feet tall, generally trellised (though some are squatly head-pruned, looking like shrunken heads fixed to the earth), and cut back to a woody nub after fall harvest. As they grow in spring, long tendrils—cordons or canes—support the end product, grapes, which mature in late summer and fall, at which point the leaves turn as multicolored as New England in October, and drop.

Plate 65. Wine grapes are influenced by hundreds of small changes that make grapes (such as these at Armida Winery in Sonoma County) good, great, or interplanetary—and much that happens later matters, too.

Irrigation is used on occasion, but there are wine grape growers who eschew any added watering, preferring that their grapes work hard to plumb and hoist from the depths any water they may obtain. When profound winter cold strikes, relict smudge pots and oil burners (their use no longer allowed) sit rusting under broad-bladed turbines that fan cold and settling winter air to keep it moving (such was once the case with oranges, too). The viticultural landscape reflects order, care, and a lot of investment. Distances between rows vary with the grape, the purpose, and the philosophy of the grower; there are those who believe that stressing grapes will produce a wine reminiscent of the soil itself—*gout de terroir*, as the French call it, or simply "terroir." (Soil scientists argue that the very concept is nonsense.) But other growers and winemakers are more benevolent. There is not an enormous amount of difference, visually, between the vineyards where raisin grapes, table grapes, and wine grapes are produced; the differences are in pruning, trellising, and harvest. This section is about wine grapes—but sizing up the end use of grapes involves a variety of steps, and sometimes a visitor's opportune question voiced to nearby workers.

Where wine grapes are concerned, California reigns supreme in the United States. There are plenty of arguments to be mounted about premium varietal or blended wines that come from grapes raised in Oregon or Washington; in fact, all but two states in the United States produce wine from grapes. What speaks loudest in California is the matter of scale: wine grapes in the state are a vast business, with wine grape sales valued at $2.4 billion (2008). All grapes, together, in California are the second largest value agricultural product (dueling with nursery and greenhouse crops); wine grapes, by themselves, would come in at seventh (in 2007, that would place them below lettuce and above strawberries). The crush was 3,674,453 tons, up 5 percent from 2006. If wine grapes are the essence of the California wine industry, they only hint at the heart and soul, which is a bigger matter. The retail value of wine shipments from California increased another 6 percent in 2007, to $18.9 billion. All these numbers suggest just a fraction of the total value of the overall wine world in California. The wine trade in 2007 was an endeavor that produced $53 billion in economic value for California, in the accounting of industry experts.

That most impressive step up in value ($2.4 billion in wine grape value to $53 billion in wine "economic value") speaks to something unique to the wine grape in California: the after-harvest markup in value generated by the fermentation, bottling, marketing, hospitality, dining, event-based entertainment (think: weddings and meetings), and tourism industry, all of which, collectively, are part and parcel of California winemaking. Wine industry sources estimate that altogether, the California wine industry generates $125 billion for the U.S. economy, including 309,000 jobs, and wine attracts nearly 20 million tourists. Those claims are at a generous scale. A lot of tourists who travel to California do end up in wineries during their visits; anyone who has seen a first-line Napa or Sonoma tasting room on a summer weekend understands the economic throughput. Nearly half of the bonded wineries in the United States are in California (2,687 of 5,958); 20 years ago, there were only a third as many (799). And direct-to-consumer sales, generally involving a winery visit, are a crucial element in California wine sales (although exports are on an upswing, too, rising 77 percent since 1997). Grapes are a relatively simple product, and California table grapes and raisin grapes yield another $1.6 billion in total value (2008).

Wine grapes are not just an agricultural product, but also the embodiment of ambition, forging and furthering a self-image of quality and prosperity that California has broadcast around the world. This description cannot begin to capture the full range and variety of the wine grape industry in California; for that, better to turn to one of the 1,200 books currently in print about one aspect or another of California wine and fine dining. Or, better yet, explore the landscape yourself, and enjoy the process. Excursions are an essential feature in the California agricultural experience, whether the destination is northern California's Napa, Sonoma, Marin, or Mendocino; the Central Coast vineyards from Monterey to Atascadero and south to the Santa Inez Valley; the ancient Zinfandel vineyards (more than 80 years old) of Lodi; or the huge wine operations in the San Joaquin Valley.

Quite contrary to the story as some might tell it, the California wine grape industry is anything but the long-standing international signpost of quality. The delightful book-length George Taber account of the so-called 1976 "Judgment of Paris," in which California chardonnay and cabernet sauvignon triumphed (though not universally) over the best of Burgundy and Bordeaux, offers a superb account of the evolution of California wines from what really was a mediocre, or at least a decidedly unimproved, product in the 1950s and 1960s through a transformation, almost by happenstance, that began in the mid-1970s. With dedication bordering at times on the fanatical, a handful of producers and landowners, and an equally few savvy amateur winemakers and professional enologists, in essence took grapes from varied parts of California and turned bits and drabs of wine into an organized and effectively reconceived industry. Strikingly, in Taber's account, it is not the wineries, their owners, or their winemakers who are the heroes. What stands out is a constant return to teachers: professors in the enology program at UC Davis and at Fresno State University, and farther away, the loose and welcoming association of winemakers around the world who generously allowed California students of viticulture into their bodegas and chateaus, whose ability and generosity aided wine and grape fanciers, teaching them what they needed to bring back, in the act kickstarted a wine revolution that is still afoot and afield.

The twists and diversions of the wine industry make simple grape-growing seem straightforward by comparison. A few details help—consumption of red wine is way up. The crush of red grapes in 2007 was 1.88 million tons; for whites, 1.37 million tons. But in 2006, red wines and white wines were sold in equal proportion (42 percent), with "blush" wines at 15 percent (down substantially from 1991, when it was 34 percent of the total). What has risen most is red wine consumption, in part because the somewhat cloying white wines of a couple of generations ago have plummeted in the public taste, and vineyards replanted to more interesting white varieties have not yet come fully online. Vineyard owners don't necessarily mind; red wine grapes bring in a higher average price: $626 a ton, as against $481 for white wine grapes.

The cultivation of wine grapes is an immensely complicated process, and the details are beyond the scope here. The production of great wine—or entirely adequate wine at reasonable prices—is wholly beyond the ambit of a field guide; for that, a deep dive into the literature, which is easily found, is indicated. Suffice it to say, the annual growth cycle of grapevines is well studied, and viticulturalists working on grapes for winemaking recognize the significance of each step of the process, and at some point or another, every element has been tested, modified, and revised. The parts of the wine grape-growing cycle are, in essence, bud break, flowering, fruit set, veraison (the color change in which individual grape berries cease being green and hard as they swiftly uptake glucose and fructose and acids start to form), harvest, and after-harvest care (usually pruning, sometimes spraying, if the vintner does those kinds of things).

Subtle differences can add up to striking variations in price. Although visually there is no difference in what is seen on the vine, Napa brought in the highest 2007 average wine grape price per ton ($3,251/ton), and Sonoma and Marin the second highest return ($2,081/ton). Mendocino and Lake viticultural districts came next, and about on the same bar, the Central Coast areas of Monterey, San Benito, San Luis Obispo, Santa Barbara, and Ventura counties. In 2007, the western foothills of the Sierra Nevada came next, and toward the less profitable end were counties in Southern California, San Joaquin Valley, and the southern Sacramento Valley. Kings and Tulare counties received the lowest prices for their wine grape crush. The categories generated from this are striking: wine below $3 a bottle is "jug wine," popular premium goes for $3–7; super-premium is $7–14; ultra-premium is wine over $14. The share of the three higher-priced grades has increased steadily since 1995. A delightful 2008 article by Rachael Goodhue and coauthors lays out these and other economic trends for California wine, but warns against complacency: competition is everywhere, and the politics of wine can be fierce. In 2005, negotiations with the European Union (EU) produced a tentative agreement not to use terms such as "Burgundy" or "Chianti" to describe U.S. wines. But Germany then launched an onslaught, waving accusatory fingers at wines from California for being "Coca-Cola" or "Frankenstein" wines, in contrast to the "bottled honesty" of "innocent" German wines. As Henrich Brunke and coauthors summed this up, "the spat . . . may have been a rear-guard action of conservative wine romantics." Or the alternative suggestion is that it was a skilled softening-up action. The agreement was signed in 2006—and with that, the German nativist cant was silenced.

California makes 90 percent of U.S. wine; it is the fourth leading wine producer after France, Italy, and Spain. Australia and China are contenders. Because some people readily buy foreign wines (and some exclusively so), it is significant that California wine accounts for two-thirds of wine sales in the United States. The acreage of wine grapes in California is always a two-stage estimate: first is the bearing acreage, which in 2008 was 482,000 acres, and second comes the non-bearing acreage—tabulating acreage in development, as it were—at 44,000 more, for a total of 526,000 acres: 0.5 percent of the physical area of the State of California is planted to wine grapes—more than 820 square miles. Keep in mind that the acreage in wine grapes in 1991 was 295,000 acres. Increased acreage over the last five years was slight, but there has been readjustment of acreage as crops shift around. Still, as word on the street brings resounding eddies of concern about "overproduction" and events such as the 2008–2009 recession hit hard and slow purchases of high-end wine, the genesis for such talk is clear.

Most wine grape harvest is by hand still, although occasional attempts to harvest mechanically are underway, using some of the same techniques and machinery being adapted to olive harvest. But to harvest mechanically requires setting up a vineyard in a special way, and that is costly. Specialization in the wine grape industry is everywhere. Grapes in California have to be on phylloxera-resistant grafted rootstock: some of the best vineyard specialists in France and Spain are now working in California to supply the market with rootstock that can then be grafted with whatever varietal a grower might wish to see. This is another of the spinoffs of California agriculture into new areas where distinct markets can be found.

Although cultivation techniques are no straightforward issue, a reality particularly interesting for someone traveling from wine district to wine district is the question of varietals. Distinguishing between them is easy only if a helpful 4-H Club has the varietal name posted next to the vineyard. Some grapes do better in particular climates and microclimates. Matters relating to slope, aspect, soil-type, parent material of the soil, humidity, and irrigation regime (or absence thereof) are all contributors to what grape is sited where.

Grapes come from diverse places, and they bring with them specific characteristics. For example, not until 2002 was it resolved—thanks to Dr. Carole Meredith of UC Davis and a strong research team—that a Croatian black grape called Crljenak Kašteljanski is identical, and parent, to both the Italian Primitivo and the Zinfandel variety of California, the wine grape so vaunted in Sonoma and other wine-growing districts. Tracking the movement of the ancestral grape from Croatia to California (and Italy) is a matter for another sleuthing expedition. As for the varietals, they are diverse indeed, from Burger and Chardonnay to Verdelho and Viognier among white wine type grapes, and from Alicante Bouschet and Barbera to Teroldego and Zinfandel (its parentage now identified) on the red. Acreage planted to each varietal depends on the preferences of winemakers, growers, and estate owners: currently, the top five reds in tons crushed are Cabernet Sauvignon, Zinfandel (not far behind), Merlot, Pinot Noir, and Syrah; for white, Chardonnay outstrips all others, with French Colombard and Sauvignon Blanc at second and third. But the acreage coming on suggests changes in the complexion of wine grape-growing. The wine industry, as a reality of California agriculture, will continue to excite . . . when it doesn't frustrate.

TABLE GRAPES (INDUSTRY INSIDERS PREFER THE TERM *FRESH MARKET GRAPES*): Viticulture came with the Franciscan friars whose cultivation made sacramental wine—and respite from the winter cold of adobe walls with wine or brandy—possible. Not until the late 1830s were table grapes really embraced, and then only thanks to William Wolfskill in Los Angeles, a pioneer in several aspects of California life. The table grape stuttered, caught, and took off, and in the Gold Rush era proved immensely popular. Acreage planted to table grapes increased, and an industry was born. Not without reason: the tables grapes have charm equaled by their profitability.

In the 1860s, William Thompson, an Englishman, planted an eastern Mediterranean grape variety known then as the Oval Kismish, near Yuba City, north of Sacramento. The Thompson Seedless grapes now leads all California fresh grape production—since "fresh grape" is the term preferred by consumers, and Thompson Seedless is that most grown by the state's 550 table grape farmers. Consumption in the United States is seven to eight pounds per person, and about a third of the crop is exported. Fresh table grapes come in generous hues: green (sometimes described as white), which includes the Perlette, Calmeria, and Thompson Seedless; reds, such as Emperor, Red Globe, Flame Seedless, and Ruby Seedless; and the blue-blacks, with Autumn Royal, Fantasy Seedless, and Marroo Seedless. Within those color classes are more than 50 table grapes in active use for production, and each has its season, its niche, and its constituency. The top table grape–producing counties are Kern, Tulare, and Fresno, along with the Coachella Valley, in eastern Riverside County.

The table grapes are pruned in different ways than wine grapes (unsurprising), and the emphasis in pruning and trellising is to give ready access to the pickers. The largest virtue of the Thompson Seedless, aside from its tastiness, is its versatility. It is a frequent choice for table grape buyers but can also be made into raisins, and it factors not insignificantly into the production of wine, being sometimes blended into wines when more sweetness is desired.

Although a substantial income producer, the table grape is by no means the dominant grape in California; both wine grapes and raisin grapes take up far more area. Table grapes grow on about 82,000 acres; hardly a small footprint, but barely 10 percent of the 789,000 acres of California sown to grapes. That said, a lack of visibility is not the same thing as insignificance: table grapes—or fresh market grapes—brought in $1.1 billion in 2008, which would locate it at the sixteenth spot in farm-gate value, firmly lodged between pistachios and cotton. Instead, all grapes are contained in "Grapes." But even if wine grapes are the more glamorous sibling, the value of table grapes is anything but insignificant.

RAISIN GRAPES: Raisins are dried grapes; how they get to that point is the essential element of the art form. Carefully produced commercial raisins have been a part of American life for decades, with brand names such as Sun-Maid (established in 1912) familiar to everyone who ever carried a bag lunch or lunch box to school. Raisins come in a variety of shades and forms, from the blond Sultana to others in green, black, blue, purple, and yellow. Most common are the dark brown or black raisin, but variations aplenty exist. Raisins are about 60 percent sugars by weight, with antioxidants roughly equivalent to their presences in prunes and apricots. Raisins can be eaten raw, or when soaked in hot water, sugars reconstitute (though the crunch of the fructose crystals is lost).

Many Californians are familiar with the simplest version of raisin production: grapes forgotten on the vine until they shrivel up and dry. Those tend not to be so delectable, since the seeds do not reduce in size, and mold or must will arrive with wetter weather. But the scientific cultivation of grapes as raisins-to-be has a hundred-year history.

More and more growers (nearing 10 percent) are using an overhead trellis system that allows the raisins to be dried on the

Plate 66. A deep blue-black cluster of raisin grapes is a frequent sight in Madera and other San Joaquin Valley counties. Rarely are raisins dried on butcher paper laid between vines; paper bins now suffice.

vine, reducing labor demands and permitting machine harvesting. The long rows of raisin bunches, set on paper trays on terraces between the vine rows, may be a thing of the past. Raisin grapes are in a modest cycle of acreage decrease, down 3.4 percent in 2008 to 225,000 acres, and nonbearing acreage—which suggests which direction the overall crop is headed—was down by one-third.

Favored raisin grape types, in acreage, are Thompson Seedless, Fiesta, and Selma Pete, with a few others like Black Corinth and Dovine coming on. Raisins can be treated to assume different looks (that is done with sulfur dioxide, to keep the yellow shade in Sultana raisins), and the Black Corinth is sun-dried to produce Zante currants, smaller than conventional raisins and with a tart flavor.

Fresno County is by far the largest raisin producer, which television watchers who were around a couple of decades ago may remember from the 1986 mini-series *Fresno*, a send-up of the then-popular winery-based series *Falcon Crest*. Starring Carol Burnett as the matriarch of a Fresno raisin-growing dynasty, it set up Dabney Coleman as her diabolical neighbor and competitor, with considerable hijinks ensuing, much of it turning on Armenian surnames, grower pretension, and agricultural humor.

The raisin is a particular favorite of California travelers, since in some quarters the picked grapes are, indeed, placed out in the sun for drying. The paper trays are a step up, or at least a difference, from the brown butcher paper that once used to be laid out between the rows of the vineyard. Quaintness aside, the raisin grape produced $422 million in revenue in 2008, which eclipses a lot of other crops, and would actually put "Grapes, Raisin" at seventeenth in the list of California farm-gate value in 2007.

KIWIFRUIT *Actinidia deliciosa*

Pl. 67

FAMILY: ACTINIDACEAE.

RANK: U.S. #1; CALIFORNIA SHARE 97%.

The kiwifruit has at least one great point of distinction: it is the only significant temperate fruit crop to be domesticated in the twentieth century. Discovered at the end of the 1800s, it was taken to New Zealand, and the movement toward domestication began. In its natural element, the lower Yangtze River valley of northern China, the kiwifruit is a woody and vigorous twining vine, with the habits of a climbing shrub. A single plant can cover a wide area if left to its own devices; the domesticated kiwifruit is usually pruned and trained into a single trunk that grows high, then spreads canes along a trellis. Chinese practice was to gather the wild fruit; in cultivation, the trick is taming the growth to allow workers to get to the fruit, oblong edible yet engagingly hirsute or fibrous berries the size of a goose egg that descend from the underside of the liana or vine. Inside, however, is a mass of black, edible seeds and a bright green or golden flesh that is scooped out with a spoon when the kiwifruit has softened, much as some people eat avocados or Bartlett pears. As an alternative, the kiwifruit can be peeled and sliced for eating.

The texture, the form of the fruit, and active and successful promotion of the kiwifruit, especially by New Zealand, as an

Plate 67. An import from New Zealand, and recently a wild plant in China, kiwifruit has a cachet among growers in excess of its domestic market, with fruit also exported internationally.

ideal crop for small-acreage growers, have made it a reasonable success in California. The other felicitous involvement was that of Frieda Caplan, a produce dealer in Los Angeles who encouraged the importation of kiwifruit in the 1960s. Because consuming in hand is the preferred way to eat kiwifruit, larger fruits are preferred. The fruit stores well; New Zealand kiwifruit is sold from May through December, and California fruit is available from November through April. There is usually fresh kiwifruit available in stores; some effort has gone into drying kiwifruit, without an ensuing surge of demand; other fruit may be processed into jam, juice, or wine, but not a lot.

The history of kiwifruit is its own minor epic. The first examples of kiwifruit seeds were brought from China to New Zealand by Mary Isabel Fraser, headmistress of the Wanganui Girls' College and a visitor to mission schools in Yichang, China, in the early 1900s. The seeds were planted in 1906 by Alexander Allison, a Wanganui nurseryman, and bore fruit four years later. The first commercial kiwifruit crop was planted about 1940. By 1986 New Zealand had 50,000 acres in kiwifruit vineyards. Significant exports ensued. Once the fruit came to California, the New Zealand cultivar "Hayward" became the standard in California, although New Zealand uses far more diverse cultivars. Other varieties are under development, including some from other species: *Actinidia arguta*, *A. kolomikta*, and *A. polygama*, each with its own attributes. The Hayward kiwifruit shows no sign of leaving the California scene, although alternative cultivars are being planted in Oregon with success.

An experimental planting of a dozen kiwifruit was established to California in 1960 (some say 1967, others earlier—the kiwifruit may actually have been introduced to California at the same time it went to New Zealand, but that is uncertain). Butte County, in the east-central Sacramento Valley, is home to the Chico Plant Introduction Station, operated by the federal Agricultural Research Service, where initial kiwifruit plantings still exist—dating, records imply, to 1934. Growers who frequented the Chico station (administered by the U.S. Forest Service since 1974) were intrigued by the kiwifruit's possibilities. As of 1971, more than 100 acres were planted in an odd distribution: Butte County and in the southern San Joaquin Valley's Kern County. As is so often the case, the geographically disjunct plantings were simply the work of two growers who fancied the crop: Judd Ingram, near Delano

(1967), and George Tanimoto, in Gridley (1968). Building on the New Zealand practices, growers evolved high trellis systems that stretched the kiwifruit above head height and allowed the fruit to dangle for harvest. The first commercial crop was packed in 1977, and production was large in the 1980s and early 1990s, although local consumption is down somewhat in the last decade.

In general, kiwifruit do well in California in the same areas where peaches succeed, so a distribution in the Sacramento (Butte, Yuba, Sutter counties) and San Joaquin (Tulare, Fresno) valleys, along the Central Coast, and in parts of Southern California is unsurprising. Other Mediterranean-type climate countries grow kiwifruit, including Italy, Spain, France, Chile, Greece, and Israel. The main investments in kiwifruit production are ongoing costs for water and labor, but more than anything else, the great expense is in establishing vines and trellises. After preparation, it takes four years for a first commercial-level crop of "Chinese gooseberries" to arrive. The overhead trellis, or pergola, can cost $3,000 an acre to install. Because kiwifruit require significant irrigation every two or three days, a permanent irrigation system (drip will work) is unavoidable, at added expense. Shading from full sun has to be provided in hotter California sites. Establishment costs, including land, can approach $12,000 an acre for kiwifruit. Such a high ante makes kiwifruit vineyard establishment, in interesting ways, more like a wine grape vineyard or an orange grove than other crops. The major ongoing cost, once the kiwi vineyard is producing, is for pruning and vine training—and irrigation water is not gratis. The only other issue is pollination, since the kiwifruit flowers are minimally attractive to bees. The preferred solution, in California, is a technique known as saturation pollination, which installs a hive or several in the kiwifruit vineyard, and in effect puts so many bees in a small area that some inevitably pollinate the kiwifruit.

Some of the early kiwifruit growers were retirees willing to take on high up-front costs with a decent probability of yielding steady income after the kiwifruit worked up to full production—a motif heard in Southern California orange culture in the early twentieth century, and among some genteel wine grape fanciers from the 1970s onward, as the wine grape craze (and acreage) built.

The acreage of kiwifruit in California was 5,300 acres in 1998, and 4,000 acres in 2007. Kiwifruit production stayed more or less steady over the 10-year period, with $33 million earned in 2008. Demand has not decreased—but a significant part of

California's production is tied up in exports. The net result is that to meet demand, imports from Chile and from New Zealand in the off-season stabilize the market.

MARIJUANA ***Cannabis sativa* (generally *indica*)**

Pls. 68, 117, Map 12

FAMILY: CANNABACEAE.

RANK: U.S. #1; CALIFORNIA SHARE >30%.

Tall and emerald-green, capable of growing anywhere in California, marijuana is a crop fundamental to the economy of the Emerald Triangle, a tricounty area in northwest California that includes Mendocino, Humboldt, and Trinity counties. In a realm of more than 10,000 square miles, there are fewer than a quarter-million people, and those are concentrated mostly in south-central Mendocino County, or in the side-by-side communities of Arcata and Eureka. Along with sparse population, prevalent winter wetness is the Emerald Triangle's distinguishing characteristic, a boon to marijuana growers, who plant their crops variously on hillslopes, in clearings, or sometimes in so-called grow-houses that are given over entirely to plants, not people. And although marijuana is inarguably the most important crop in upper coastal California, seizure statistics mapped by the federal government in 2006 show marijuana grown in all but seven of the state's 54 counties.

Beginning with a 1983 attempt to exterminate marijuana-growing, the California state government created CAMP, the Campaign Against Marijuana Planting. CAMP has met with limited local success, but at nearly unrestrained cost (billions of dollars), and without making any perceptible dent in steadily rising annual production. Furthermore, with increasingly skilled cultivation, the quality of the marijuana buds—the female plant's flowers, which are dried and consumed by inhalation or eating—has risen swiftly and profitably. Announcements by both federal and California government in 2009 suggest that enforcement of antimarijuana edicts could well abate, and there is talk of legalizing marijuana to tax it. Just what would happen to the possibly thousands of felons in prison in California for sundry marijuana possession and distribution charges (prisons being another huge money sink in California) is an interesting ancillary question, but the reality is clear: marijuana in California is big business,

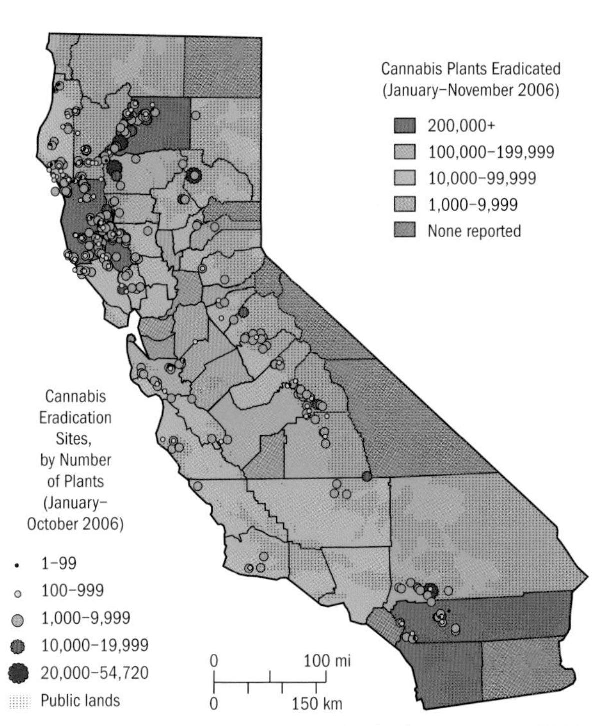

Map 12. Outdoor marijuana plants seized in California by county, 2006 (based on DEA Domestic Cannabis Eradication-Suppression Program map, 2007).

likely the largest value crop (by far) in the state's lineup, and it is perhaps the single largest commodity produced in California, including tourism. No one, though, knows for sure.

The inclusion of marijuana in a field guide to agriculture might seem destined to court controversy. Marijuana is a potent variant of hemp—variously marihuana, ganja, pot, kief, weed, or a dozen other names—and is known in its refined form as hashish. In the official agricultural statistics of California, marijuana is unlisted, because until recently its cultivation and possession was illegal—a criminalizing step first embraced in significant parts of the world in the early 1900s. But cultivated it is. Marijuana's criminality in California was eased in 1996, with passage of Prop. 215, a state initiative with ambiguities somewhat cleaned up in 2003 by Senate Bill 420. The result is that sections of the California legal code, as enhanced by state referenda, make medical marijuana cultivation legal, under specific conditions, as are sales from a marijuana dispensary to someone with a prescription for medical marijuana from a physician.

The avid consumer has a fair amount of spiritual help, with the state's voters making it clear that they would as soon see law enforcement officials redirect efforts toward other forms of crime. Just how much marijuana is cultivated in California, its worth, and how much marijuana contributes (or detracts) from the state economy, are variables in a complicated equation made more difficult because almost everything is done from best guesses. But here is an estimate: marijuana growing in California in 2008 produced a crop worth at least $19.1 billion at wholesale value (to the producer) but that might easily be valued at $40+ billion.

Plate 68. Illegal in varying degrees in all states, marijuana is a stunningly beautiful plant that grows abundantly in California and that is without a doubt the most valuable "agricultural" crop in the state.

Either of these figures is sure to produce hesitation, and adjustments may boost total value by various multipliers: 2? 3? 10? But here's how one estimate is arrived at: federal figures claim 3.8 million marijuana plants were confiscated in California in 2006, six times more than the second-place state, Kentucky. There are oddities in the details of this number that suggest the completely opportunistic nature of marijuana seizures under federal or state auspices. First of all, the largest number of outside plants seized came in Lake County, hardly the state's hotbed of pot cultivation; Riverside and San Diego counties had just a third less marijuana seized than Lake County. Federal enforcement efforts are strongly directed now toward suppressing cultivation in the national forests rather than on private land. The largest county for indoor seizures was Humboldt—but Alameda County, with no outdoor plants confiscated, has nearly the same number of indoor plants seized.

Yearly, these county-by-county figures for seizures and enforcement vary enormously. California's CAMP program released in 2008 its seizure figures from 542 raids (not including the federal marijuana take), which in a remarkable coincidence total 211 more plants confiscated (2,905,232) than the year before (2,905,021). The CAMP-estimated wholesale value—somewhat unimaginably—is $11.6 billion. And that is just from one agency: it does not include federal raids. According to California Department of Justice figures, 2008 saw a total of 5.3 million plants seized by all interdiction efforts: federal, state, and local. Alas, the best maps are for 2006—and not updated.

But for this one entry, perhaps it pays to let the math play out. If 10 percent of the crop were seized in raids, that would mean the total crop cultivated in 2008 would be 53 million plants, indoors and out. Now, only the most halcyon estimates claim 10 percent of marijuana grown in California is seized; most authorities offer 5 percent, and many say less. Policy expert Jon Gettman guessed at 17 million plants grown in California in 2006. That, however, would imply that the 2006 seizure of 3.8 million plants intercepted nearly 25 percent of all marijuana plants grown, or 2.5 times the most wildly optimistic law enforcement figure. A further bone of contention is how much "bud" comes from each plant. In California, the marijuana grown tends to be high-quality and a reliable producer. For what it is worth, medical marijuana growing provisions in Senate Bill 420, and guidelines issued from Humboldt and Mendocino counties, suggest that three pounds of marijuana can reliably be raised from six mature plants (all female) in a 100 square foot area (the "six plants" is an unlikely number driven more by politics than reality). Claims of a half-pound-per-plant notwithstanding, four to 32 ounces of clean bud per plant, trimmed and dried, is realistic and within conventional expert and insider estimates.

Using 2008 seizure figures, and the most conservative figure of a quarter-pound (4 ounces) of trimmed sinsemilla per plant, the total yield within California would be 13.25 million pounds (a quarter-pound from 53 million plants). And from there, the estimate is easy: current retail prices are $430 per ounce ($15 per gram). Let's say that a price of $100 per ounce to the grower is used. The total value would be $21.2 billion (subtract, then, the 10 percent confiscated, $2.1 billion) for a net product of $19.1 billion. But bear in mind, that figure is income to the grower, not street or retail price. For retail price, multiply by four.

Of course, if the 5.3 million plants seized in 2008 were only 5 percent (not 10 percent) of the total, then total wholesale production would be $40.3 billion (that's minus the 5 percent seized). And that is entirely realistic: for all the marijuana confiscated, more and more venues for growing cannabis are expressly off-limits to the roving bands of law enforcement. "Seizure" totals are therefore coming from an increasingly small geographical area (in general, public lands), which means private pot growing is untapped—and quite unmeasured, since there is now no way to track it.

The $19–40+ billion range of marijuana "farm-gate value" is an eye-opener. Honesty requires mention that drug policy experts dish up far smaller numbers. Used are odd, impractical, and inverted metrics that, among other things, ignore that California is producing at least three times as much marijuana as is likely consumed in the state—the rest goes elsewhere. Many policy experts ground "production" figures in a broad estimate of demand—attempting to calculate how many marijuana users there are, they reckon how much users might smoke or consume in an average day, and from that deliver a seat-of-the-pants suggestion of how much marijuana is consumed. It's a remarkably bottoms-up calculation of production. In a world of sophisticated supply–demand curves, that scheme is a galley slave pulling an oar on the demand side.

Thankfully, in a field guide, the questions instead are simply: how much marijuana is actually produced? and what is it worth? In California, marijuana production, and prices, are supply-driven. The pot grower's manta for the "field of dreams" is, "If I grow it, it'll sell." Too much supply could drive down the price, but what's grown is grown, and the best measure of how much is grown is what is actually tracked in the field. Since "amount grown" is unknown, using confiscations or "interdictions" is about the best measure there is—keeping in mind that medical marijuana plots are no longer being aggressively sought in the Emerald Triangle region, nor are sheriff's departments, the State of California, or—since February 2009—the federal Department of Justice going after smaller producers. Being sought are the big commercial growers, especially those with ties to gang members or foreign-directed operations, or those growing heavily in national forests or regional or state parks.

Play with the numbers: if $19 billion in production is accurate, then marijuana as a commodity is worth a half of the output of all other agricultural products in California. If $40+ billion is accurate, then the value is greater than the collective worth of everything else that California agriculture produces. And the fact is, nobody really knows just how much is out there. Of course, were marijuana legalized, as some California state senators suggest, the value per ounce could plummet. But marijuana is a highly specialized crop produced by skilled (and furtive) growers, and the dimensions of the entire industry now and in the future can only be surmised by best assumptions.

Marijuana is grown in varied circumstances. For obvious reasons, unlike most California crops it is rarely visible, whether in backyards or grow-houses or on remote acreage or (especially illegally) in the national forests dotting the state. With the rise of growers tapping into the medical marijuana market, the era of booby-trapping growing sites and a pervasive paranoia about being ripped off has dissipated, somewhat. But travelers within the Emerald Triangle at harvest time know to expect frequent traffic stops by police and highway patrol, and for a five- to six-week period each fall tension rules. The sales of packaging materials rockets upward, and scissors for trimming leaves from the buds become impossible to find. An aura of resin scents the air. Predictably, not everyone is a fan of the pot-growing culture, although irrigation equipment, lights for indoor growing, and an enormous under-the-table economy based on cash are features of these northern California counties. Stories spun about pot cultivation are an adventure in their own right. Among the best is Thomas Pynchon's dense *Vineland*, a novel about the chaotic understory of Mendocino County life that captures its character with a postmodern appetite for sharing the bizarre and paranoid local sense of place. But anyone who has traveled in Mendocino or Humboldt county in harvest season recognizes a radiant aura of persecution that even the breathless pro-weed adulation of magazines such as *High Times* cannot quite banish.

The cultivation of marijuana can take place almost anywhere: the plant is weedy and undiscriminating in its choice of locales; Iowa teenagers are regularly arrested for loading "ditch weed" into car trunks—"weed" that is, most definitely, low-grade marijuana growing wild on the road verge. But that feral crop is a far piece from what is grown across California. Generally favoring sunny hillside locations, marijuana thrives with careful cultivation. To produce the highest-grade (and the most potent) product, all male plants are removed before they pollinate female plants. Seed production is eliminated, and the result is a female-only planting that can grow six to eight feet or taller, with inflorescences (buds) loaded with an active ingredient that purchasers seek: Δ^9-tetrahydrocannabinol (THC, or delta-9-tetrahydrocannabinol). Potency increases in the marijuana grown in California as a result of improved agronomic techniques, including sinsemilla (Spanish, "without seed") which keys on cultivating the flowers or buds of female plants, absent males. Breeding techniques produce a crop with the active ingredient, THC, ranging from 1 percent to 22 percent, with an average nationwide above 8 percent, according to U.S. Department of Justice data.

Processing involves drying the plants, which concentrates active ingredients, removing the buds, and trimming any leaves. The wizened flowers have tripled in potency since the 1980s, and although there is argument about whether there is really any difference between the named varieties of marijuana, each has a loyal following and pricing to match—just like wine varietals or luxury wristwatches. The plants require irrigation in all but the wettest sites, and widely available books describe the optimal cultivation techniques—with instructions anything but consistent. As with wine grapes, the process of learning how to grow the best marijuana is iterative—learn as you go—and at the risk of editorializing, even a small medical marijuana plantation backed by a blue sky is a thing of beauty.

Any discussion of the future of marijuana growing in California is necessarily speculative, since its legal status seems likely to remain veiled for some time ahead. There can be no doubt, however, that marijuana will continue to grow in California, and that profits will enrich the Emerald Triangle economy for decades to come. A degree of decriminalization and relaxing of federal enforcement statutes will decrease violence associated with the hidden product. But for now, the profit motive has taken over, and—like marijuana—it is potent stuff.

MELONS

Cantaloupe — ***Cucumis melo* var. *reticulatus***

Honeydew or mixed melon cultivation — ***Cucumis melo***

Watermelon — ***Citrullus lanatus*, Inodorus group**

Pl. 69

FAMILY: CUCURBITACEAE.

NOTES: *CUCUMIS MELO* INCLUDES CRENSHAW, CASABA, PERSIAN, WINTER, AND OTHERS.

RANK: U.S. #1, CANTALOUPE (CALIFORNIA SHARE 55%); U.S. #1, HONEYDEW OR MIXED MELON CULTIVATION (CALIFORNIA SHARE 72%); U.S. #4, WATERMELON (CALIFORNIA SHARE 16%).

Melons are a serious business, and they do best in sustained heat and in relatively sandy and undeveloped soils, where the earth is essentially a neutral medium for the creeping vines to do their thing. Little wonder that melons in California are found in two broad areas: the central San Joaquin Valley, where harvest takes place from late June into October, and the Imperial Valley and eastern Riverside County (where the Coachella Valley is a magnet for crops that include dates, winter lettuce, carrots, and winter onions that, collectively, can take the desert heat) with their complementary May–July harvest. With cooperation from the weather, melons can be in markets from early May through October.

The gourd family, which includes the cantaloupe (and the cucumber), is robust, with 90 genera and 750 species, and is well represented in North American agriculture. While many melons are grown in California, the primary contributors are cantaloupe, honeydew, and watermelon (actually from another genus). California produces about 60 percent of the U.S. crop of melons. Evidence in historical records traces the movement of the muskmelon from India and Iran to Europe, with evidence of cultivation in Egypt as early at 2400 BC. Melons moved to the Americas with the later voyages of Columbus, and Native Americans soon put it to use, until the muskmelon was known from Florida to New England by the mid-1600s. Melons moved west with American colonization efforts, and commercial supplies have come predominantly from Arizona and California.

In time, other melon varieties besides the muskmelon arrived, and they are now important contributors to agricultural production, although watermelons are the least significant in California. Melons are improved with genetic manipulation, especially at the F_1 cross, which produces vigorous hybrid seed that avoids some of the problems of the open-pollinated varieties. In particular, hybrid seed tends to produce determinate crops that mature at the same time, and since it is difficult to harvest just a few melons at one time, there are benefits to taking the entire crop at a single entry into the field.

CANTALOUPE: Most of the cantaloupes in the United States are now grown in California and Arizona, where aridity and well-drained soils produce attractive melons. The characteristic feature that identifies the cantaloupe is the netting on its skin. In 1881, W. Atlee Burpee Co. introduced the "Netted Gem" cultivar, thought to be the clearest antecedent of the today's cantaloupe. There are some naming peculiarities, since what is called "cantaloupe" in the United States is actually the true muskmelon; the true cantaloupe are actually members of the Inodorus group, which includes the honeydew and sundry other melons. For here and now, the American cantaloupe will remain what we call it; the warning reinforces the treachery of common names.

Cantaloupe has a familiar raised-rind netting over a yellow-tan background, and can be produced from either open-pollinated or hybrid varieties. To produce the earliest fruit, a variety of techniques are used—including lots of plastic, to accelerate solar warming. A drive by one of these fields will leave no doubt about what is being done. The cantaloupe is a warm-season vining crop that spreads out to a substantial volume with vines and roots, a pattern common to other melons. For melons that are raised nonorganically (and melons tend to be troubled by pests), black plastic and soil fumigation are sometimes used, though growers are attempting to move away from fumigation. Melon acreage has changed drastically in California through the years, rising and falling but without any clear trend. In 1920, 26,500 acres were harvested; the peak came in 1990, with 90,200 acres. There are still 46,100 acres in cantaloupe in 2007—a little above the median acreage of those intervening 87 years, with production worth $149 million in 2007.

Fresno County is the giant in melon production, for all variants. Imperial County is not far off in its production. All three of the main melon varieties are down, in bearing acreage and in total value of harvest, from 1998 to 2008. But then again, here is a classic California story. The acreage for honeydews dropped 13 percent from 19,000 to 16,600 acres; the value of production decline was half that, at less than 7 percent. So although acreage was down significantly, income was down far less. Efficiencies of scale and improvements in yield pay off.

HONEYDEW, CRENSHAW, CASABA, JUAN CANARY, SANTA CLAUS, AND PERSIAN: The "mixed melon" family invokes another subspecies of melon, and can be grown somewhat farther north than the cantaloupe, including Sacramento and Stanislaus counties, essentially flanking the Delta Region north and south. These substantial producers require some sophisticated seeding techniques. Spring plantings offer no certainty of emergence, so slanted-bed sowing

Plate 69. Melons, which have an ancient history as cultivated plants, fill out swiftly and do well in California, as with this still-green honeydew in Sutter County, in the Sacramento Valley.

is used for honeydew and its companions, as is sometimes done with cantaloupe. Seed beds are prepared at a tilt, to increase soil temperature by orienting the south face of the bed toward the winter sun. The best germination comes when the sun's rays hit the soil surface at a nearly perpendicular angle, so the ideal orientation is a bed angled at 35 to 37 degrees from the horizontal. The seed line has to be located with care, because after germination, growers will rework the bed to remove the tilt, shaving away soil in several treatments until the entire field surface is essentially flat, with the germinated melons 80 inches apart. Once the leveling is complete, cultivators return to the field to cut new furrows for irrigation. All this is hardly a process for the faint of heart. The San Joaquin Valley offers a more robust (read: warmer) environment, so less highfaluting techniques are called for, although hybrid varieties are popular, and the honeydew variety brought in $56 million in 2007. Acreage is down from historic highs in the late 1980s, but only from 21,300 acres then to 16,000 now: well within the usual range of acreage, and yield per acre is near the all-time high.

WATERMELON: Watermelon production in California follows a geographic pattern similar to that of the cantaloupe: some watermelons grown around the Delta, but most production in the southern San Joaquin (Kern and Tulare) and in the southern desert regions of Imperial and Riverside. Yields are substantial—when watermelon is in demand—but when the market falters, watermelons will simply be left in the field, because weight makes them not worth transporting. It is easy to recognize watermelon, although the proliferation of seedless, or triploid, varieties is definitely on the upswing, despite a nearly stratospheric seed cost that leads most growers to use transplants rather than field-seeding. Also on the rise, if slowly, is the taste of those seed-free variants. The seeded watermelons tend to loom largest in the world of flavor, with the deep taste of the Calsweet a favorite; Sangria and Fiesta are hybrids but have seeds, as does the Royal Sweet. About 50 varieties exist, but they vary in size, flesh, color, and whether they are seeded or not.

NOPAL

Prickly pear pods and fruit ***Opuntia ficus-indica***

Pl. 70, Part 4 opener photo

FAMILY: CACTACEAE.

UNRANKED—BUT CALIFORNIA IS PRINCIPAL U.S. PRODUCER.

There are only a few sources of information about the nopal, many of them in Spanish or within the brain space of prickly pear connoisseurs. Part of the charm of California agriculture is that some crops are not going to make it onto the statistical charts, and squarely in this category is the nopal, the elegant fruit of the prickly pear cactus.

The nopal, from the Nah'uatl word *nopalli*, which describes the pads of the *Opuntia* cactus, is a vegetable made from the young segments of the prickly pear cactus, the pads known in Spanish as penca, and in botanical terminology referred to as the cladode. Cleared of spines, peeled, and carefully sliced, nopales are commonly sauteed with onion, or used in salad, or with eggs (huevos con nopales), and are important ingredients in the evolving cuisine of New Mexico. The pad is prepared with a firm grip. Hold its base, scrape the skin on both sides with a blunt knife until the spines are removed; then peel the pads and cut them in shoestring strips, or dice them according to whatever recipe is being followed. Nopales taste somewhat like green beans, but with a mucilaginous texture that is slightly unnerving, until the eater realizes that they are sweet and delicious.

Alternatively, nopales can be used as cattle feed. The prickly pear is intimately tied to the life of the cochineal, an insect that

Plate 70. Although the prickly pear cactus has its pads harvested to make *nopalito* and bears fruit (*tuna*) that can be roasted, fermented into wine, or eaten fresh, it is not a common sight; here seen in Monterey County.

infests the prickly pear cactus. Cochineal, and therefore the cactus, provides the carmine (deep red) hues that fetched peak prices in colonial Europe. The fruit of the prickly pear is also harvested, and depending on the route a driver takes through California, there is a good chance that around one corner, a field of cacti will appear: there are the prickly pear and their nopales. Along River Road, in the exquisitely beautiful benchland west of Soledad, California, in the Salinas Valley, is a great field of prickly pears, a reminder of agricultural innovation. As Hispanic populations rise in the Southwest and California, nopales gain in popularity.

There are over 100 species of nopal in Mexico, reports Yvonne Savio, a master gardener at UC Davis, but only a half-dozen of them can be considered common. Cacti are cultivated over more than a quarter-million acres of Mexico for their nopales. Tolerant of varied soils, temperatures, and moisture levels, the prickly pear does best in well-drained sandy loam where there is protection from cold winter winds. The secondary product, and only from spring through late fall (depending on the variety), is the fruit of the prickly pear (*tuna*), which can be eaten raw, cooked into jams or preserves, or reduced in cooking to a syrup. A rare, but significant, use is conversion of the ultimately reduced syrup into a paste that is then used to ferment "coloncha," a drink of considerable potency; the relationship to tequila or mescal is in plant origin and the distillate's vigor.

SQUASH AND PUMPKINS — ***Cucurbita* var. species**
Summer or winter squash — ***C. pepo, C. mixta, C. moschata***
Pumpkins — ***Cucurbita pepo* or *C. mixta***

Pls. 23, 71, 107

FAMILY: CUCURBITACEAE.

VARIETIES (SQUASH): ACORN, GEM, PATTYPAN, CROOKNECK, SPAGHETTI, BUTTERNUT, YELLOW CROOKNECK, ZUCCHINI, BANANA, BUTTERCUP, KABOCHA, HUBBARD, CANDYROASTER.

NOTES: *C. MAXIMA* IS THE CLASSIC HALLOWEEN PUMPKIN THAT ALSO PRODUCES PEPITAS.

RANK: U.S. #2, SQUASH (CALIFORNIA SHARE 19%); U.S. #3, PUMPKINS (CALIFORNIA SHARE 11%).

Squash and pumpkins are closely enough related that they should be catalogued together, although there are slight morphological differences. In essence, the pumpkin has a more rigid and squarer stem than squash, whose stem will be softer and more rounded. Aside from that, there is more than enough variability in the "squash" category to encompass pumpkins. As one of the authoritative taxonomic guides remarks, "there is much confusion also between the terms pumpkin and squash." In use, history, origin, and significance, they share a role. The Cucurbitaceae family of plants may have more species used as human food than any other; *Cucurbita* is the most important within that. Squash in 2008 brought in $31 million to the California agricultural economy; the less seasonally gifted pumpkin earned another $17 million—the take for squash was up substantially, pumpkins down, from 2008.

Plate 71. A pumpkin patch may be the sole contact urban youth have with agriculture; Uesugi Farms (based in Gilroy, here in Morgan Hill) has featured a Halloween pumpkin patch for more than 20 years.

In U.S. usage, the main difference is the pumpkin's ascendancy as the seasonal indicator of Halloween, and the limitation of a highly useful crop to a narrow seasonal window. But squash shares the virtues of pumpkin: both pumpkins and squash varieties have formidable abilities to sustain storage well into the winter and spring months. Such a benefit may seem insignificant in an era in which produce can be flown across an ocean or the length of a hemisphere to meet culinary whim and provide fresh fruit or other produce. But not long ago, there was no global market in foodstuffs—the world was not "flat," in Thomas Friedman's phrase—and it was considered something magnificent to have at hand a spaghetti or acorn or Hubbard or butternut squash to diversify reconstituted dried grains or pasta. And then there are the summer squashes—zucchini, pattypan, crookneck. Technically, the squash fruit is a *pepo* in the view of botanists, a special sort of berry with a thick outer wall. To the cooking world (not the most accurate taxonomists) pumpkin remains a vegetable. And it is delicious, with preparations including butter and brown sugar. There is a universe of pies made from the pie pumpkin, an especially sweet version of the traditional Halloween pumpkin that is raised for its workable endocarp—the flesh that becomes pumpkin pie filling.

Over and above the flesh that forms the innards of squash or pumpkins, there are seeds inside those shells (the exocarp). Seeds can be roasted, ground into a paste, or pressed to produce vegetable oil. Squash blossoms are not only edible, but also form a model for sculpture and for some of the finest silver work in Native American artisanal tradition. Squash and pumpkins are, in other words, nothing to be trifled with; they are an important part of traditional world diet, and we think of them as table centerpiece decorations only at great risk to our health and sense of history. Squash for centuries offered much of the diversity in the human diet through winter.

PUMPKINS: If a celebrated fact of American agriculture, pumpkins are a decided urban presence in California. Pumpkins are associated with much revelry and an amount of acting out: carving pumpkins into frightening faces is a legacy of All Hallows' Eve, and feral high school youths (usually male) in some communities seize carved pumpkins and break them—it's no coincidence that Billy Corgan's rock group named itself the

Smashing Pumpkins. There is a broader tradition that makes pumpkins highly visible, and actually part of a cycle of agricultural tourism. "Pumpkin patches" tend to appear in towns and cities during the month of October. In many cases, there were no pumpkins there before—instead, an industrious entrepreneur purchases a few truckloads of pumpkins, sets them strategically in a leased field, watches children scatter and pick their own, and sells pumpkins to the masses ahead of Halloween and Thanksgiving, the two holidays most linked to the pumpkin. If there are few oddities to this, the first is that going to a pumpkin patch is increasingly one of the few contacts that urban residents, especially children, have with production agriculture. The second peculiarity is that only in very few cases are the "pumpkin patches" real—which is to say, actually the site where pumpkins are grown. Usually, the pumpkins are imports.

Taxonomically, the name "pumpkin" means "large melon," and with cause. Most pumpkins are raised to a modest size, and the smallest among them are pie pumpkins, which have large amounts of flesh, grow to a six- or eight-inch diameter, and often become pie filling. If as much for show as anything else, the pumpkin has an interesting distinction: it is one of the few crops that children reliably interact with when it isn't on their plate. Pumpkins are carved to make jack-o'-lanterns, but they have far broader uses, and are a respectable contributor to California agricultural account books. That said, Halloween boosts demand for pumpkins. Seasonal agricultural crops can be spooky in their own right: grocers worry about a huge surfeit of pumpkins after Halloween almost as much as Christmas tree lots fear unsold trees.

Agricultural tourism is a varied activity with festivals dotting the calendar—often at the behest of local chambers of commerce, seeking an upswing in visitors. Pumpkin lots are a mainstay of fall harvest. California's most lurid pumpkin show is an annual Half Moon Bay pumpkin festival at which topping the scale grants the winner title to that year's "The World Championship Pumpkin Weigh-Off." The winner receives $6 a pound—which brought in nearly $10,000 in 2009, when the prize went to a 1,658-pound Atlantic Giant that gave Don Young the winner's check.

Pumpkins likely originated in North America, and share the same classification and biology as squash. They can range in size from one pound to over 1,700 pounds. In general, they weigh 9 to 20 pounds, in highly varied shapes and forms. Although most pumpkins are yellow or orange, they can be dark green, gray, white, or striped. Pumpkin flesh is loaded with beta-carotene, which generates vitamin A in the body, and is another vital influence on health. The uses of pumpkins varies widely—in some districts, they are fed to livestock, used as ornamentals, or eaten as pie. Usually planted in July, pumpkins are popular crops started in summer, and grown to full size by fall—in time for Halloween. They require warm soil temperature and soil that holds water well.

Pumpkins are grown commercially in eight counties, but San Joaquin outstrips all comers by a factor of 10, with $17 million in pumpkin income in 2008. Ventura, San Mateo, Santa Clara, and Sacramento counties all produce relatively small quantities of pumpkins for the urban market; Fresno is the champ. Sweet pumpkins or pie pumpkins are widely used in cooking or incorporated into soup, as a vegetable akin to mashed potatoes, used in sweet dishes, or in tempura. Pie remains a favorite, with homemade pumpkin puree. The pepita, or pumpkin seed, is a flat and green seed, best roasted, and high in protein and tryptophan, an essential amino acid obtainable only in food.

SQUASH: Squash, with pumpkins, is an ancient crop with origins likely in Mesoamerica, likely 8,000–10,000 years ago. With corn and beans, squash was one of the so-called "Three Sisters" of Native American agriculture. The dual forms of squash—summer and winter—make it a notable food resource. Although often used as vegetables, squash and pumpkins are actually fruits, being receptacles for a plant's seeds. Squash can be served in a multitude of different ways, but in general are cooked.

Unlike pumpkins, which in California come significantly from the San Joaquin Valley, squash are regionally more diverse. Sizable squash fields, often home to mixed varieties, are in Santa Barbara, Monterey, and San Diego counties, within shouting distance of the coast. Growing in low vines sprawling across the landscape, squash will essentially harvest themselves when fully grown by drying their stems and severing connection to the mother vine. Squash may be left for some time in the field and allowed to dry more thoroughly, but should not be rained on. Storage in a dark, dry area can then preserve squash for months. When they emerge from storage to go to retail markets in December or later, the variety never ceases to surprise: the turban squash, the Hubbard, the acorn. All that awaits the clever cook in winter is some inspiration.

RAISINS

SEE: GRAPES.

TOMATOES

Solanum lycopersicum,* with synonyms *Lycopersicon lucopersicum* and *Lycopersicon esculentum

Fresh market and processing

Pls. 33, 72, 115, Map 13

FAMILY: SOLANACEAE.

RANK: U.S. #2, FRESH MARKET (CALIFORNIA SHARE 33%); U.S. #1, PROCESSING (CALIFORNIA SHARE 95%).

Tomatoes are a billion-dollar California crop, two-thirds of the value attributable to processing tomatoes (canned, peeled, diced, and paste) the other third to fresh market tomatoes. Interest in lycopene (technically, an antioxidant carotenoid phytochemical) has boosted the tomato's role in American eating—this despite the belief 300 years ago, among early American colonists, that the tomato was poisonous. Technically a perennial, most tomatoes are grown as an annual. From roadside, tomatoes will display the characteristic dark-green shade, grown in long rows. Processing tomatoes are determinate varieties that are harvested in one fell swoop, rather than as individuals or a few at a time. The varieties raised in California can provide some surprises: the processing tomato is produced in a low-growth form, as a sprawling plant, rather than as the climbing vine more familiar to home gardeners. California produces a third of the U.S. fresh market tomatoes, and nearly every processing tomato is California-born.

According to the USDA, tomatoes are the fourth most popular fresh vegetable, with just potatoes, lettuce, and onions consumed in greater quantity. Only lettuce brings in more revenue among these four "pillars" of American cuisine. First domesticated in Mexico, likely originating from wild plants found in the Andes of Peru and Ecuador, tomatoes do

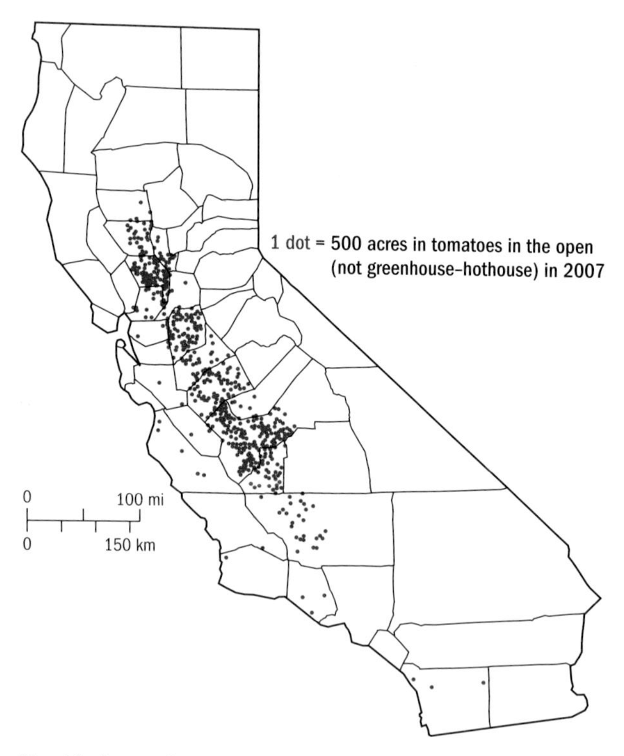

Map 13. Acreage in tomatoes (non-greenhouse), 2007.

remarkably well in California. When Europeans arrived in the western hemisphere, tomatoes were in Mexico, and cultivated. Spanish explorers in the sixteenth century changed the Nah'uatl (Aztec) word *tomatl* to *tomate*, still today the Spanish given name for the fruit. The tomatoes are the reproductive body (ovary) of a plant. Technically, the tomato is a fruit—a berry, although in 1893 the U.S. Supreme Court declared the tomato a vegetable in the sort of quixotic decision that certainly endears the "Supremes" to the American public. Technically, there is no fixed botanical definition of a vegetable, so a tomato can be both fruit and vegetable at the same time. There are more than 7,500 varieties, and heirloom strains are becoming more popular, although rarely among commercial growers.

However, in the world of farmers markets, the heirloom is a significant asset during summer months, and prices in excess of $4 a pound are not uncommon leading up to the heavy production of summer.

An outbreak of salmonella in 2008 caused fresh market tomatoes to be removed from stores in the United States and Canada, but the exact culprit was never determined. That was by no means the only linking of tomatoes to salmonella, and food safety concerns are never far from the mind of producers and technical consultants. Tomatoes are often cooked down and canned as sauce or as whole tomatoes, so the nature and philosophies of home-canning are a frequent home economics class theme.

PROCESSING TOMATOES: All processing tomatoes are mechanically picked and sold under contract. Processing tomatoes may lack some of the attractiveness of their fresh market brethren, but in tonnage, they are huge, and they provide an enormous variety of foodstuffs: tomato paste, sauces, ketchup, and "other products" (which include juice, whole tomatoes, and puree). The firms that buy processing tomatoes work generally with paste, which can then be retasked into sauces, ketchup, and whatever else is needed. Since the early 1990s, the United States has exported tomato products, with tomato sauce and paste accounting for much of the exports.

A major innovation in the processing tomato world of the 1960s was development of a mechanical tomato harvester. Designed at UC Davis and commercialized by Blackwelder Manufacturing Company in Rio Vista, California, the UC-Blackwelder tomato harvester was released in 1964, and five years later, most processing tomatoes were harvested by machine. To work efficiently, that required creation of an entirely new type of tomato, with tougher skin and easily detached from the vine, and these had to take the somewhat rougher handling that machinery-harvest required. Initial varieties were decried as "square tomatoes," but since then there has been considerable improvement. Harvesting costs dropped by half, and ended a world of stoop labor requiring tomato harvest workers to bend and stand for hours on end. A smaller crew staffed the harvester as it crossed the field, providing quality control. The original UC-Blackwelder machine in 2005 was dedicated by the American Society of Agricultural and Biological Engineering (ASABE) as an Engineering Historical Landmark. Big agriculture will respect and honor its innovations.

FRESH MARKET TOMATOES: Commercial production of tomatoes began in the mid-1800s. Although processing tomatoes outweigh fresh market tomatoes by 6:1, the fresh market tomato brings in a much higher price by the pound. Production shifts around California according to favorable climate, but there is an essential split: San Diego has the spring and fall market, and the San Joaquin Valley (Fresno, Merced, San Joaquin, Stanislaus counties) control production during summer. Imports from Baja California are cutting into the natural market for San Diego–grown tomatoes, but urbanization in North County San Diego (from La Jolla, north) has also cut into land where tomatoes used to be the sole crop. A relatively small number of growers control the fresh market tomato crop, and vertical integration of production is the rule, with a few exceptions to keep things interesting. But demand is on the upswing, rising from 12 to 19 pounds of fresh tomatoes purchased per person (1981–2007).

Fresh market tomatoes are intriguing for another reason: they are the vegetable most often grown in home gardens,

Plate 72. Giant trailers used for transporting tomatoes start moving through the Sacramento and San Joaquin valleys to processing plants in August; a mountain of tomatoes is no uncommon sight in Yolo County.

eclipsing even lettuce. Since Americans consume an average of almost 90 pounds of tomatoes and tomato products each year, it can truly be considered a staple. While store-bought tomatoes are seen as occupying a separate food reality from the successful homegrown crop, commercial tomatoes earn a lot of money: in 2006, the state's fresh market tomatoes broke $500 million, from 41,000 acres in California.

The main drawback with fresh market tomatoes is the time involved in harvest and shipping. They can be raised either on poles or stakes, or as bushes, absent any support. If pole-supported, the tomatoes are picked when pink, usually in several passes over several weeks, and are marketed as "vine-ripened." The bush varieties are harvested in one or two picks, with the fruit at the mature but green stage. The cluster, hothouse, cherry, and heirloom tomatoes are a different cultural (and marketing) practice, although they are getting penetration into the major grocery chain market. Farmers markets and specialty groceries are the more common outlet for those less usual variations, which do not appear in the USDA standard statistics. The pole-grown varieties are preferred in Southern California and are often grown in the ground under plastic row covers, with half-tents set up to protect plantings. When you spot a wide area of plastic in March or April, the crop is most likely strawberries or tomatoes (special varieties of tomatoes are chosen for pole cultivation). Costs for pole production tomatoes generally exceed those of bush production, depending on the quality of the harvest.

The alternative is bush production, and the varieties chosen are always determinate (producing and ripening all at one time); young plants are generally transplanted onto the growing sites. Machinery dominates: the seedlings are transplanted mechanically, at the dense rate of 5,300–5,800 plants an acre. Picking starts when 10 to 15 percent of the field is red, and much of the field must be at least mature, if green—the precise level is "mature green 2 (MG2)." Fruits are packed into cartons, and then head to market. The mature green fruits will often be cooled quickly and shipped to cold storage facilities, where they can be ripened slowly. This is a popular solution with fast-food restaurants, which look for consistent tomatoes that they can treat with ethylene for one or two days to generate quick ripening. It is nevertheless startling to see a double-bin truck completely loaded with green tomatoes—leaving behind ton upon ton of red tomatoes in the field. Such are the markets; these are not processing tomatoes (they will eventually be served "fresh"), but they are most definitely subject to handling.

At least 10 types of fresh market tomatoes make it to the U.S. market; some are grown by conventional commercial means, others are organic, and still others (in colder seasons) are raised in greenhouses. Most tomatoes purchased at retail will benefit from a few days at room temperature; they should not be refrigerated unless overripe.

Keep in mind that although there is a significant genetic heritage supporting mass-produced unprocessed tomatoes eaten in hand (or in salad or pasta primavera), in the last decade California has seen growth in heirloom or heritage tomatoes, grown from seed saved season to season, and therefore an open-pollinated variety, not a hybrid that has to be purchased yearly. If heirlooms tend to be less disease-resistant and a somewhat less orderly crop, fans argue that the flavor and often dramatic colors of heirloom tomatoes excuse any of the plants' disadvantages. The names alone can be inspiring: Aunt Ruby's German Green, Big Rainbow, Box Car Willie, German Johnson, Magnum Beefsteak, Cherokee Purple, Paul Robeson, Kellogg's Breakfast, Druzba, Amish Paste, Stupice, White Wonder, Black Krim, and Brandywine (Lantis Valley). Heirloom fresh market tomatoes have a strong fan base, but no discussion of tomatoes would be complete without mention of the famous Guy Clark song, "Homegrown Tomatoes," which sets an all but unattainable standard for the commercial fresh market tomato to have to equal: "there's only two things / that money can't buy / and that's true love / and homegrown tomatoes."

Field, Root, and Row Crops

Three generalizations can be made about most of the crops that appear in this next section. Nearly every crop is at least in part of its production a "fresh market vegetable"—a product that goes to market fresh, slated for quick consumption. To make that succeed requires a sophisticated distribution network and diligence in maintaining it. Second, California dominates the production of 16 of these 33 crops, producing more than 50 percent of each of those (a half dozen more are most likely majority-produced in California—but the USDA doesn't track them). Finally, these are foodstuffs that hugely diversify the U.S. diet—and, in some cases, that sustain variety in the diet of several or many countries around the world. Their presence in the California cornucopia make all of our lives better, and potentially more healthy—at least if you assidnously eat your vegetables.

California sometimes pats itself on the back for producing such varied fare, reflected in crop reports and annual summations of each county's products that are provided by each county's Agricultural Commissioner. But the fact is, that varied fare is real, and it makes eating in California (and across the United States) an infinitely more diverse and interesting activity, at least in prospect.

Vegetables are a sizable contributor to the agricultural economy and a major player in regional land use. Often there are two forms of these crops: one for processing, the other for fresh market (the two categories traditionally broken out for statistical study). Fresh market vegetables go out onto long tables or into sectioned bins at the grocery emporium, farmers market, or health food store. They are fresh and raised to go to market, there to be selected and brought back to someone's home or restaurant for eating, sometimes cooked and sometimes not. Processing production involves varieties raised to go straight to canning, jarring, or other preservation techniques. Sometimes, as with processing tomatoes, which in overwhelming quantity are initially processed into paste, a crop will not only be processed, but later (sometimes months later) may be reconstituted, with seasoning or spicing added, and made into sauce or other products: the techniques are sophisticated, and well tested.

The important thing to grasp is California's statistical dominance in fresh market and processing vegetable production. In a January 2009 USDA report, California was the leading fresh market state. It had 44 percent of the harvested area, 49 percent of the production, and 50 percent of the value, which is tracked for 24 selected vegetables in the USDA roster. Those fresh market choices include artichokes, asparagus, carrots, cauliflower, chili peppers, pumpkins, spinach, and squash. Processing vegetables are less ruled by California, but the state still has 24 percent of the area planted to eight selected vegetables (beans to tomatoes), handles 68 percent of production, and generates 51 percent of the value. In short, when you think of the vegetables you are likely to eat, chances are California grows about half of what you buy.

The discussions here are primarily for the commercial versions of these crops. Many crops, however, have organic or biodynamic equivalents. The organic movement is evolving, as witnessed by major supermarkets adding store-brand organically certified fruits and vegetables (and milk, butter, and cheese) to their produce sections. The industry has developed. Although for some of us raised in the 1960s, "organic" means something raised in a small farm plot with chickens scratching to find bugs, it must be recognized that by 2010 there are plenty of big—even massive—organic producers. If a shopper wants to eat and feed a family organically, then large organic producers will grow and deliver produce that meet government-set standards. This is, however, a remarkably complicated issue, and food policy debates on organic standards, and especially those promulgated by the USDA, have been argued for at least two decades. The standards and practices for raising organic produce in California for years were strict and enforced; the imposition of a federal standard that was much weaker is still regarded in many organic circles as a betrayal of the faith and a usurpation of the local right to a more stringent definition of what constitutes organic food and produce.

For those who believe, as E. F. Schumacher once put it, that "small is beautiful," there may be issues. The federal and California labeling rules for organic food are no simple matter and will likely be subject to much refinement. Small or large, growers believe they are providing food at a higher and healthier standard, or at least one absent pesticides and inorganic fertilizers. Finding a source for vegetables raised to that standard is a personal search, and certainly a triumph when attained.

ASPARAGUS *Asparagus officinalis* L.

Pls. 32, 73, Map 14

FAMILY: ASPARAGACEAE; SOMETIMES, LILIACEAE.

RANK: U.S. #1; CALIFORNIA SHARE 52%.

An attractive and eccentric crop, asparagus is another of the utterly memorable agricultural products for which California delivers better than half the total United States production. With 20,000 acres generating $70 million in revenue (2007), California asparagus production is ahead of Michigan and Washington, which grow the rest. In appearance and growing requirements, asparagus is quite distinctive; for those who love it, asparagus

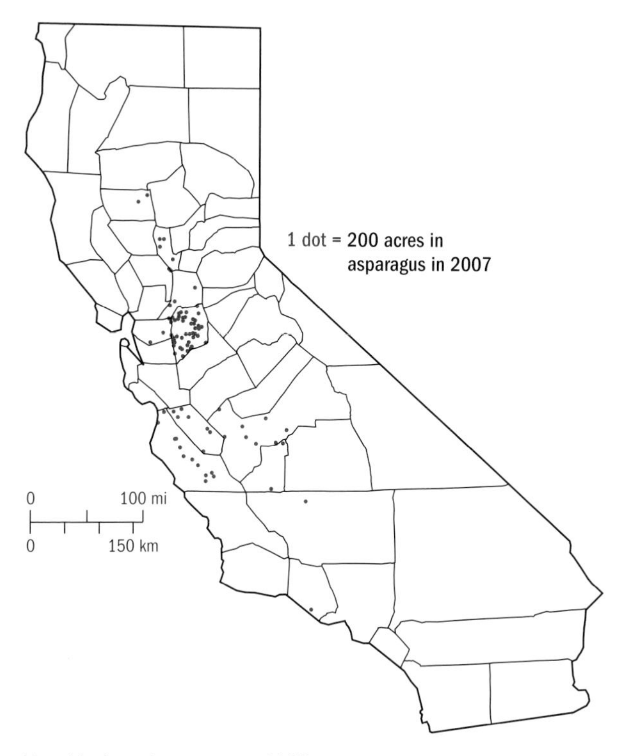

Map 14. Acres in asparagus, 2007.

is no less than a culinary wonder. It has a distinctive taste and offers a flavor and crispness in salad or stir fry, or plain, that is to some utterly addictive. But, like artichokes, another notable and substantially Californian product, asparagus has its initiates, and then among the masses are those who cannot imagine what to do with it. They may never know what they are missing. Asparagus production is declining in acreage and in the value of the entire crop, a pullback attributable to competition from alternative crops (vineyards in Contra Costa County, and lettuce in the Salinas Valley), and edges of the Sacramento–San Joaquin Delta are particularly susceptible to the effects of regional urbanization.

Asparagus is a member of the Lily family and is native to the eastern Mediterranean region, where it was first cultivated in Antiquity. In Greek and Roman society, asparagus might be eaten fresh or dried for winter consumption, and *De re coquinaria*, the oldest extant book of recipes by the Roman Apicius, offers a recipe for preparing asparagus in Book III. Roman texts comment on the curious influence of asparagus on the digestive and excretory system. Asparagus has pronounced effects on the human body, sufficiently so that in some cultures asparagus is regarded as particularly healthgiving. Thanks to high levels of asparagusic acid in asparagus, volatile organic compounds, including forms of sulfur, are metabolized and then pass very quickly—in as little as 15 minutes—into the urine. In no way is this a physiological ailment, but the effect is a source of some humor and will surprise those untutored in the ways of the asparagus. Wild asparagus is still much prized in Europe (Spain, Portugal, England, Russia, Italy, Germany), where the fine spears are chopped and commonly cooked with eggs. When asparagus is shipped to Europe, the market for green asparagus prefers thinner stalks, likely because those are closer to the wild asparagus form. In the United States, the thicker stalks retain more sugar, although the woodier base should be broken off before cooking.

Asparagus is now grown in much of the world. The edible asparagus is related to the asparagus fern (*A. plumosus*) and smilax (*A. asparagoides*), which are used for decoration. Asparagus is planted as a small stalk, or sown from seed, and grows into an underground crown that then sends forth shoots that are known as stalks, or to growers as "grass," which grow quickly and are then harvested and eaten. When the culinary asparagus plant finishes its yearly growth cycle, losing vigor and producing fewer spears, it bolts, and the remnant spears become tall fern-like fronds with bright-red berries, which then dry and are chopped or windrowed to return the material to the soil. That rambunctious final gasp is, ironically, when an asparagus field is at its most visually distinctive. During the earlier months-long stages in the life of an asparagus field, spears emerge overnight from long built-up rows, and workers harvest those spears daily. When field workers are done each day, the field again appears to be long rows of relatively fine dirt, occasionally darkened by irrigation waters.

The history of asparagus in California is tightly tied to the Sacramento–San Joaquin Delta region, where it was first planted in 1852. The market was largely local until around 1900, when the first trainload went east. Asparagus is not an especially robust

Plate 73. Asparagus is a crop almost unique to the Delta, a fast-growing perennial that produces "grass" from spring to early summer, and one that must be packed and processed carefully for best results.

traveler, since it requires substantial cooling to arrest decay when harvested, and if kept around too long and not prepared fresh, the spears can seem woody and less than tasty. Interest in asparagus started in the nineteenth century. A substantial market for asparagus developed in California, and eventually farther afield, and in 1919, Thomas Foon Chew built the Bayside Cannery in Isleton, which began packing green asparagus, shipping, at peak production, some 600,000 cases a year. In the Delta, horse-drawn plows worked the ground, which was so light that draft animals would wear large-area "peat shoes" specially designed for use in the Delta. Animal traction was replaced in the 1930s and 1940s by track-laying caterpillar tractors, created in Stockton specifically to provide flotation on Delta peat soils, keeping the machinery from sinking into or excessively compacting the ground.

Asparagus is a perennial plant, grown in soils that are loose, easily worked, and relatively uniform, which facilitates the timing and duration of irrigation. The Sacramento–San Joaquin Delta region is ideal asparagus ground, with slightly saline soils and a high degree of organic matter, although the earth cannot be allowed to grow too wet. The Coachella–Imperial Valley area and the Central Coast, particularly the Salinas Valley but also Santa Barbara County, can be added as prime growing areas; Fresno is a secondary producer. A crucial element in asparagus production is caution about heat: when the temperature grows too hot—optimum is 65 to 85 degrees Fahrenheit—the spear-tips feather or splay quickly, making for a rather less attractive product. Springlike temperatures are benign, and a mature stand of asparagus—a cleanly kept field can be harvested for eight to 10 years or even more—will produce for up to 30 weeks, yielding 50 to 80 cartons of asparagus per acre. In the Delta, harvest from mid-February to the end of June is common; hotter areas are earlier, while cooler areas can continue a bit later. To keep the spears growing and cool, 15 irrigations a year are ideal in the low desert valleys; drip irrigation line, laid into the rows, is not uncommon.

A major shift in asparagus preference took place in the 1960s. Before then, sentiment in asparagus-eating in the United States was dominated by immigrants from Europe, where the preferred presentation was as white asparagus, which is still available, carefully canned, in European grocery stores. White asparagus carries a high price, and is produced by mounding adjacent soft soil over asparagus rows as the spears prepare to emerge, building it to a height of 12 inches or so. Because the spears are covered, they develop without chlorophyll, and remain white; if exposed to the sun, they turn green. As the tip of the spear starts to emerge from the mounded soil, the harvester reaches in with an asparagus knife—razor sharp at its curved broad tip—and cuts off the spear at its base. White asparagus is correspondingly a very labor-intensive crop, but a profitable one. In Europe, growers will now sometimes shield the spears under black plastic; in California only one grower still produces white asparagus, essentially for family consumption, and it rarely goes to market. Although the varieties favored in California vary somewhat by location, there is a general standardization on the UC157 F_1 and F_2 hybrids. The older Mary and Martha Washington varieties are rarely seen anymore, although there are limited acreages of Apollo, Atlas, and Purple Passion (a specialty line).

The picking of asparagus is labor-intensive activity; it is done by crews of laborers who, in California, are generally Mexican, although some Delta growers will employ a specialty crew of Hmong workers. Sprouting from a mature subsurface crown, individual spears grow inches per night, so a field has to be harvested each day. As it emerges, the grass is either hand-cut with an asparagus knife or snapped off. Snapping takes only the edible upper part, but most growers prefer to remove the entire spear, since that adds weight and a degree of protection, even though the added portion should be removed by the consumers before eating. There are storage advantages, too. A part of the butt is sawn off at the packing plant in processing, and the fibrous butt of the spear will preserve moisture during transit. When properly trimmed and cooled, asparagus can travel for up to three weeks.

Production is on a downswing for the last five years, in both acreage and value, with about half the area harvested (14,500 acres) in 2008 as in 1995, and the commodity value of asparagus in 2008 was forty-eighth among state agricultural products. Out-of-season imports come from Peru and Chile, and competition from Mexico is an ongoing concern. Many California crops demand labor for short periods of time. Asparagus, however, is a long-term commitment; a four- or five-month harvest is at particular risk of labor shortages or escalating pay scales. Delta producers, who predominate in production, consequently provide worker housing that is conspicuously better than the rule, and many of the asparagus workers are year-round employees.

Traditionally, packing of asparagus used to be done in distinctive wooden crates, substantially wider at the base than at the top, and 30-pound crates are still sometimes used for shipping to Japan, a premium market. Most grass destined for domestic consumption goes out in waxed cardboard or fiberboard boxes. A booming market for asparagus crate labels on eBay suggests the degree to which nostalgia has inserted itself into some of the California agricultural markets. Asparagus may be a crop seeing some decline in total crop production value—but even with the reduction in annual output (or perhaps especially because of it), announcements are often heard in February or March on the more food-conscious radio stations about the early arrival of asparagus in better grocery stores.

BOK CHOY — ***Brassica rapa* L.**
Asian vegetables — **subspecies *chinensis***
Pl. 74

FAMILY: BRASSICACEAE (FOR BOK CHOY AND MOST OTHERS IN THIS LIST).

NOTES: ASIAN VEGETABLES INCLUDE (IN CANTONESE TRANSLITERATION) *HINN CHOY* (EDIBLE AMARANTH), *FU GUA* (BITTER MELON), *NGAO PONG* (EDIBLE BURDOCK), *GAI LAN* (CHINESE BROCCOLI OR KALE), *SI GUA* (CHINESE OKRA), *DAI GAI CHOY* (BIG LEAF MUSTARD), *NAM GUA* (CHINESE SQUASH), *TONG HO CHOY* (CHRYSANTHEMUM GREENS), *NAI YOW CHOY* (FALL-FLOWERING CHINESE LEAF CABBAGE), *YOW CHOY* (FLOWERING EDIBLE RAPE), *CHOY SUM* (FLOWERING CHINESE LEAF CABBAGE), *SHANGHAI CHOY SUM* (SHANGHAI BOK CHOY), *SIEW CHOY* OR *WON BOK* (NAPA CABBAGE), *LO BOK* (ASIAN RADISH), *GAI CHOY* (RED AND GREEN MUSTARD), *DOW MIU* (SNOW PEA SHOOTS), AND *TUNG CHOY* (WATER SPINACH).

UNRANKED—BUT CALIFORNIA HAS MAJORITY SHARE.

Asian or Chinese vegetables are a constant presence in markets that cater to an Asian clientele, but are less often seen in open-field situations. Many Asian vegetables prefer a climate more subtropical than temperate, and do best in greenhouse situations, which are common at the urban fringe of Riverside, San Bernardino, San Diego, and Santa Clara counties (the edges of the major cities of San Diego, greater Los Angeles, and the San Francisco Bay Area). That said, the variety of Asian vegetables is stunning, and Asian cooking (especially the many regional Chinese and Southeast Asian cuisines) makes full use of this variety. Beyond the California Agricultural Commissioners' category of "Vegetables, Oriental, All," there is little official tabulation of the crops. That category, in the 2008 report, was worth $55 million for the counties that chose to report it—San Luis Obispo, Riverside, and Fresno were the main beneficiaries. No doubt "Vegetables, Unspecified" is in much the same dilemma of lack of specificity, and that category in California generated a total value of $985 million, in 2008, nothing to sneer at.

So bok choy is a stand-in for the much greater variety that could be "captured" with a much longer entry and good bit of added research. It merits mention that Cooperative Extension

Plate 74. Although the Asian vegetable category is a complex one, bok choy is easily recognized by anyone, so Bryan Brown was right to claim in the 1986 film *F/X,* "Bok choy has great texture, like alien flesh."

agents in many parts of California, including the San Joaquin Valley and the Santa Clara Valley, actively work with first- and second-generation growers from East and Southeast Asia, providing them with advice, sometimes translation, and often a degree of guidance in dealing with the produce-marketing establishment, which has its own sometimes arcane ways of doing business.

Bok choy and other choys (choy sum, yu choy, gai choy, tai cai) are nonheading types of cabbage. The variations are subtle in appearance, but far less so in taste. The best known of the choys is bok choy (bai tsai, or white vegetable or cabbage, as transliterated into Mandarin Chinese), with its tall branches resembling those of a romaine lettuce (cos) head, with a broad central mid-rib. Bok choy comes in many variants, and can be harvested at any point along its life cycle, from baby bok choy to a fully mature plant.

Members in the "bok" category are issues of the family (Cruciferae) that includes mustard, cauliflower, mustard, cabbage, and broccoli. Among the many charms of bok choy is its spatial thrift: it can grow in a small area, and when harvested, all parts can be used. Generally cool-seasoned plants, bok choy and other Asian vegetables are inevitably cooked. Many of the Asian vegetables offer a distinctive, slightly mustardy taste that is considered highly desirable.

Fresno County has the largest number of Asian farm operators in California, but numbers are also significant in Tulare, Merced, and San Joaquin counties. In a 2001 report, two Cooperative Extension experts in Fresno reported that there are over 2,000 acres of Asian vegetables in Fresno, and 62 percent of the Asian farmers were Hmong from the mountain regions of Laos, with another 30 percent Lao, from the Laotian lowlands. In 2007, 1,290 of Fresno County's 6,081 farmers—21 percent—were Asian (another 25 percent were Hispanic). Many of the Asian growers run intensive operations, and many are self-funded operations, growing on parcels of leased land and using micropayments or small business loans. Although the size of operation and the ethnicity or race of a farm operator cannot be linked in Census data, it is notable that in Fresno County, which produces the largest dollar value of agricultural products in the United States, 53 percent of farms had sales of less than $50,000 in the 2007 Agricultural Census. Even in the biggest ag county in the United States, there is room for small production and diversity.

BROCCOLI *Brassica oleracea,* Italica group

Pls. 75, 113

FAMILY: BRASSICACEAE OR CRUCIFERAE, ITALICA GROUP.

RANK: U.S. #1; CALIFORNIA SHARE 93%.

Broccoli has a serious mojo charm in the field: the distinctive bluish-green shade of a broccoli plantation is matched in coolness only by cabbage. The broccoli plant has stature. Its stalks show an intriguing fractal form, and a field sown to broccoli gives a dense and cluttered feel. As broccoli grows, the flowery heads are arranged on the plant like a tree canopy rising above the firm stalk. Broccoli fields, like cabbage, actually show up on Google Earth as a distinctly separate shade of green when looking at the Salinas Valley from on high.

Broccoli has had less than 100 years as an important vegetable in the United States, but is well appreciated elsewhere. The name is from *piccoli bracci,* which means "little arms," and like cauliflower, which it somewhat resembles, broccoli is a Brassica, a

Plate 75. Broccoli, the crown prince of cruciferous vegetables, is a sizable producer in farm-gate value, nearing $700 million (a total exceeded only by lettuce and tomatoes). Monterey County is the biggest producer.

member of the mustard family. Broccoli and related members of the Cruciferae family are so-called because the four-part flower, looked at in cross-section, shows the shape of a cross. That said, the taxonomy of broccoli, as with many brassicas, gives one pause. A half dozen distinct plants are in the same genus and species as broccoli—*Brassica oleracea.* Among them are cauliflower (*Brassica oleracea,* Botrytis group); kohlrabi (Gongylodes group), brussels sprouts (Gemmifera group), Chinese broccoli (Alboglabra group), kale, and collard greens (Acephala group). Rapini (*Brassica rape*) is a broccoli relative also grown in California, but in relatively small amounts. Such variety indicates noteworthy plasticity in nomenclature, and represents a world of fear for the student of cladistics (the study of taxonomic systems). Although each of those plants appears quite different from the other, all the changes in physical appearance were introduced over thousands of years by human cultivation and propagation. *Brassica oleracea* has proved an adaptable and amenable crop—and one hugely useful to humans.

First grown in Calabria, Italy, broccoli was originally known as Calabrese, and while there are distinct varieties, the Italian green or sprouting variety is most popular among growers (and buyers). Broccoli is grown in many states, though California is by far the largest fresh market producer in the United States. Specialization is increasingly common in the California vegetable growing trade, and the broccoli industry is highly concentrated in a small number of large farms that, in turn, control a great deal of the production; two-thirds of the farms producing broccoli are 500 acres or larger. In a move that displays a great deal of understanding of the higher characteristics of the two-income household and shared cooking responsibilities, broccoli growers have latched onto the value-added fresh vegetable sector and are providing broccoli split from the stem as florets, cello-bagged and essentially ready to cook.

A cool-weather plant, broccoli does best in places where the average daily temperature is 65 to 75 degrees Fahrenheit, which accounts for its largely coastal presence in California. The big broccoli counties are Monterey, Santa Barbara, and San Luis Obispo; each has a climate moderated by coastal fog. Broccoli is also grown in the desert valleys and the San Joaquin Valley in winter months. More than a dozen distinct cultivars are raised in Monterey, 14 more in the desert, and 10 in the San Joaquin

Valley. Each cultivar has its specialty, its own niche and timeline that growers aim to nail. With broccoli, the kickers in terms of influence are facts of area, volume, and value. The California broccoli harvest in 2007 brought in $669 million, and broccoli was the thirteenth largest commodity in California agriculture, down one spot from 2006 but up three spots from 2005. That is a lot of fresh market broccoli—since relatively little is processed.

The acreage in California planted to broccoli has not risen much in the last decade, but broccoli is essentially a year-round crop, harvested somewhere in California every month of the year. Broccoli is usually seeded in double rows on flat 38- to 42-inch raised beds, and often transplanted from seedlings. Sprinkler or furrow irrigation is used, and sometimes both, starting with sprinkler irrigation and shifting to furrow irrigation as the plants are established. At 127,000 acres, the extent of broccoli is actually down a few thousand acres from 1999. The value, however, has risen steadily, as the price paid for good-quality broccoli rises. Broccoli, and the cruciferous vegetables in general, have undergone a vast rehabilitation in reputation and are now considered one of the important plants for providing vital minerals, phytonutrients, and vitamins—so long as the broccoli is not overcooked. Broccoli is sold as a full stalk, as crowns, and as just florets. In all forms, it finds takers.

BRUSSELS SPROUTS *Brassica oleracea,* Gemmifera group

Pl. 76

FAMILY: BRASSICACEAE OR CRUCIFERAE, GEMMIFERA GROUP.

RANK: U.S. #1; CALIFORNIA SHARE 98%.

Generations of children were once known to turn slightly pale at the sign of brussels sprouts, which are not at their best when profoundly overcooked; they acquire a slightly sulfurous smell and emit an aura of the indelicate afterworld. But properly cooked, brussels sprouts are entirely another story. As adults, many learned to appreciate the small yet delicate member of the Brassica family, a cruciferous vegetable with sprouts that resemble miniature cabbages. They (we) discovered that, cooked properly, brussels sprouts are a delight—although it takes some of us until age 50 to learn that. Such are the benefits of growing up. And too bad that the children could not

Plate 76. The curious growth pattern of brussels sprouts, with the sprouts forming on a tall stalk, endear it more to some eaters than do the tiny cabbage-like sprouts, although those are delicious when lightly cooked.

see brussels sprouts (widely popular in Belgium, starting in the seventeenth century) in their natural element, where they look very much like something that would appear among the alien barflies in *Star Wars*.

The sprouts grow in a spiral array around a strong stalk up to four feet long; the sprouts can be plucked by hand and dropped into baskets, or the entire stalk can be picked, which is a treat to see in retail establishments, where full stalks loaded with sprouts are increasingly common. Hand labor is the traditional first route for harvest. More and more brussels sprouts are harvested mechanically, if a plant is to be taken all at once. For machine harvesting, cultivars that are determinate, ripening all at once, are now in play. The preferred technique for one-harvest cultivation requires popping off the top of the plant (technically, the apical meristem) weeks in advance of anticipated harvest.

That stops growth of the plant, forcing the subordinate sprouts to mature.

Translocation of brussels sprouts from Europe to the United States began relatively late, with the installation of French settlers in Louisiana. Plantings came to California around the late 1920s, mostly in Monterey County and north along the Pacific Coast, where they mostly are concentrated today. Brussels sprouts are relatively heavy users of water, but standing water or a high water table can kill off a crop, so saturated soils need to be avoided. They do well in Santa Cruz County, in a small strip of land along the Pacific Coast, and in San Mateo County, altogether growing on about 2,000 acres that in 2008 produced some $15 million in revenue.

Brussels sprouts look like golf ball–sized knobs, and are a favorite vegetable for winter, when they offer a variation on the usual holiday foods. The bitter taste that once characterized brussels sprouts is mostly gone, thanks to improved breeding techniques and new varieties brought into play, including the Jade Cross, Oliver, Content, and the late season varieties, Genius and Rampart. The crop remains a relatively small financial contributor to the agricultural rolls, but an interesting and respected one, in California agriculture. A sizable proportion of the crop goes to the frozen food market; the rest is eaten fresh.

CABBAGE *Brassica oleracea linne,* Capitata group

Pl. 77

FAMILY: BRASSICACEAE OR CRUCIFERAE, CAPITATA GROUP.

RANK: U.S. #1; CALIFORNIA SHARE 22%.

Although the form of cabbage familiar to most of us is green and leafy, there are purple and red variations that make it a perpetually interesting crop to spot in the field, and it is harvested year 'round in California, so the cabbage fanatic can always find a fix. The heart of the cabbage is a dense core, familiar to anyone who has sliced open a cabbage as a prelude to making coleslaw, but the outer leaves, which are abandoned in the field when a cabbage is harvested, are what provide much of the energy for growth of the cabbage heart. These external leaves will spread for up to three feet, surrounding the head. A cabbage field, spotted

Plate 77. Cabbage, with its profusion of leaves and densely packed head, can blanket a field as far as the eye can see, as here in San Benito County.

from the roadside, can look for all the world like a giant field of leaves. It may take a sharp eye to note the heads. That the leaves can be a dusky green, purplish-red, or flat-out purple, makes identification all the more interesting.

And can there be any crop quite so diverse in its culinary presentation as cabbage? There are distinctive, even signature, preparations of cabbage in many parts of the world, and especially in Europe. Cabbage soup is a staple in central and eastern Europe, cabbage is often an ingredient in borscht, and stuffed cabbage is the essence of the Yiddish dish *holishkes*. *Garbure* is a Gascon, Béarn, and Basque dish made of cabbage with bacon, meat preserved as confit, and often beans; cabbage is common to dishes in India; and the leaf used in the delectable Greek delicacy dolma is a cabbage leaf softened by parboiling or being placed in the freezer. Fermented cabbage is an alternative to cooking, and has long offered a means of preserving food for winter. Not only is sauerkraut familiar to many American hot dog fanciers, both Chinese and Korean cuisine have dishes made from fermented cabbage—suan cai and the often-fiery kimchi.

This is not to say that cabbage is a wonder food—but it is most definitely a solid producer for California agriculture, and one that brought in $98 million in 2008. Some cabbage is grown under contract with fast-food outlets and coleslaw manufacturers, with varieties and growing specifications ordained by the contractor. Despite occasional see-sawing in cabbage market prices, which can lead to cabbages being left in the field when prices drop, or to significant windfall profits when shortages occur, almost exactly the same acreage is in cabbage in 2008 as in 1998. There is some slight buffer, for cabbages can be processed (into kimchi or sauerkraut), but much goes to the fresh market, and if there is a cabbage surplus, the varieties that are processed are not the same as those favored for grocery store fresh sale.

Head cabbage is grown in 10 counties, largely along the coast: Monterey and Ventura, San Luis Obispo, Santa Barbara, and, somewhat in the interior, San Benito. A winter crop is grown in the Imperial Valley. There are variations on the straight-up head cabbage, with diverse growth forms, and these brought in almost a third of "cabbage" production value in 2006. Chinese cabbage is both tall and white, and the tighter, more compact head of a Savoy cabbage is familiar in specialty markets. Oxheart has an oval head; the drumhead cabbage is flattened, with a squatter look. Cabbage benefits from a cooler climate, as its coastal growing habits should suggest. Often cabbage is grown on 42-inch raised planting beds in two lines, and this nearly four-foot-wide surface is the launching board for cabbage that can either be seeded or transplanted.

CARROTS — *Daucus carota,* var. *sativus*

Pl. 78

FAMILY: APIACEAE.

RANK: U.S. #1; CALIFORNIA SHARE 66%.

Half a billion dollars, that's what carrots were worth to California in 2007. The entire value of the state's carrot crop was $495 million, including both fresh market and processing carrots, which come out of the ground year-round for the eating public. Carrots are a curious crop, a root vegetable that by standard consumer preference is orange, although the color scheme extends to yellow, red, white, or purple. Carrots are slow to develop, so they are seeded as soon as possible, and they are impossible to transplant, because the fruiting body of the carrot is the taproot. That long taproot makes carrots a bit tricky to

Plate 78. Among the most successful of fresh market root crops, carrots spend much of their growth sending up leaves and stalks, and at a crucial point pour energy into the root growth that is eaten.

identify in the field for someone who has not raised them in a garden setting, because the part of the carrot that has commercial value is hidden away underground. But carrot greens are a signature feature of the crop—green fronds rising on multiple stalks from the carrot root, which is what we eat. In the days of greengrocers, it was common practice to buy a bunch of a dozen or more carrots, and, unbidden, the grocer would take off the tops, or leaves, with a practiced twist of the wrist.

Carrots are botanically a biennial crop grown as an annual, and if left alone will produce a flower head in the second year. Few carrots, of course, survive that long—what is eaten is, in effect, a youthful manifestation of what the carrot plant can become (the life-form gives new meaning to the phrase "eat the young"). The flowering stage is commercially undesirable; as the carrot plant starts to flower, it pulls energy and water from the root (botanically, the hypocotyl crown), which shrinks. The Nantes-type carrot grown in gardens is a rarity in commercial production. Hybrid varieties are generally direct-seeded into the field, and their names rally around martial themes: Avenger, Legend, Dominator, Apache, Navajo, Comanche, Flame, Blaze, and Neptune. The main open-pollinated variety is Long Imperator 58, which sounds like a character in a film starring Arnold Schwarzenegger, though the nomenclature is all the carrot industry's own. Usually carrots are mechanically harvested with self-propelled machines that can process 1,000 tons of carrots a day; a two-bed harvester will dig two beds at the same time, and is indeed something to see in action: an example of the specialization of agricultural technology that makes California a superb competitor in industrial agribusiness.

A brawny economic producer within the state agricultural economy, carrots are cherished mainly for their adaptability and diversity. They can be cooked in dozens of different ways, and it is a tribute to carrots that they are considered a staple of both sides of the human age spectrum: infancy and senescence. Stewed or run through a ricer, carrots are a near-perfect starter food for infants broadening their dietary investigations from breast milk to other fare. But for those who are of what Europeans call the "third age," who we might describe as "seniors," carrots return to the fold, processed as juice, stewed, or mashed so that they are not a challenge to teeth that may or may not be there. Carrots are a crop for any age; they are a huge benefit, and their dollar value reflects that. Somewhat processed carrots (peeled and bagged), and baby carrots (likewise, bought in bags) are now readily found—baby peeled carrots account for 35 percent or more of the California production.

In their commercial life, carrots are a curiosity. Although with a little patience they are easily grown in a household garden, carrots go overwhelmingly into fresh market consumption—only 1 percent of carrots go to processing. The California carrotscape is arranged around seasons and heating cycles. Carrots are not at their best grown in high temperatures, because the roots start to shed color and the foliage loses its vigor—plus, the carrot root accumulates carbohydrates more effectively with cool nights. The commercial carrot is almost always orange; multihued carrots that are quite popular in farmers markets or specialty grocery stores are an unusual sight commercially, although yellow and purple carrots were the most common varieties in Europe until the orange varieties were developed in Holland. Yellow carrots came to North America with early settlers but were replaced by orange carrots in the early nineteenth century. To meet winter demand, Imperial County is a sizable carrot producer (worth $82 million in 2006). Grown on 73,300 acres (more than 114 square miles) in 2007, carrots were in Kern, Imperial, Monterey, San Luis Obispo, Santa Barbara, Fresno, and Riverside counties—but none in the Sacramento Valley. The scale of carrot operations is vast; increasingly, carrot production is done with sophisticated tractors equipped with multiple GPS units that keep the machine on a straight line, accurate to within a half-inch, so the driver can focus on engine and equipment performance.

CAULIFLOWER *Brassica oleracea,* Botrytis group

Pl. 79

FAMILY: BRASSICACEAE OR CRUCIFERAE, BOTRYTIS GROUP.

RANK: U.S. #1; CALIFORNIA SHARE 86%.

Cauliflower is brassica, a cruciferous vegetable of exactly the same family as broccoli, chard, brussels sprouts, and cabbage,

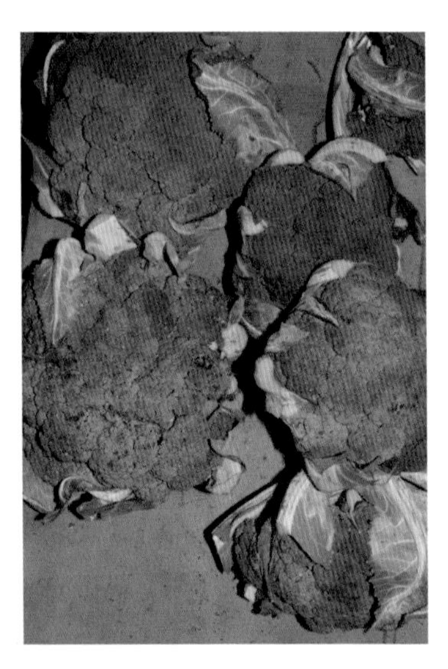

Plate 79. Quite variable in appearance and in coloration, harvested cauliflower is delicate, requiring careful handling that sometimes includes the use of a plastic film to avoid bruising the "curd"—the flowers that are eaten.

although in a different assemblage, the Botrytis group. At the early growth stages, it bears an uncanny resemblance to broccoli, which is its closest relative. As cauliflower matures, a flower grows from the plant, and that "curd" is what is eaten.

Although the familiar cauliflower is white, there are variants grown in saffron, magenta, emerald, plum, and cream-colored: a vegetal coat of many colors. Each of the other distinct shades of cauliflower brings fractionally different nutrients to the table—although everything about cauliflower is healthy, some parts of the crop are more beneficial than others. The curd—the edible immature flower buds—of the cauliflower is eaten, whereas the stalks and leaves are typically abandoned. That the curd is consumed makes cauliflower an oddity in its family of plants, since usually it is the leaves of brassicas that are eaten—besides cauliflower, the other exception (more or less) is broccoli, where it is the green flower buds that are consumed.

There is a good bit of variation in the explanations of just where cauliflower originated and just when it was recognized as a valuable contributor to human foodways. Some argue that domestication came in Asia Minor over 2,000 years ago; there are suggestions that origin in Cyprus or elsewhere on the Mediterranean fringe is possible—what is known is that cauliflower appeared as a domesticated plant some 15 centuries later than cabbage. The naming, with its reference to the caul (an item of headgear or an artifact of childbirth), is sometimes attributed to monks living in thirteenth-century southern France. Regardless of the specifics, cauliflower is a staple of the crudités plate, where the curds are broken up, with some careful paring knife work, as an appropriate receptacle for dip accumulation.

The vast majority of U.S.-grown cauliflower is from California, with much of the crop coming from Monterey, Santa Barbara, and San Luis Obispo counties, where it can be produced across the calendar. Fresh market cauliflower is a $100 million crop in Monterey County alone. For the months that are too cold, cauliflower is grown in eastern Riverside and Imperial counties. The acreage remains essentially unchanged from 1998, at 34,000 acres. And the value is about the same, too, at $198 million in 2007.

Fields are picked three or four times, with those cauliflower heads that show mature curds taken each time, clearing space for sunlight to reach the remaining plants. Lots of sun is good for everything except the curd, which to be kept white has to be shaded, so leaves are rubber-banded together to cover the edible part. That process, called blanching, inhibits photosynthesis, but has to be done manually and is relatively skilled labor. Harvest is also done by hand, and with some care—if the curds touch the ground or a work surface, they will decay or show browning. Once trimmed, the cauliflower heads are placed in plastic bags, sealed with tape, and packed in cartons according to size. Although a relatively recent arrival on the world plant scene, cauliflower is much appreciated, and a welcome addition to the California plant roster.

CELERY — *Apium graveolens*

Pl. 80

FAMILY: APIACEAE.

RANK: U.S. #1; CALIFORNIA SHARE 95%.

Known for its satisfying crunch, celery is loaded with fiber, has almost no calories, and delivers vitamin C at a great rate. Variously used for its stalk (the petiole) or its root, celery is a biennial plant that, like the carrot, is planted as an annual. California produces almost all of the U.S. crop, and has for some time. Overwhelmingly, celery is a fresh market crop, shipped widely around the country and widely used as an amendment to salads, eaten alone, or used as a transportation device to the mouth for dip, blue cheese, peanut butter, black olives with cream cheese, or almost any other intriguing spreadable foodstuff that it would be considered déclassé to break out a utensil and spoon up. Somehow, having edible "silverware" is considered far more genteel.

Traditionally, celery was considered a medicinal plant, rather than an edible one. Celery still grows wild throughout wet places. It began to be considered edible or even delectable in the late seventeenth century, and was eaten as a food by the eighteenth century, arriving in the United States before the American Revolution, although uncommercialized until the late 1800s, when Dutch farmers in Michigan started marketing celery. Considering the relatively recent wild origins of celery, it is interesting that in the United States wealthier consumers tend to eat far more celery than the population at large. Celery is routinely cooked or eaten raw; the celery root,

Plate 80. The tall stalks of maturing celery are familiar to produce shelf habitués who welcome the crispness; in growth form, the crop starts low and grows fast, as here in Santa Paula, Ventura County.

a solid block, can be trimmed, is sometimes cooked, and is more often consumed abroad than in the United States.

The crop tends to grow out swiftly from greenhouse-grown transplants, and is a remarkably even spreader through a field, which means that the entire area can be harvested in one fell swoop instead of incrementally by repeated field entries. Such simplification is never unpopular with growers. An area planted to celery can produce up to 2.5 full crops per year, a sizable production. Some growers try to rotate celery with lettuce or strawberries, but with consumer resistance to pesticides building, the advantages of moving into a field with previously inoculated soil will decrease. If transplanted, celery usually has a two-month start (up to 70 days), and then is transplanted, double-rowed, onto the familiar 40-inch wide raised beds. As a crop used to a high water table in its wild form, celery does well in saturated soils.

Celery is grown throughout the year in California, and marketed around the country and abroad for export. All the counties where celery is grown in California reflect a strong coastal influence, with Ventura leading the pack (46 percent of the crop), followed by Monterey (32 percent), then Santa Barbara, San Luis Obispo, and San Benito. Celery is a gainer in the world of crops, up 10 percent in area harvested from 1998 at 26,500 acres. It is a matter of token amusement that although celery counts as a "Minor Crop" in the records of the California Minor Crops Council, it earned $401 million in 2007. Minor, indeed, especially for crudités.

CHARD — ***Beta vulgaris*, Cicla group**

Swiss chard

Pl. 81

FAMILY: AMARANTHACEAE.

UNRANKED—BUT CALIFORNIA PRINCIPAL PRODUCER.

Chard is a creature of many colors, forms, and names. In essence, it is a beet grown for its edible leaves, and by the roadside it is a stunner: a plant 18 to 24 inches tall, with glossy, crisp leaves that can grow up to 15 inches long and 10 inches wide, marked by a strong internal structure. The notable features are stems and ribs that range from white to red to yellow. Though commonly spoken of in California as chard, the plant was called Swiss Chard starting in the nineteenth century to distinguish it from a variety of French spinach; the first varieties are traced back to Sicily. Chard can be used for salads, stir-fried, or wilted and added to soup. Along with kale, mustard greens, and collard greens, chard is one of the several leafy vegetables referred to generically as greens.

Genetically, Swiss chard is the same as sugar beets, or table beets, although visual recognition of that would be difficult; its visual resemblance is much closer to spinach. Baby chard and mature chard are sold widely, but relatively little production is at a commercial scale, and the USDA stopped tracking chard production in the United States in 2002. A pity—the health literature is paying consistently more attention to Swiss chard and other greens with suggested healthful qualities; aside from being high in vitamin A, with a high ratio of beta-carotene content, qualities as a carotenoid may have anticarcinogenic qualities. In short, a food to keep an eye out for.

Plate 81. Chard may be the most colorful and dramatic of vegetables with its distinct colored ribs and a glossy flesh—a soup- or salad-maker's favorite.

Planted in early spring or fall, Swiss chard's stems make it recognizable, for they tend to be in colors so lively as to seem lurid. In the field, it is rare to see a huge planting of Swiss chard all in one variety. Smaller aggregations, by color, tend to dot a field, and Swiss chard is sold in bunches, rather than by the pound. However, bags of mixed greens are increasingly sold bagged, washed, and ready to cook. Edward Wickson noted in 1917 that "chard is not largely grown in California because conditions are so favorable for continual supplies of spinach, which is preferred, and yet many find it desirable." A voice in the wilderness, then: but now the market is growing, and local consumption is rising.

CHILI PEPPERS ***Capsicum*** **var. species**
Tobasco, chiles de arbol, malagueta ***C. frutescens***
Naga, habanero, Datin, Scotch bonnet ***C. chinense***
S. Am. rocoto ***C. pubescens***
S. Am. aji ***C. baccatum***
Pl. 82

FAMILY: SOLANACEAE.

RANK: U.S. #2; CALIFORNIA SHARE 43%.

The pepper plant is a familiar presence in gardens, but less so as a significant agricultural crop. Chili pepper plantations in California are something of a rarity—under 6,000 acres are grown—and they are scattered in different production areas, and more than 350 varieties exist; all derive from peppers originally grown in the Americas. Chili peppers show no single morphology or look—they are highly varied in appearance, and there is no relationship of size to hotness, except that many of the very hottest chili peppers are quite tiny. Peppers may be eaten plain, or dried, or ground. The common ingredient present in chili peppers is capsaicin, a colorless, odorless alkaloid that is concentrated in the placental tissue of the pepper. Capsaicin gives the pepper its punch—or makes it as mild as bell peppers, which are a member of the same species (but are distinct in crop record keeping and in use—so they have their own entry). Animals are kept away by capsaicin, which targets pain receptors specific to mammals. Birds are unaffected, likely because they act as distributors of seeds.

Peppers are grown in California with nowhere near the research—or, for that matter, the variety—of chili peppers mustered by other states. The leader is New Mexico, which produced chili peppers on 12,000 acres in 2007, and which hosts the Chili Pepper Institute at New Mexico State University in Las Cruces. Where California does lead is in agricultural technology associated with the mass production of peppers and their incorporation into more complex foods, such as salsas. The jalapeño, for example, has a stem that is difficult to remove, but growers at the southern edge of the San Francisco Bay Area have evolved equipment that, in a major labor savings, pop the tops off jalapeños, which are then pickled whole or sliced for an avid hot pepper market in which pepper consumption has picked up from 12 pounds per person in 2001 to 14 pounds per person in 2006—amounting to an avalanche of interest, in the food-growing world.

Like eggplant, tomatoes, and potatoes, the chili pepper is a member of the nightshade family (Solanaceae). The chili pepper comes in diverse shapes, colors, shades—and varying levels of hotness judged by Scoville heat units (SHU), which is the number of times a chili extract must be diluted in water to lose its heat. Bell peppers come in at 0 SHU, jalapeños about 4,500 SHU, and habaneros at 300,000 SHU. In India, the Naga Jolokia pepper is rated at 1,000,000 SHU, the world's hottest—smeared on fences in northeastern India, the residue is said to

Plate 82. Peppers have a justly earned fame for their varying hotness, judged by Scoville heat units, which range from zero (for the bell pepper) to an exotic pepper from Assam, India, that clocked in at a million SHUs.

keep wild elephants at a distance. Unsurprisingly, it has gained popularity as an import to the United States.

From origins in the New World, the chili pepper has been incorporated into national or regional cuisine in almost every corner of the world, from paprika in Hungary and pimenton in Spain to the Tunisian harissa or Indonesian sambal. Of course, there is a universe of indigenous chilis known and eaten in Mexico and Central America. Chili peppers may be dried in long strings (ristras), or incorporated into decorations. There are studies underway examining the effect of chili-eating on endorphin production in the human body. Some analgesic creams use capsaicin to assist with pain management, which appears to work without any reduced effect from prolonged use. And pepper spray really does employ capsaicin, which comes from chili peppers. Let it be agreed: the plant is nothing if not versatile.

The varieties of chili peppers in California change from year to year, with changing fashions. The "standards" are paprika, jalapeño, Anaheim (or New Mexican), wax, Fresno, Thai, poblano, habanero, Scotch bonnet, and chileno peppers. The California acreage in chili peppers has varied little through 80 years of record keeping: there were 3,350 acres harvested in 1929, and a peak of 7,100 acres in 1946 (and 7,020 in 1981), but in 2007, 5,800 acres were harvested. Although there was never a marked leap in acreage, a whole lot more peppers are grown on each acre of land: yield was 1,500 lbs/acre in 1929, but 30,000 lbs/acre in 2007—an increase of 20 times. Little wonder that the value of the chili pepper harvest in 2007, $62 million, was a record. Production areas include the Desert Valleys of Imperial and Riverside, the Central Coast (Monterey, San Benito, San Luis Obispo, Ventura), San Diego and Orange counties, and Tulare, Fresno, and San Joaquin counties in the San Joaquin Valley.

CILANTRO — *Coriandrum sativum*
Coriander
Mexican parsley
Chinese parsley

Pl. 83

FAMILY: APIACEAE (ALT. UMBELLIFERAE).

UNRANKED—BUT CALIFORNIA IS MAJOR PRODUCER.

Cilantro is both a spice and an herb, with appropriately diverse uses. In leaf form, it is known as Mexican parsley, or simply as cilantro, and is widely used as a garnish to Mexican cooking, and as an essential ingredient in some relatively exotic sauces, such as the Canary Islands standard meal accompaniment, mojo verde. This leafy cilantro has an exceptionally strong (but never unpleasant) aroma, and an equally distinctive taste that is either sought after or avoided, and those who dislike cilantro detour from it as they would a listing plague ship.

In the field, cilantro is not particularly easy to recognize; it looks a great deal like Italian parsley, and the only way to be sure is to pick a leaf and breathe deeply: the smell will tell. Fresh cilantro (it can also be referred to as fresh coriander) is chopped and added to food or, in the case of mojo, is chopped finely, mixed with garlic and high-quality olive oil, with some vinegar, and spooned over

Plate 83. The singular taste and even more distinctive smell of cilantro mean that although some avid eaters search out what some call "Mexican parsley," others shun it entirely.

food. But the reality of cilantro is that it is a crop known throughout the world, with uses that are remarkably varied—fresh or dried, whole seed or ground. Fresh leaves are put to use in chutneys, in Chinese dishes (sometimes including the roots of the plant, which have an even stronger taste than the leaves), and in Mexican salsas and guacamole. Ceviche properly has several pinches of cilantro in it to moderate the taste of lime.

The mature form of cilantro, generated when the leaves bolt and go to seed, develops to coriander seeds, themselves widely used in cooking. Ground, the seeds are a crucial component of many Indian curries, but coriander can also be purchased pre-ground, or in seed form at most larger supermarkets or specialty stores.

The use of coriander seeds is purported to go back to 5000 BC, which would make it among the oldest known spices. Native to both European and Middle Eastern regions, coriander seeds were known in Asia, cultivated in Egypt, and mentioned in the Old Testament as one of the bitter herbs ordained to be eaten at the Passover. The Romans used coriander seeds, appropriately ground in mortar and pestle, to spice bread and preserve meat; when coriander seeds traveled to Britain, they were used in Tudor times as a wedding libation called hippocras—a highly spiced wine. The seeds, crushed or chewed, are used to flavor gin, liqueurs, chewing gum, cigarettes, and a Belgian brew called witbier. Coriander is still used widely in Mexican, North African, and Near Eastern cooking.

Relatively little cilantro (coriander) is harvested for seed; the herb is fresh market ready at 40 to 45 days, and the key growing areas are in the Central Coast from Ventura north through Santa Barbara, San Luis Obispo, and Monterey counties. There is limited production in Santa Cruz, San Benito, and Santa Clara counties, but winter production in the Coachella Valley and San Joaquin Valley (Fresno and Stanislaus) is common. Cilantro is often used in rotation with other crops, and some growers will produce multiple crops in a single year. In 2008, cilantro generated $30 million from 4,572 acres.

CORN, SWEET — ***Zea mays* var. *rugosa***

Pl. 97, Part 2 opener photo

FAMILY: POACEAE.

RANK: U.S. #2; CALIFORNIA SHARE 16%.

SEE ALSO: CORN, FEED.

Anyone can recognize a cornstalk, but not everyone ends up figuring out what is growing inside it. This is about sweet market corn, which is for the table, and quite separate from feed, or field, corn. Both are raised in California, and in the eye of the passerby, they look a lot alike. But there is a world of difference; although both are of the same species, and each is a grass, sweet corn is raised to be eaten fresh. Feed corn is considered a grain, and is allowed to dry in the field. Then the kernels are mechanically removed from the husk and transported to a grain elevator, where they are again dried to a common standard. Further confusion is possible, because what is confidently called "corn" in the United States is referred to in British English as maize. Maize comes from the Taino word for the plant, which is *maíz*. For the British, "corn" has always been wheat, and to make matters still more complicated, Europeans frequently learn English in the British Isles and not the United States, so this confusion about just what is being presented in the market or on the table endures. For non-Americans, corn is maize, and this entry is restricted to the sweet table variety, which is a staple crop of huge significance. If, as some authors argue, Middle Eastern civilizations were based on wheat and Chinese civilization on millet and rice, then Anahuac civilization in the Americas was grounded in the cultivation of corn.

Sweet corn varieties differ in sweetness, time to maturity, and color. If a field seems divided into sections, with one maturing, another mid-height, a third just emerging from the planting beds, and one section apparently fallow, that is the normal sequence of successive plantings to ensure a continuous harvest. Sweet corn is the result of a recessive gene that controls conversion of sugar to starch. In growing sweet corn, hybrid seeds are used, and although sweet corn is the largest component of human-directed corn, popcorn and ornamental (blue or Indian corn) can be grown in California, and those must be planted at least a quarter-mile away from sweet corn, or the pollen from those other varieties will carry by wind and affect the sweet corn. Production of sweet corn follows a planting and yield cycle similar to a number of California field and row crops. Winter plantings in the southern desert valleys (eastern Riverside and Imperial counties) capture the early market, with harvest from May through June, and then again from November through early December. The rest of the California crop is planted from February through July, for harvest in July, continuing through October, and goes into the ground in the south coast (San Diego and Orange county, with additional planting in Santa Barbara), in the warmer regions of the Bay Area (San Benito, Santa Clara, and Alameda counties), and in the San Joaquin Valley, including eastern Contra Costa County, on the edge of the Delta.

Production is substantial, earning $108 million in 2007, with the big agricultural counties in the lead (Fresno, Imperial, San Joaquin, Riverside's Coachella Valley), but Contra Costa County remains a respectable sweet corn producer, largely to move sweet corn to a greater San Francisco Bay Area market of seven-plus million people.

The history of sweet corn is somewhat obscure. Although Native American populations began the process of corn domestication 7,000 to 12,000 years ago in the Americas, sweet corn is a relatively recent arrival. The first recorded example was a gift from the Iroquois to Europeans in the eighteenth century. It caught on, and became widely available in the next 150 years. Sweet corn is a staple in the cuisine of Mexico, and is widely used in North America by many Native Americans, although sweet corn did not arrive in California until European incursions in the late eighteenth century, for reasons that remain complicated.

The mechanics of sweet corn production are perhaps familiar, but merit a quick mention. Seeding beds are used more commonly now than rows, 33 to 40 inches wide, with one or two rows of plants for each raised bed. Sweet corn has to be irrigated, and although furrow irrigation is still common, in some parts of California, drip irrigation tape is buried in a permanent bed. Hybrid seed is almost always used, although varieties are changing. The three genetic classes of sweet corn are standard endosperm (SU), sugar enhanced (SE), and supersweet (SH_2). Which variety was planted is usually evident in the eating. Some varieties are raised for processing—think corn niblets, or frozen corn—but others are shown a quick bath in boiling water and then head to the table. Americans eat an average of 26 pounds of sweet corn a year; more than 9 pounds of that is fresh sweet corn, the rest processed. Organic sweet corn is also produced; the varieties are different, and care has to be used not to intermix pollen from different corns—and pest management issues are never trivial.

CUCUMBERS *Cucumis sativus*

Pl. 84

FAMILY: CUCURBITACEAE.

RANK: U.S. #3; CALIFORNIA SHARE 10%, BUT PRODUCES MAJORITY OF FRESH MARKET CUCUMBERS.

The cucurbit family includes cantaloupes, squash, pumpkins, watermelons—and cucumbers. They are vine crops, sometimes grown on poles or trellises to suspend the fruit, which is what is eaten. Cucumbers are grown in three varieties. Two of them can be seen on California's 5,000 acres of cucumber land: a processing cucumber used for pickling, and a fresh market cucumber for slicing. The third variant is hothouse cucumbers raised for slicing, and they are grown in greenhouses and therefore all but invisible, although greenhouse structures may themselves be in evidence.

Cucumbers in the field may be on poles or trellised, or may run in rows that allow the plants to spread, vinelike, along the ground. Seed goes directly into the field, sown two to three inches deep, and vines emerge soon after if the ground is sufficiently warm. The numbers of plants involved is impressive, especially for pickling cucumbers, which can go in at 40,000 to 90,000 plants an acre; some growers will plant 150,000/acre, and machine harvest—most commonly in Michigan, which leads the nation in pickling cucumber production. Whether slated for pickling or otherwise, bush and vine cucumbers alike generally spread to all the available area, if they are not trellised or tied up on a pole. Cucumber sex is a difficult process, with vastly more male flowers produced than female, which makes the current frontier of cucumber genetics the production of seed for a cucumber that reliably generates more female flowers (which alone bear fruit). Success so far is mixed, but that accounts for a lot of flowers and not so many cucumbers; by 10–20:1, the flowers are typically male. Pumpkins have much the same problem, with a preponderance of males over females—and of course, they are related to the cucumber.

The cucumber fruit is harvested at a variety of stages, from quite young to mature, but before the seeds inside reach their

Plate 84. A modest but high-producing crop, cucumber can go to many purposes, from salads and sandwiches to bread-and-butter pickles and gherkins.

final stage, when they enlarge and harden. Any size of cucumber, short of the overmature, can be tasty. They are quite variable in presentation, from the small pickling cucumber up to the long and slender slicers, but apple or lemon cucumbers are nearly spherical, and yellow. There are seedless (sometimes called burpless) cucumbers that are English or European varieties, grown in both fields and greenhouses.

Acreage in California is down by 50 percent over the last decade, and farm-gate value was $20 million in 2007. With the exception of San Benito County, where there is a small pocket of cucumber-growing, the crop takes advantage of the generally warmer climate of the San Joaquin Valley and Southern California, with San Diego the largest dollar producer, and with additional coastal production in Ventura County. The largest acreage sown to cucumbers is in San Joaquin and Tulare counties. The high price for Southland cucumbers comes because of the row covers that are set up to protect the vines from wind and boost their temperature. More delicate than peppers or tomatoes, early crop cucumbers will be planted outside with tunnel row covers that look like a plastic half-circle tent atop the plants. That reserves both water and heat from the soil, which would otherwise vent to the sky; even in Southern California, the late winter and early spring months can bring a chill. The advantage is the far higher prices that early cucumbers offer—to the benefit of north-county San Diego and Ventura.

EGGPLANT — *Solanum melongena*

FAMILY: SOLANACEAE (NIGHTSHADE).

RANK: U.S. #3; CALIFORNIA SHARE 18%.

Eggplant are not a prime value producer in California, bringing in $10 million in 2006, more in 2007, and a little less in 2008. There are a few oddities about eggplant production, however, that give it added interest. First, nearly two-thirds of the $12 million in eggplant produced in 2006 was in Fresno—a remarkable concentration, and a figure that rose to $9.5 million in Fresno County production in 2007, before dropping 30 percent in 2008. Second is that within Fresno, a good bit of the eggplant produced was grown by farmers within the Hmong community, who have made a specialty of producing intensive eggplant crops on limited acreages, including Chinese and Japanese eggplants and the traditional elongated or teardrop-shaped aubergine. While farming in the agricultural niches, Fresno's small farmers (50 percent of Fresno County growers gross less than $50,000) obviously muster some phenomenal productivity, and the Cooperative Extension experts in Fresno are attentive to specialty markets and ethnic farmer needs.

The eggplant generates a somewhat disconcerting shape, with a strong central trunk and a scattering of relatively large leaves. The flowers are often fairly spectacular, depending on the variety grown.

The regional concentration of eggplant in California matches the plant's affection for a moderate climate: the desert valleys of Riverside and San Bernardino, Orange County along the southern coast, and Stanislaus and Fresno in the San Joaquin Valley. With careful planting on the 1,106 acres sown to eggplant in 2008, fresh market eggplant is available from April through October. Eggplant is a voracious rooter if planted in sandy or loose soils, with roots three to four feet deep. Self-fertile, eggplant is content within a wide range of temperatures, but heat above 95 degrees Fahrenheit causes the flowers to drop.

Thomas Jefferson is thought to have introduced eggplant to the United States, growing it in his Monticello garden. Considered a vegetable, eggplant is another fruit with the morphology of a berry, with numerous small, soft, edible seeds—and those fruitlike qualities link it to the bell pepper, tomato, and potato. When cooked, the sometimes bitter eggplant taste mellows. With a remarkable ability to absorb cooking fats and sauces, eggplant is a boon to anything that involves complex sauces, including Thai dishes, the French ratatouille, the Italian eggplant parmigiana, Greek moussaka, and the Arabic dish baba ghanoush. Eggplant is stuffed, fried, baked, and grilled, and in Indian cuisine it is noted as the "King of Vegetables." The United States is a minor producer, but California stands among the top three eggplant-growing states.

FENNEL — *Foeniculum vulgare*

Anise

Pl. 85

FAMILY: APIACEAE (FORMERLY UMBELLIFERAE).

RANK: U.S. #1; CALIFORNIA SHARE 99%.

Plate 85. Two aromatic crops, leeks (gray, on left) and fennel (airy leaves, right), grow side-by-side in San Juan Bautista, where the superior soils near the Salinas Valley provide prime growing ground.

Fennel is a hearty crop prized for its root, which is a large and solid block of white flesh, much like the subterranean bulb of celery or the Mexican jicama (*Pachyrhizus erosus*). The fennel bulb can weigh up to two pounds and has strong stalks that radiate from the root or bulb and that look much like celery stalks (the plants are related). The fronds that appear above the ground and that provide the energy for root growth reflect fennel's status within the Apiaceae family, which used to be known as an Umbelliferae—and the plant displays a dramatic splay of leaves, seeds, and fine flowers, reminiscent of the last yearly growth of asparagus. Fennel is not a widespread crop in California, but is routinely found in stores, where the neatly trimmed roots, oval in cross-section but also ribbed, are placed among the produce.

Used as a vegetable, fennel is a crop with distant origins; it was known as a vegetable with useful properties in Roman times. Particularly when roasted or thinly sliced, fennel has a distinct and most pleasant taste of anise, and it has, among its uses, a role as one of the principal ingredients of absinthe, the legendary distilled spirits drink that was a darling of European society through the nineteenth century, but then was banned for much of the next hundred years because of the alleged psychotoxic effects of wormwood, another ingredient. With those effects disproven by twenty-first century chemistry and spectrometry, anise (and fennel) have come back as a darling of adventurous society. And, indeed, fennel has the same distinct, yet slightly cloying taste as the absinthe to which it contributes so much flavor. In Greek mythology, Prometheus supposedly used a stalk of fennel to steal fire from the gods, making it a crop of diverse uses.

A perennial herb, fennel distributes taste through its roots, its leaves, and its seeds. The aniseed flavor comes from anethole, also in anise and star anise. California saw 786 acres harvested in 2006, down from a peak of 1,063 acres in 1999, yielding 12,003 tons of fennel—enough for a lot of salads.

GARLIC — *Allium sativum* L.

Pl. 86

FAMILY: ALLIACEAE.

RANK: U.S. #1; CALIFORNIA SHARE 86%.

They are tall and elegant tufts, much like those of leeks, onions, or scallions that launch skyward. The stalks that ascend from growing garlic are neither particularly dramatic, nor are they telltale: the fact is, it's not easy to tell exactly which crop is under there. But the general conformation speaks "root crop," which garlic is. Along with onions, garlic is certainly the foremost root bulb product used reliably in U.S. cuisine. California is a sizable producer, although Nevada and Arizona each have substantial production. Much of the processing of garlic—which takes the characteristic heads in their white papery wrappings, breaks them down into their 4 to 15 compound cloves, and chops, minces, grinds, or dehydrates the garlic into nonperishable form—is done in either northern Nevada, near Fallon, or in the town of Gilroy, California, at the southern end of Santa Clara County, which has for 40 years been a garlic stronghold. Fresh market garlic is a prominent product,

Plate 86. Garlic is a mainstay of Mediterranean cooking. Although Gilroy lays claim to the "Garlic Capital" title, Fresno County grows 500 times more than does Gilroy. The Gilroy Garlic Festival is held the last full weekend of July.

too, and a major moneymaker; processed and delivered in jars, in a suspension of oil, garlic is now a staple of American eating.

Valedictory speeches on the virtue of garlic date well back through time to the Bible, to ancient Chinese scrolls, to the writings of Dante, Shakespeare, and Sir Francis Bacon. French chef Marcel Boulestin once wrote, "It is not really an exaggeration to say that peace and happiness begin, geographically, where garlic is used in cooking." An Arab legend has it that "when Satan stepped out from the Garden of Eden after the Fall of Man, Garlick sprang up from the spot where he placed his left foot, and Onion from that where his right foot touched." Since it is universally known from the popular literature that garlic repels vampires, a garland of braided garlic with the long strands attached is at once a decorative statement, a supply for the leaner months, and (who would have guessed it) a home defense mechanism against the undead (but not zombies, so it definitely pays not to get cocky).

Gilroy, California, prominently bills itself as the Garlic Capital of the World, even though the area immediately around Gilroy produced barely a million dollars' worth of garlic, and at that on 297 acres, the second lowest county total reported in 2008 agricultural production statistics, ahead of only tiny

Mono County, on the eastern side of the Sierra Nevada. But Gilroy wasn't, and isn't, known for garlic, it's known for the Gilroy Garlic Festival, which celebrates garlic in all its magical manifestations and improbable possibilities. Should that claim seem over the top, consider garlic ice cream, garlic fondue, garlic hand lotion, and garlic massage oil: in Gilroy, the unthinkable becomes an everyday experience, so long as it involves garlic. Gilroy in fact deserves credit for ambition in seizing upon the marketing opportunities of garlic, and the now-defunct radio station KFAT, which lasted through 1983 and broadcast until the very end from the "top of the Gilroy Hotel," was a stalwart in the Garlic Festival support group. The Festival originated, however, when the President of Gavilan College in Gilroy read in 1978 about a garlic festival in Arleux, France, that drew 80,000 people to a community with just a couple of thousand native residents. The first Gilroy Festival was inaugurated in 1979, and continues today—one of the dozens of such festivals in California designed to draw in wellwishers, and the hungry (and those with well-padded wallets), to sample the comestible fare. Christopher Ranch remains the main Gilroy-based producer, and has been so for more than 50 years.

In fact, garlic hit its peak production in California in 1999, with 40,000 acres, but the value of the "stinking rose" was still $203 million in 2007, a highly respectable total, and enough to boost garlic from thirty-fourth to twenty-seventh in the list of farm-gate value crops. But to embrace garlic is more than supporting a crop—garlic is a lifestyle. Garlic establishes an enduring imprint on those who make it a regular part of their lives, and it counts as a significant part of the Mediterranean-type diet. Historically, garlic was thought to have benefit as a medicine; now it is cherished more for its flavoring.

As a USDA commodity profile puts it, "no other vegetable has exhibited such strong sustained demand growth in the United States as garlic." Per capita consumption of garlic rose from 1.3 pounds in 1990 to a record 3.3 pounds in 1999. That was a decade ago, and consumption has dropped fractionally, but has stabilized at about three pounds a person (perhaps 15 heads of garlic per person per year). Unabashed celebrity chefs pitching the virtues of garlic have made this an essential California crop. Garlic shipments abroad from California were worth $25 million in 2007. Although China is a far larger producer, the United States ranks fourth worldwide; within the country, however, demand exceeds even supply, so there is a net import of garlic from outside the country.

California production is an elaborate process. The crop has a nine-month growing cycle, which makes it susceptible to weeds or disease, but garlic is fairly hearty, as befits a crop with origins in the mountains of central Asia in what is now Kazakhstan, Uzbekistan, and Turkmenistan. There are dozens of different cultivars at use in California, chosen for soil type, planned irrigation, and end use. Within California, Fresno County by far is the largest producer, with more than half the total state production. Total acreage within California is 24,900 acres—an altogether respectable area. Kern County is the second-ranked producer. For Santa Clara County, garlic was a small player in county agricultural income in 2008; the giant was mushrooms—garlic produced just 2 percent of the $57 million in revenue that Santa Clara mushroom production earned in 2007. But it's the thought that counts, and Gilroy (and California) loves its garlic.

HORSERADISH *Armoracia rusticana*

Pl. 118

FAMILY: BRASSICACEAE.

NOTES: SYNONYM, *COCHLEARIA ARMORACIA.*

RANK: U.S. #1; CALIFORNIA SHARE: MAJOR.

Horseradish is grown only in the northern section of California and is tightly focused on 975 acres around the small town of Tulelake (pop. 1,200), in Siskiyou County, along the northern edge of the state—although neighboring Modoc County claims another 700 acres in production. Statistics on horseradish yield are not up to date, but these estimates are realistic.

Another member of the brassica family, which includes broccoli, cauliflower, and cabbages, horseradish when eaten is a potent force, dramatic and pungent in formulation. Few American adults have failed to overindulge, at some point, and then had to live with the consequences: watering eyes and a grab for a refreshing adult beverage. There are industrial uses for horseradish, with Calzyme Laboratories in Tulelake looking at peroxidase production from the root.

A cultivated plant since 1500 BC, it is the root of the horseradish that is cherished. The horseradish root is unprepossessing, and can be bought raw in your better farmers markets. But much of the production is distributed in 1,200-pound pallets to processors who grate the root to release the volatile oils that make horseradish all that it is. The main problem with horseradish production in California is finding enough root cuttings to do spring planting. The cuttings, a little more than half an inch in diameter and 8 to 14 inches long, are inserted at a 45-degree angle with just a bit of the top visible above the soil. Planted when the ground can be worked, usually in early April, the crop takes about another year to grow; a spring harvest makes it possible to obtain more cuttings for replanting.

The inelegant root with a dramatic effect is used for the creation of culinary horseradish that when grated releases enzymes from damaged cells that produce a volatile and potent mustard oil (allyl isothiocyanate) that irritates the sinuses and the eyes. Grated or ground, horseradish is variously a condiment, an herb, and a stimulant—a potent bit of work, and one that is not always easy to pick out while in the field: as a root crop, it is the leaves that are seen, not the root. Although horseradish can be peeled and then ground up in a food processor, it's best to do that outdoors, or at least with a lot of fans running; the effects of grinding straight horseradish on the eyes and mucous membranes make onion tears seem as insignificant as dew.

Mixed with distilled vinegar, horseradish stabilizes, and the vigor of the root is preserved. Some horseradish is combined with other ingredients to make cream horseradish, or "prepared horseradish," mustards, or dips—but the essentials are vinegar and horseradish. A generally amicable duel is ongoing between Collinsville, Illinois, which hosts the International Horseradish Festival each May, and Tulelake, California, which bills itself as the "Horseradish Capital of the Universe." There is, perhaps, a kind of escalation at hand, but laying claim to a signature crop obviously can be good business, as garlic growers will also attest.

KALE ***Brassica oleracea,* Acephala group**

Cover photo

FAMILY: BRASSICACEAE OR CRUCIFERAE, ACEPHALA GROUP.

UNRANKED—CALIFORNIA IS MAJOR PRODUCER.

Sometime after the domestication of plants began, humans selecting what they wanted to eat began to create variations among brassicas in the wild. Some brassicas were selected for leaf area, other for stems. With time, one group of plants chosen and cultivated had larger stems that were robust, could travel, and were suitable for cooking. This was kale, which has the Latin botanical name *Brassica oleracea* variety *acephala*, which means "vegetable without a head." And, indeed, that is kale—a variation on cabbage, lacking the compact-leaved head.

Kale is distinct within the Brassica group. Kale is readily identified by its great leaf spread, although it can be curly, or plain-leaved, leaf-and-spear, cavolo nero, or the singular category known as rape kale lutes, colloquially known as Hungry Gap kale (because it was available when no other crops were available). The fact is, there are variations enough to make identification interesting, and a downright challenge.

With origins in Antiquity, kale is considered closer to the wild cabbage than many other forms of brassica, and manifests a robustness that few, if any, other vegetables can match. It is raised for its leaves, which though edible raw (in salad, usually) are generally cooked, often in soup. Kale is cherished in part for its slightly bitter taste, but also for recognized antioxidant properties. Medical literature is exploring kale's anti-inflammatory qualities. Distinguishing between varieties of kale is not easy, but identification generally turns on leaf types and the length of stems. Kale is used in two dozen national dishes, including feijoada, the favored dish of Brazil, where kale (known as couve) accompanies the dish. Kale can be used in decoration, where it is often misidentified as "ornamental cabbage." Kale is not a huge producer in California, with $24 million in production in 2008. Of that population, 80 percent was in Monterey County; Ventura County, with fields on the Oxnard Plain, was the second largest producer. The growth of kale for farmers markets is common, and there is a significant organic crop that is not always captured in the statistical record, which had 2,618 acres harvested in 2007.

KOHLRABI *Brassica oleracea,* Gongylodes group

Pl. 87

FAMILY: BRASSICACEAE OR CRUCIFERAE, GONGYLODES GROUP.

RANK: U.S. #1; CALIFORNIA SHARE: MAJORITY.

Plate 87. Looking for all the world like an alien spaceship tinged light green or purple by long travel, kohlrabi has a delicious and sweet bulb that can be used like celery root.

An oddly shaped crop with stems that jut from the plant, kohlrabi looks uncannily like a grounded Sputnik or, as others put it, like an octopus, appearing in shades of purple, cream, or green. Kohlrabi is not a huge producer in the California economy, worth only a few hundred thousand dollars. However, this vibrant brassica is growing significantly in production. Kohlrabi literally translates as "cabbage turnip," and its role in the eastern European foodchain was crucial: sometimes cooked thoroughly, it could as readily be eaten without cooking, and after prolonged storage. Eventually, kohlrabi was replaced by the potato.

When eaten, kohlrabi has something of a bite to it, like radish or cabbage heart, but sweeter. What is eaten is a bulbous swelling known as the corm, which forms a single kohlrabi on each of the plant's stems. Varieties include the White Vienna, Purple Vienna, Grande Duke, Gigante, Purple Danube, and White Danube. Kohlrabi grows well in coastal California, and is commonly marketed through farmers markets (and, increasingly, in larger supermarkets). It can be eaten raw, or cooked first (which makes removing the skin easier). Or it can be grated, and julienned into salads. Perhaps the most impressive thing about kohlrabi is size—it can grow to 40 pounds, though younger examples are sweeter and more tender, when they are about the size of a tennis ball.

LETTUCE, SALAD GREENS, LEAFY VEGETABLES *Lactuca sativa*

Lettuce, plus salad greens, radicchio, endive (escarole and frisée)

Pls. 4, 15, 88, 104, 113, Preface photo, Part 3 opener photo, Map 15

FAMILY (ALL): ASTERACEAE.

NOTES: EACH VARIETY IS DIFFERENT SUBSPECIES; LEAFY VEGETABLES CAN BE OTHER GENUS AND SPECIES.

RANK: U.S. #1, HEAD, LEAF, AND ROMAINE VARIETIES; CALIFORNIA SHARE IS HEAD 76%, LEAF 90%, ROMAINE 83%.

The lettuce category is a catch-all, since "salad" is itself such an inclusive term. Salad greens often grow nearby fields of head,

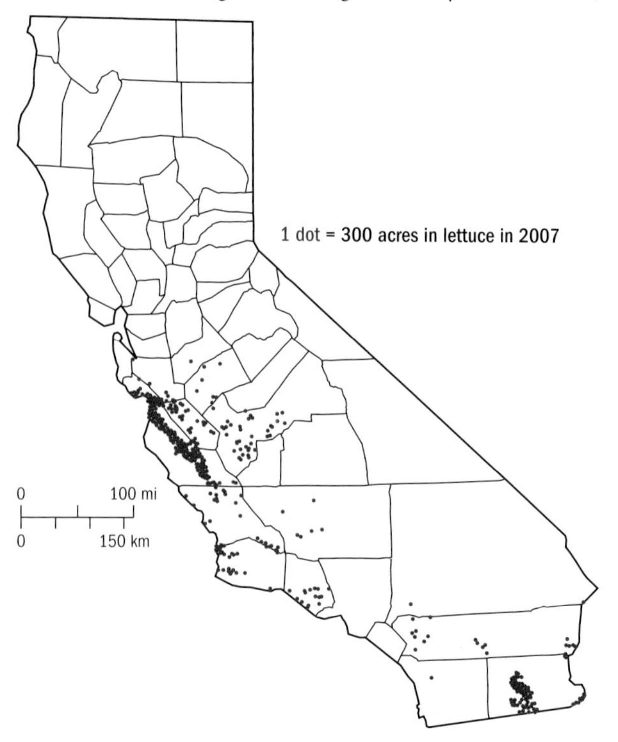

Map 15. Acreage in lettuce (all), 2007.

leaf, or romaine lettuce. An area good for lettuce is suitable also for growing so-called boutique vegetables, salad greens, or leaf vegetables. But make no mistake: lettuce and salad greens constitute a large, and growing, category in California agricultural production. Lettuce and salad greens were worth $2.2 billion to the agricultural economy in 2007, enough to boost lettuce up one notch to the fourth spot in farm-gate production value—below "nursery and greenhouse" and above almonds.

Lettuce and leafy vegetables can be grown in many different ways, from small boutique organic crops for specialty markets to raised beds where romaine (known as cos), head, or leaf lettuce stretch literally into the horizon, irrigated by furrow or sprinkler and tended by an army of workers, generally driven to the field in retired school buses or minivans. Acres of laser-leveled fields sown to salad greens are harvested by rotary mowers that collect hundreds of soon-to-be bags of greens in a single pass along a raised bed, cool and sanitize the gleanings on the move, and have the crop ready to go into hydrocooling, drying, and packaging before leaving the field. A couple of weeks later, the greens have grown out to five

Plate 88. Lettuce is close to an omnipotent force in California agriculture, especially since Monterey County so dominates production. More than a $2 billion crop, lettuce in all forms says "California."

to seven inches and are harvested again. Such is California's lettuce world.

Fields of green and red welcome travelers southbound through Monterey County's Salinas Valley. There indeed is the "Salad Bowl of the World," as one grower announces to traffic headed toward the city of Salinas, a sprawling and fast-growing community at the head of the Valley. Monterey County leads all comers in the main varieties of lettuce that are tracked, and so much so that it raises a question: is there possibly any major-value crop that a single county so dominates as Monterey in its variations on a theme of lettuce? "Bulk salad products" in 2008 were worth $105 million to Monterey, "Head Lettuce" another $461 million, "Leaf Lettuce" a neat $223 million, and "Lettuce, Romaine" another $423 million—all this within Monterey County. Monterey produced nearly 64 percent of the lettuce in California. Add salad greens (another $172 million) and spinach ($130 million in 2008), and you have, at the very least, a remarkable market share.

The only counties that really come close to such dominant production for a single significant crop are Fresno with garlic (88 percent), Merced with sweet potatoes (96 percent), Monterey with artichokes (68 percent), and then Monterey with spinach, again (63 percent). On greens, Monterey is in truth in a class by itself. The production of lettuce in 2008 upheld the previous year's standard. Head lettuce statewide earned $815 million on 116,000 acres, an area unchanged for a decade from 1998. Leaf lettuce brought in $468 million, down from the previous year thanks to a health scare, but the extent, at 59,000 acres, was up 20,000 acres from 1998. And the acreage of romaine showed a similar increase, from 25,000 (1998) to 57,000 acres in 2008, and romaine was worth more than a half-billion dollars. The main producers of lettuce products in California, aside from Monterey, were Fresno, Imperial, Santa Barbara, and San Benito counties.

Lettuce was known and eaten in Roman times, with an origin around the Mediterranean Basin. Brought by Columbus as seeds to the Americas, it was widely raised, but only garden by garden; in California, the missions found it an easily raised crop. Commercialization began in the nineteenth century and picked up speed rapidly with the development of fast-cooling and shipping techniques that finally made it possible to transport lettuce long distances (a theme of considerable importance to the Salinas Valley growers, and a motif in John Steinbeck's *East of Eden*, as in the Elia Kazan movie of the same title). The mechanical world associated with lettuce production is infinitely more sophisticated now than in the 1930s, and it has to be: if efficiency is paramount, not far behind are food safety concerns. Field, harvest, and packaging technology could hardly be more high-tech, and for a reason: consumer confidence is paramount. After several scares since 2004 involving spinach and lettuce contamination, the California Department of Food and Agriculture implemented the California Leafy Green Products Handler Marketing Agreement (LGMA), which covers 99 percent of the volume of lettuce, spinach, and leafy greens in the state. Such are the costs, and the imperatives, of consumer confidence.

Lettuce comes in many varieties, but the three foremost are familiar to most Americans: crisphead and butterhead lettuce (iceberg is the best known, but in fact is a 1940s creation), leaf types—which do not form a head, but stay as an open rosette, and romaine or cos (named after the Greek Island of Kos in the Aegean), which has upright leaves forming from a long loose head. Although head lettuce has remained relatively unchanged in earnings over the last decade, California leaf and romaine has gone through the roof in acreage, production, and crop value.

Another front in the leaf vegetable world involves salad amendments, generally from the same family as lettuce (Asteraceae), but with differences. There are literally dozens of leaf vegetables, which, if anathema to some salad eaters, are dear to others. There is radicchio (*Cichorium intyus*), a leaf chicory with white-veined red leaves that has a bitter but distinctive taste, and that is often used to add color. Endive (*Cichorium endivia*) is a common salad addition as either escarole (broad-leaved endive) or curly endive—increasingly known as frisée—with its curly outer leaves, and sometimes covered with white caps to maintain a pleasing shape. Various mustard greens are common additions to salad mix, and the prom queen is unmistakably arugula (*Eruca vesicaria* ssp. *sativa*), which is loved or hated, sometimes for political reasons, and which adds a peppery taste to salads. Finally, many groceries are starting to carry mesclun mix, a Provençal term for an assortment of very young lettuces and greens, including bitter greens. All are available commercially, all are grown predominantly in California, and all are harvested with a remarkable degree of care. These brought in another $21 million in 2008.

MUSHROOMS *Agaricus campestris*

Agaricus, button mushrooms

Pl. 89

FAMILY: AGARICACEAE.

RANK: U.S. #2; CALIFORNIA SHARE 14%.

The *Agaricus* family starts its life as "spawn," spends its life in the dark, sprouts pasty-white with its feet deep in a heady mix of composted horse manure and agricultural by-products, grows at a phenomenal rate, and then is summarily severed at the base, rinsed, and shipped off to be eaten. And sales of *Agaricus* brought in $177 million to California in 2008. Not bad for the life of a button mushroom. There are two dozen different mushrooms and fungi raised commercially in California, with *Agaricus campestris*, the white button mushroom, by far the most popular. But that is only a small part of the overall pool; consumers who gather their own can be rather elitist in their mushroom-seeking, and Mendocino County alone is home to more than 3,000 mushroom varieties. Little wonder that mushrooms are sometimes referred to as "nature's hidden treasures."

Mushroom production in California is second in the United States only to that of Pennsylvania (who would have thought?). Sixty-nine growers in Pennsylvania earned some four times the income from mushroom cultivation as the 17 growers in California. For mushrooms, "bearing acreage" is not an issue, since they grow in huge bins, beds, or trays that are moved around by forklift, and advance from one pitch-black room to another as they grow.

Because the button mushroom, like most of its family, is raised in darkness, the *Agaricus* is not something likely to be seen from the roadside, except as rather large warehouse-like buildings closed off to light, usually with massive concrete pads lying behind them where long, 20-foot-high lines of compost are steaming and dribbling a by-product of deep brown liquid, since the compost piles must be kept suitably wetted down to avoid too much internal heat—and spontaneous combustion.

Plate 89. *Agaricus* is just one of the varied cultivated mushrooms, but the button variety is known to more Americans than any other. Its production is in deep darkness, which keeps the results white, absent chlorophyll. Most mushrooms are grown within easy travel distance of horse racing tracks and large stables, which provide a growing medium.

The creation of mushroom compost is a process even more carefully monitored by growers than by garden hobbyists who care for their own personal compost bins. The crucial ingredient in mushroom compost is nutrients, which generally come from horse stables, and especially race tracks, which tend to feed their animals exceptionally high-quality feed and collect manure with white-glove care. This manure is manna to the compost maker, since it is high quality and of known origin. Turned and stirred as the composting occurs, the compost is completed after 15 to 25 days, at which point the compost mix is pasteurized to remove any pests or remnant seeds present in the compost. The result is loaded in bins, inoculated with spores, and covered with peat moss. It takes a month for the first mushrooms to appear, and as the leaders are picked, other mushrooms are emerging for six to 10 weeks, at which point the compost is exhausted and the room steam-cleaned, and the cycle begins again with new compost. The productivity is remarkable.

NURSERY AND GREENHOUSE CROPS, FLOWERS AND FOLIAGE — Various species

Pls. 90, 91

RANK: U.S. #1, ALL; U.S. #1, NURSERY AND GREENHOUSE (CALIFORNIA SHARE: UNKNOWN); U.S. #1, FLOWERS AND FOLIAGE (CALIFORNIA SHARE, UNKNOWN).

Without a moment's hesitation, it can be said that nursery and greenhouse crops are by far the most complicated category in this guide. Nurseries are everywhere; so are greenhouses. However, there are distinct classes of nurseries and greenhouses—some nurseries raise crops that other nurseries then buy to distribute: there is a wholesaler–retailer step to the process. But there is a simple and direct beauty to understanding nurseries and greenhouses as well. They are near where the larger populations of people are—in every case except for just a few, which get their own mention. The nursery-strong counties (and they bring in an

Plate 90. Nursery and greenhouse crops join floriculture as a hugely profitable business in California, and none more lucrative than the mature shrub business, here with Bougainvillea Purple Queen as a star.

enormous amount of money) are an exception to the rule that agriculture is often at a discrete remove from urban centers.

Nursery products are raised in two categories: for landscaping, and to produce "starts" for the California agricultural industry. Example of starts include strawberries ($86 million in 2008) and fruit-nut-and-vine products ($188 million in 2008). Those go to industry. But the rest largely have the household garden market as a final destination.

The big nursery and greenhouse product counties are, to put it simply, those at the edges of big cities with lots of people: for the Southland, that is San Diego, Orange, Ventura, Riverside, San Bernardino, and Santa Barbara. Equally, for northern California, there are San Mateo, Contra Costa, Santa Clara, Alameda, Monterey, and Santa Cruz counties, surrounding the San Francisco Bay Area. The interesting outliers appear in a few categories in the cumulated reports from the County Ag Commissioners: Lassen, Shasta, and Siskiyou counties are the largest producers of strawberry starts, in the 2008 data—worth $86 million (*see* Strawberries for an explanation). Humboldt County is the second-largest producer of "nursery products (misc.)," and the category of "nursery, fruit/nut/vine, nonbearing" is headed by Stanislaus, Kern, Tulare, and Sutter counties, with their huge nut and stone-fruit starts, including olives, peaches, almonds, and pistachios (the entire category was worth $188 million in 2008, and the dollar figure will rise). But the biggest category of all, at $1.2 billion in 2008, is "nursery, woody ornamentals." With San Diego producing $304 million, Riverside $190 million, Orange $129 million, and Ventura $183 million, that adds up to a whole lot of trees. Within that exceptionally rich category are included, for example, the palm trees that are either raised from starts or purchased from landowners who are tired of the ornamental palm's prodigious mess of fronds and fruit, or who are simply enticed by a check. Huge four-tined excavators come, dig up a palm, and deposit it on the bed of a flatbed truck, upon which the palm is wrapped and secured and transported to a nursery, where it can bring top dollar for the homeowner who wants a palm and wants it immediately.

Greenhouse plants and floriculture are related, but not identical. Suffice it to say (and for those who are interested, there are plenty of details) that a lot of money moves through floriculture and greenhouses. In 2008, potted plants were

Plate 91. Palms are prosperity—for nursery owners. Growers will adopt, or raise, young palms; others excavate palms fully grown, then box and transplant them for those willing to pay.

worth $200 million, cut flowers $318 million, and bedding or garden plants $359 million. And these are just for operations with more than $100,000 in sales. Landscaping is a huge business in California, and it is agriculture to the nines. Although this guide leaves out Christmas trees (we also leave out horses, emus, jojoba, and cherimoya), there is no doubt that although the human relationship to nature may be somewhat attenuated by asphalt and gated communities, the need to make revisions to landscaping that is rife with conspicuous consumption is, apparently, forever.

Without being too dollar-oriented about this, it helps to understand just how large the total category is. Nursery crops in California were worth $3 billion, and flowers and foliage another $1 billion in 2007, which, when added together, total $4.07 billion. That total amount ranks just below the category of "milk and cream" as the largest-value agricultural product in

California. Every time a suburban homeowner takes up a shovel to transplant a tree or plants a tomato or chili pepper start that is bought from a local nursery, or a realtor calls a trusted supplier to come install appropriate outdoor and indoor plants to "stage" a house for sale, it is the California nursery and greenhouse plants industries that breathe an audible, if muted, "Thank you!"

ONIONS *Allium cepa* L. var. *cepa*

Includes leeks, shallots, and scallions

Pls. 85, 92

FAMILY: ALLIACEAE, ALSO LILIACEAE.

RANK: U.S. #1; CALIFORNIA SHARE 38%.

Well known to anyone who has bitten into a hamburger, a bowl of chili, an onion ring, or a well-prepared stew, the onion is an ancient cultivar of humanity. There are thousands of ways that onions are incorporated into dishes, and although onions are often cooked, they need not be. There are people around who will grasp an onion with the outer skin removed, remove a slice from it, lay that over a piece of apple, and bite hard. The onion is a remarkably plastic plant, raised for an underground root that is removed, allowed to dry, stored, and eaten. The bulb of the onion is actually a food-storage device for the plant itself: garlic, fennel, and carrots share the same technique for storing carbohydrates underground for use by the plant as it goes to seed. Leeks (another *Allium*) are not significantly different in morphology, and neither are shallots (which are dried) or scallions (green onions, eaten fresh)—each also of the genus *Allium*. Some cuisines favor one form of these spicy, layered root crops over others, but most parts of the world make use of them. Certainly California does.

The onion field looks not so different from a field of green onions, leeks, or shallots: multiple stalks rising from the ground in a gray-green hue, with distinctive hollow leaves. As the onion grows, it may actually raise itself partly out of the ground. The bulb of the onion, which is dried and eaten, was considered in Ancient Egypt to be a symbol of eternal life because of its spherical shape and concentric rings and was venerated as an expression of the human relationship to an encompassing universe. Not many onion devotees in the United States would go quite that

Plate 92. Regarded as a sacred crop in Antiquity because of its rings of flesh and internal symmetry, the onion is a modern-day favorite for its ability to withstand storage.

far, but the onion is known to have arrived with Columbus in his 1492 expedition, and it immediately went into cultivation. A main eccentricity of the onion is that it is a two-year crop, requiring added time to fill out its bulb before the plant prepares to go to seed, although it must be harvested before that final step.

Although onions can be seeded, they are more commonly started from sets begun the year before, and sown densely so that they produce small bulbs, which are then planted. Even though onions are not a huge-value crop in California, 49,000 acres of onions brought in $259 million in 2008, with Fresno the dominant county, followed by Imperial, Kern, and Monterey counties. Farm-gate value of shallots was another $30 million (mostly in Monterey), and leeks earned $5.4 million, with most of that hailing from Monterey, and the rest from San Mateo County. (As is often the case, there can be problems with ag statistics—2006 data has Fresno County producing $233 million in onions; the 2007 report says $123 million, an unlikely shift.) Onions come in diverse forms, and are grown for particular traits: there are spring onions, summer storage onions, summer nonstorage onions, and processing onions, and each has its niche.

A particular pleasure of an onion field is the aroma, especially when the field is freshly irrigated. Aside from the cleanly earthen smell of wet dirt in the early morning, the field-side pause in still air, can definitely produce a whiff of the characteristic onion smell.

PARSLEY — *Petroselinum crispum*

Pl. 93

FAMILY: APIACEAE.

UNRANKED—CALIFORNIA IS MAJOR PRODUCER.

Like cilantro, to which it is related, parsley is a handsome crop, a bright green herb largely used as a spice. Parsley was once the default decoration of the American chef, obligatory on a baked potato or anything to be improved with a sprig of color. Those days are largely gone, but the role of parsley in adding something to the spice of life has not entirely departed, and parsley is one striking example of a crop that has proved to have staying power, thanks in part to its widespread presence in Middle Eastern, European, and American cooking.

An herb that does not thrive on excessive heat, parsley is grown largely in Monterey and Ventura counties, two areas that manifest a climate cooled by the Pacific Ocean, at least for the first 20 miles or so inland, which takes in major growing areas for both counties. It is the leaf of parsley that is employed in culinary use, although there are two varieties: the common curly leaf, and the slightly less common Italian, or flat leaf, parsley (*P. neapolitanum*). The two varieties brought in $20 million in 2006, with a quarter of that hailing from Monterey, and another $3 million from Ventura County. Perhaps the most remarkable detail about parsley is its extent: 2006 production was on 3,807 acres—just a shade under six square miles of parsley, surely an aromatic sea.

Either form of parsley, whether curly leaf or Italian, is used as an accent. In gardens parsley is considered a companion plant, because it attracts beneficial insects, many of them predators,

Plate 93. Acre upon acre of parsley may seem an unlikely presence, but the crop grows plentifully in eastern Riverside County, and serves to garnish millions of baked potatoes each year.

that protect nearby plants from outside incursion. Within the kitchen, parsley may be dried and used in soups or sauces.

PEAS *Pisum sativum*

Snow, sugar, pod, green, chickpea, dry edible

FAMILY: FABACEAE.

UNRANKED—CALIFORNIA SHARE IS MODEST.

Peas are not the most glamorous of California crops. They come in both vining and bush form, like beans, to which they are related. The general form of either bush or vine is readily recognizable, but what is trickier is deciding which crop is being grown—a more accurate verdict may not be possible without stopping to walk closer. The variety of peas raised in California is significant, but the acreage produced and value is relatively modest: $35 million in 2008, with Monterey the dominant county ($30 million), Santa Barbara second at $3 million, and San Luis Obispo's considerable snow pea production at $2 million. The Sierra Nevada foothill counties are minor producers, especially of chickpeas (garbanzo beans). Snow peas grew on 653 acres in 2008, and green peas on 2,059 more; acreage of other pea varieties is difficult to estimate.

An advantage of peas is that they mature in a relatively short time, 60 to 75 days, which makes them a highly useful crop for northern-latitude environments, where summers are relatively short—and peas make good use of longer day length. That is less an issue in California, although peas are cultivated most often in coastal environments where the heating regime is far milder than in the Sacramento and San Joaquin valleys, which can resemble a blast furnace during the late summer months. Many peas are plantable in winter or in the early spring, and can grow well up to summertime; fields are sometimes run through multiple fresh plantings in a single year. That said, peas have a remarkably strong constituency and are a regular item in fresh market vegetable bins in the greengrocer's aisle.

Like numerous other items that appear among field crops, peas are in fact a fruit, but are generally classed as a vegetable in cooking. Snow peas and green peas are classic California crops, in the sense that they fit the needs of the state's remarkably broad ethnic constituency. What would Chinese cooking be without snow peas, or Thai food without pea pods? Dry beans were a staple in Europe of the Middle Ages, and peas—generally also dried—were leavening in an otherwise relatively monotonous diet. Peas include, of course, the lentil, split pea, black-eye pea, and sweet pea, so in addition to the highly significant categories of green peas and snow peas, the dry peas are also contributors to California agriculture. Peas are growing somewhere in California every month of the year; the art is in finding them, although distribution channels are multitudinous.

Many of the edible-pod peas are harvested with the pod still fully edible and the pea inside at an immature phase. Among the charming aspects of edible-pod peas are the names of the cultivars or varieties: a reliable California snow pea is Mammoth Melting Sugar; smaller acreages of Oregon Sugar Pod II similarly exist, if they can be found. Among the sugar peas,

Cascadia and Sugar Snap are popular. The main drawback to edible pod peas in California is the need to trellis them—the varieties grow tall, and quickly, so support is required. Chickpeas are particularly prized when they are green, and therefore are a field crop rather than a dried bean—and reports travel fast among food fans when the chickpea (garbanzo) harvest is up and running. Humus fans rejoice . . .

PEPPERS, BELL

Sweet peppers **_Capsicum annuum_, with various cultivars**

Pls. 36, 94, Part 1 opener photo

FAMILY: SOLANACEAE.

RANK: U.S. #1; CALIFORNIA SHARE 48%.

A cultivar of *Capsicum annuum*, the same species that gives us many of the chili peppers (*see also* Chili Pepper), the bell pepper is the very mildest member of the family, clocking in at exactly 0 Scoville heat units (SHU), which means that bell peppers have taste but no heat to them. They do, however, offer a remarkably tasty addition to salads or cooked foods, and are regarded as an essential addition to the many base preparations, such as sofrito, that begin by sautéing a mix of onions, garlic, bell peppers, and sometimes tomatoes, in a style akin to the French mirepoix. Bell peppers come in red, green, yellow, orange, white, aqua, and brown, and some varieties are blue, shading toward black. Packagers now routinely offer a selection of different-colored bell peppers to the larger grocery markets.

The bell pepper plant is little different, visually, from its spicier cousin, the chili pepper. A relatively stiff stalk with branching arms supports the large fruit—three to four inches long, usually with four or three lobes. About 200 varieties are available; the Sweet Bell, Bell Boy, California Wonder, Golden Bell, and Jupiter are popular examples.

Bell peppers do well in the same regions where tomatoes can be grown, but are finicky starters: they prefer warm temperatures, and seed will start very slowly in cooler weather. On the other hand, like chili peppers, bells produce well into autumn. If allowed to mature on the plant, the flesh will be sweeter than if picked and ripened inside.

Plate 94. A staple of cuisines in much of the world, and ever popular for their lively colors, robust flavor, and absence of "heat," bell peppers are relatives of the chili pepper that are highly marketed, if somewhat delicate to ship.

California bell peppers brought in $225 million on 18,180 acres in 2008. Produced in many parts of the state, they grow in exactly the same areas as chili peppers: the desert, San Diego and Orange counties in the Southland, all along the Central Coast, and in the San Joaquin Valley. A good bit of the production hails from eastern Riverside and Imperial counties, where they are grown in the winter and spring months before the extreme heat comes on—but also before other areas can start growing the heat-admiring plants.

PEPPERS, CHILI

SEE: CHILI PEPPERS.

POTATOES

SEE IN: GRAIN, HAY, AND PASTURE CROPS.

RADISHES *Raphanus sativus*

Half-title page photo

FAMILY: BRASSICACEAE.

NOTES: DAIKON IS ALSO A RADISH.

Millions of children get their start in gardening growing radishes. The seeds all but start themselves, with quick and rewarding results. There is an enduring radish crop in California, mostly issuing from Ventura, Monterey, and eastern Riverside—a small niche production on just 1,800 acres in 2008. The total value of radish production in 2008 was $9.3 million, with the pungent Scarlet Globe the preferred variety.

SALAD GREENS

SEE: LETTUCE, SALAD GREENS, LEAFY VEGETABLES.

SPINACH *Spinacia oleracea*

Fresh market and processing

Pl. 95

FAMILY: AMARANTHACEAE (FORMERLY CHENOPODACEAE).

RANK: U.S. #1; CALIFORNIA SHARE: FRESH 83%, PROCESSING 63%.

SEE ALSO: LETTUCE, SALAD GREENS, LEAFY VEGETABLES.

A remarkably resilient leafy plant, respected for at least a couple millennia for its tasty leaves, spinach is a member of the same botanical family as the sugar beet, kale, mustard, and collard greens. In that roster, spinach is the likely star. Its culinary uses are varied, from soups to stews to salads; the vintage comic book hero, Popeye the Sailorman, knew spinach, and introduced generations of American (and world) children to its nutritional advantages, which include a robust level of iron and calcium.

A native of central and southwestern Asia, spinach is a hearty annual plant that grows to a height of twelve inches or so. Spinach had a prickly form until a smooth-seeded variant was noted and bred up by plant breeders in the sixteenth century. Spinach arrived in the United States by the early nineteenth century but was not widely popularized until much later. California remains the dominant producer, with more than 80 percent of the fresh market supply. Spinach is classified by leaf type as savory (wrinkled), semisavory, and smooth (flat). Picking the right variety of spinach is important, since a variety will bolt if exposed to heat that exceeds the plant's threshold. The upside of this is the tolerance of spinach

Plate 95. The fresh market spinach crop was an early winner in the packaged-vegetable trade in the 1990s, and the scale of operations is readily seen from this single field near Salinas, in Monterey County.

to winter cold, and as a cold-season herb, it can weather severe frosts. Spinach does particularly well in cooler parts of California, principally in Monterey County, but also accepts upland sites, and grows in the desert counties (Imperial, eastern Riverside), where spinach reaches harvestable size before hot weather arrives. There is modest production in the central San Joaquin Valley (Stanislaus and Tulare counties). In the coastal valleys, spinach is a year-round crop, direct-seeded onto three- to four-foot-wide raised planting beds. Sprinkler irrigation is used to start the crop, and then is usually replaced by furrow irrigation as the plant matures. Monterey can often produce double crops of spinach in a year: the first for fresh market, the second for processing.

Production and total value have remained relatively constant for the last decade, with the exception of a crash in 2006 after the discovery of *E. coli* O157:H7 in 21 states, an outbreak that ultimately caused 141 hospitalizations and three deaths. After an interval of several months in which spinach and some lettuce were nearly shut out of the food supply, the FDA tracked contamination back to an organic spinach field grown by Mission Organics in Paicines, California, in San Benito County east of the Salinas Valley. The assumption was that wild swine breaking through fences had contaminated water supplies used for irrigation with feces, and that the *E. coli* spread through irrigation water. Three years later, the dense impenetrability of fences around lettuce and spinach fields looks like a testimonial to some divine power that rules by barbed wire. The state of California's Leafy Greens Handler Marketing Agreement is designed to lessen the likelihood of further contamination, and to accelerate response should there be another outbreak.

Fresh market spinach is yet another crop for which the Salinas Valley and Monterey County dominate U.S. production, with $88 million of the 2008 crop hailing from Monterey. The total spinach yield for California in 2007 was $169 million, with another $7 million coming from processed spinach. Crop production figures are still recovering from the 2006 recall, which affected consumer confidence in the leafy vegetable food supply.

STRAWBERRIES *Fragaria x ananassa*

Pls. 10, 96, 110, Preface photo, Map 16

FAMILY: ROSACEAE.

RANK: U.S. #1; CALIFORNIA SHARE: FRESH MARKET 86%, PROCESSED 93%.

To think of a fruit, a perishable berry, coming in at the eighth spot in farm-gate value in California is only slightly less surprising than having lettuce in the fourth spot. Strawberries were a $1.6 billion California crop in 2008, and they brighten the lives of a lot of people who elect to eat them regularly. Strawberries share in an interesting two-part history that is characteristic of the ingenious adaptability at times evident in California agriculture. In terms of strawberries, the United States is by far the largest world producer—Russia grows just a quarter of the American yield. And within the United States, California raises 86 percent of the fresh market strawberries, and more than 90 percent of the processed strawberries. The California

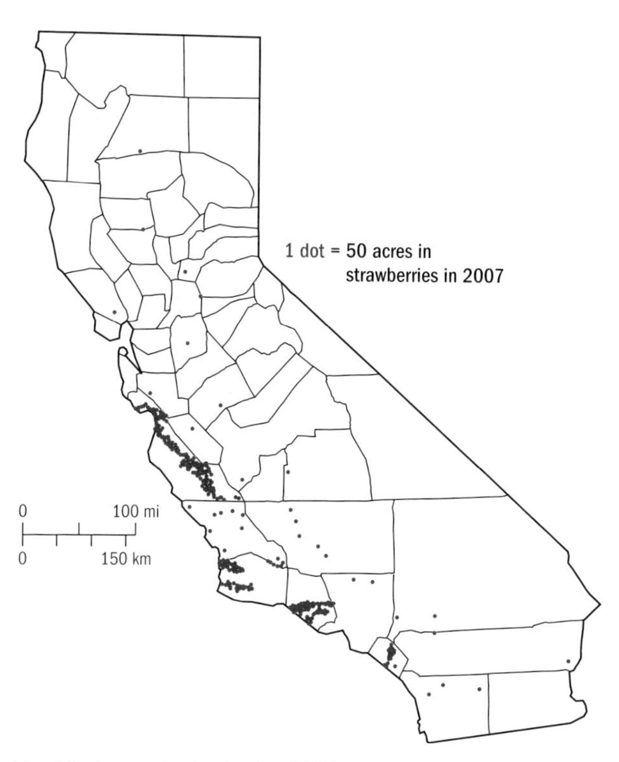

Map 16. Acreage in strawberries, 2007.

strawberry narrative is, then, more than just a casual yarn: it is the story of an industry leader that produces nearly 20 percent of the world's total commercial strawberry production.

A member of the Rose family, strawberries are robust crops that by nature propagate with runners, as is common in plants that favor sandy soil or bottom land. When it is time to reproduce, strawberries sprout from their main stems, send out runners with long tendrils (botanically, the runner is a stolon), and reach to a new area where the tip of the stolon can root, in effect leapfrogging from one spot to another. Strawberries that are raised for fruit production have their runners removed so that they will concentrate on providing fruit. Strawberries have quite varied cultivars, and their fruit differs in sweetness, in ability to handle shipping, and in resistance to

Plate 96.
An abundance of stories attach to strawberry production, including high prices to the grower, the strawberry's preeminence among labor-demanding crops, and significant changes in the last 20 years that include wholly new cultivation strategies.

pests and weeds. Genetic engineering wrought on strawberries has produced both results and controversy. Strawberries are hand-harvested by farm workers who travel up and down the rows, generally daily, picking the ripe fruit and wheeling the crop in bags that look amusingly like the roller bags that are now a staple of airline travel. The work is unforgiving, involving constant stooping and standing, and is another of the realms in which immigrant labor completely dominates a major California crop.

The California strawberry story is a curious one, and involves a narrative that helps to explain a curiosity in the "Nursery Crops" entry. Toward the end of the nineteenth century, Thomas J. Loftus, a Shasta County rancher, moved his family south to Sweet Briar, just north of Redding, California. Loftus noted a particularly sweet strawberry on his ranch, a remnant from a previous owner's cultivation attempts. He marked the crop and in the next several years saved enough runners to start a quarter-acre patch of what he named the Sweet Briar. When word spread of the extraordinary strawberries being produced, Loftus was contacted by growers in the Pajaro Valley, in Monterey County, who sought to propagate the Sweet Briar strawberry (subsequently renamed) in the Pajaro Valley. But reproduction failed, with plants transferred to Watsonville unable to send out the runners essential to reproduction. A plan evolved: runners would be brought down from Redding to the Pajaro Valley by train, the Sacramento Northern Railroad, and then transplanted. This established a practice maintained today within the industry—separating plant production from fruit production. In 2008, $86 million worth of "Nursery Plants: Strawberry" were raised in Siskiyou, Lassen, and Shasta counties, in northern California. They are the basis for a crop approaching $1.6 billion in value (2008) that is brought to fruition overwhelmingly in the Central Coast counties of Monterey, Ventura, Santa Barbara, and San Luis Obispo. Processing strawberries are overwhelmingly a product of Ventura, Santa Barbara, and Santa Cruz counties. But the ties to the colder north remain.

Strawberry production is carefully monitored by the California Strawberry Commission, which does yeoman work in support of the crop. There are controversies associated with strawberry production, however. Many of the fruiting strawberries traditionally have been raised under a chemical-intensive regime, with soils fumigated to prevent competing weeds, with irrigation tubing added along the crown of the row then covered with black plastic, and with strawberry runners inserted through holes cut in the plastic. Methyl bromide, a fumigant, was phased out in 2005, and growers are trying out alternatives, including organic production.

Grain, Hay, and Pasture Crops

The domestication and development of grains and pasture crops are the greatest inroad humanity has made against hunger. How curious then that staples—what the writer E. J. Kahn once memorably called "the staffs of life"—are not a decisive element in California agriculture. The acreage planted to rice, wheat, feed corn, and even cotton is not insignificant at 1.5 million acres. But that is less than the area in hay crops, which by and large go to feed livestock and dairy animals that convert otherwise indigestible roughage into protein—a process we humans cannot undertake efficiently on our own. And 1.5 million acres is just 8 percent of the area in private rangeland, used for grazing. The oddity of California's food staple–bereft story is not entirely surprising. The state was famously described by crusading journalist Carey McWilliams as *The Great Exception*, a title that fits California agriculture better than might any other tag.

Only a few of the staples crops grown in California are raised in quantities sufficient to make barking at the tires of the federal farm bill subsidy wagon worthwhile—rice and cotton are essentially the only contenders. California produces just 1 percent of the U.S. barley supply, 8 percent of cotton lint, 6 percent of alfalfa and alfalfa hay, 2 percent of oats, 3 percent of potatoes, and 1 percent of wheat. It is twenty-seventh in feed corn production, and seventeenth in sorghum yield—not even enough to earn California a spot in the quarter-finals of sorghum production. In the oil seed crops sweepstakes (including safflower and sunflower), California produces a smidgen—nowhere near its national "share," considering that 12 percent of the U.S. population calls the state of California home. Besides, the state is an increasingly adept producer of olive oil (*see* Olives).

The resolution of this tale is in part explainable through a maze of federal agricultural subsidies. There California is curiously absent from the most flamboyant, malevolent, and abusive accounts. Instead, with a couple of dramatic exceptions (well-publicized), California is in essence a sort of hero of unsubsidized agriculture. Or, at least, that can be said about staple production subsidies; there has been, in the past, quite epic support that came in the form of dams, irrigation infrastructure, and a federal aqueduct system.

The explanation for why California raises relatively little in the way of grains and silage is partly climatic but mostly financial. Complex economic analyses regularly establish how much net income farmers are actually making for their food, feed and hay, and grain crops. USDA figures suggest that growers of everything but cotton are making more for their crops now than in 1990. But—and this is the difficulty—when the prices farmers received are compared to cost to farmers for farm wages, interest, taxes, and prices paid, the resulting so-called "parity ratio" shows that farmers of those crops in 2009 (cotton included) earn just 78 percent of what they would have earned in 1990.

Economics, then, suggests that the story is actually relatively simple: California's agricultural niche is in "minor" or specialty crops that diversify the diet of the state, the nation, and—truth be told—the world. Specialty crops are far more likely to earn money than grain production, considering the dismal way in which domestic grain, hay, and cotton prices have failed to keep pace with rising costs. Only in 2008 did specialty crops become eligible for $1.2 billion in federal subsidies. California growers concluded early on that Mediterranean-type ecosystems had certain strengths that could be manipulated to advantage. The weakness is an absence of summer rainfall that makes raising grain without irrigation infeasible—you can choose to irrigate staple crops, but that is hardly free.

The strength is that summer-dry regions can grow crops that peoples living on the edges of the Mediterranean Sea (north, south, east, and west) have cultivated for thousands or even tens of thousands of years. Those crops grow swiftly, and often well, in California, especially when innovative agricultural technology is brought to bear. The discovery, adoption, genetic improvement, and perfection of crops from the Americas—and from the rest of the world—through crop introduction and adaptation make the advantage of California all the greater. And let us be realistic: growing grain—or even cotton—is not a big moneymaker compared to carrots, lettuce, almonds, pistachios, or tomatoes (processing *or* fresh market). Cotton growers have abandoned traditional cotton and are turning to long staple American-Pima cotton, some raised organically—and major purchasers are flocking to them, to take advantage of the political and ecological coup implied by such a crop.

The staple crops, it might be argued (and somewhat unfairly), are for the faint of heart. They may feed us, form the basis of our bread and pasta, and (mixed with hops or fermented to a fare-thee-well) put a foolish grin on our faces. But they are not symbols, or even remotely representative, of what is going on in California agriculture. For a state built upon experimentation and edginess, the great grain and field crops are a bit of a yawn (rice excepted), however important they may be in a global sense. Of that California farmers are quite certain, and they will continue flocking to their specialty crops, thank you very much.

ALFALFA

SEE: HAY (ALFALFA AND OTHER).

BARLEY *Hordeum vulgare*

FAMILY: TRITICEAE.

RANK: U.S. #12; CALIFORNIA SHARE 1%.

Although the share of barley produced in California is trivial, by U.S. or global standards, let there be no disrespect intended for the barley growers in California, who in 2008 sold more than $18 million in barley on 55,000 acres (although barley and other staple grains see a considerable yearly variation in acreage and value). Barley has been a classic crop through human history, if used largely at this point as animal feed, for health food, and in the production of libations—both beer and whiskey.

Barley, like so many large-headed grains a member of the grass family, is considered in some circles the oldest of all domesticated grains; wheat and millet in China, and rice in Southeast Asia, provide the competition. Certainly it was among the first domesticated grains in the Near East, and the origin of barley domestication dates back more than 19,000 years (17,000 BC). Part of the confusion is clearly related to eurocentrism; after all, not only are beer and whiskey European additions to the human consumable vocabulary, they are deeply lodged in early forms of English. The word "barn" originally meant "the building where barley is lodged," or "barley-house," in Old English. Without argument, then, barley is an ancient addition to human diet and comportment.

An extraordinarily handsome plant, wild barley is instantly recognizable because of its spike—a brittle head that facilitates dispersal of seed. The domesticated version has either two rows or six rows, and has varying uses. Barley was a steppingstone for early humans, the origin point for the earliest bread. Through human history barley has been considered fairly rough fare, something to graduate from rather than a dietary end point to be sought out and cherished.

The role of barley in California is modest; barley beer brewers bring in their fare. The main advantage of barley is that it survives in conditions that would wipe out other grain crops, so it is raised where dryland farming is essentially the last option—or, in many cases, where cold climates preclude the growing of anything else. The major producing counties in California were Fresno and Siskiyou, followed by Merced, Monterey (in the eastern county foothills), and San Bernardino. Acreage planted is up in 2008, to 55,000 acres—the yield has yet to be reported. But the characteristic barley "beard" should still be recognizable when sighted, and keep in mind that you are looking human origins, and the elimination of early want and hunger, straight in the eye.

CORN, FEED *Zea mays*

Feeder corn, field corn, seed corn, silage corn

Pls. 11, 97, Part 2 opener photo

FAMILY: POACEAE.

PRODUCTS: GRAIN OR SILAGE.

RANK: U.S. #30; CALIFORNIA: MINOR SHARE.

SEE ALSO: CORN, SWEET.

Corn varieties are taxonomically identical, but with genetic differences sufficient for corn to fit within two completely different plant categories. Sweet corn is harvested after pollination but before the starch has formed, usually in late summer or early autumn. Feed, or field, corn is left in stalks in the open, to allow the corn ears to dry. Although feed corn still has ears and kernels, husks and silks, it is treated altogether differently. The crucial role of corn is twofold in California (and in other parts of the United States). Feed corn kernels, dried on the stalk, are run through a combine, where the kernels are knocked free, stored, and delivered to the "elevator" as future cattle, hog, or chicken feed, after various forms of processing.

Feed corn may also be harvested much earlier, chopped whole in sophisticated machinery (including stalks, leaves, and ears), and delivered on-site to dairies (usually) as silage—a moist, fermented feed that is already somewhat preprocessed (a by-product of the fermentation) and that is easily digested by the livestock to which it is fed.

Silage operations are a magnificent sight in California, and they come in several telltale forms: in trenches where silage is chopped and piled, in silos such as the characteristic blue Harvestore towers that were once the telltale signature of nearly every dairy farm with a favorable sense of itself. The most common system in 2009 is to prepare silage in elongated 25- to 40-foot-high mounds of feed, typically covered with white 9-mil plastic, with tires cut in half and wired together atop the plastic to hold all in place. The plastic sheeting not only protects silage from drying out, but it also provides a shelter from storms. One final storage vessel involves a further paean to plastic film, and in this scheme, what appear are stout white tubes up to 25 feet high that

Plate 97. Both sweet corn and feed corn are raised in California, and this native product of the Americas is a sufficiently iconic summertime crop to merit a sign at a fruit stand along Hwy. 99.

fully enclose their contents: clearly, they are stuffed like a Cuban cigar with something. Inside is silage, and that proprietary system, with long white silage worms that look like an outtake from the film *Tremors*, is seen outside a quarter of all the dairies or feedlots in California. Other crops besides corn can be used (there is oatlage for oats, haylage for alfalfa), but sorghum, feed corn, and grass are common alternatives. Corn for silage was harvested in California from 495,000 acres in 2008, and earned more than twice the revenue of corn for grain (*see also* Hay).

Corn can be made into a world of products: a variety of alcoholic drinks (including corn and bourbon whiskey in the United States), dog food, corn flakes—a breakfast cereal—biofuels (a deservedly controversial use, because there is argument about whether the energy put into raising corn exceeds any energy that would come out at the end of the process). Corn cobs can be burned in stoves; biogas is produced from corn, corn whiskey is of course a time-honored use, and there are even ornamental uses of corn (sweet or feed, it doesn't matter . . .) in edifices such as the epic corn palaces of Mitchell, South Dakota, a tribute to a crop that still makes the deepest sort of mark in the Midwest and North American Great Plains.

Grain, or feed, corn is not a huge product in California of late. But then again, on 170,000 acres, "grain corn" was worth $172 million in 2008. The big producers are predictable in looking at the dairy and feedlot trade: San Joaquin, Sacramento, Tulare, Merced, and Glenn counties. When there have been past agricultural subsidies in California, they have tended to go to these sorts of crops: grains used for feedlot (confined animal feeding) operations.

COTTON

Upland cotton — ***Gossypium hirsutum***

Pima, or American-Pima, cotton — ***Gossypium barbadense***

Pls. 26, 38, 98, Map 17

FAMILY: MALVACEAE.

NOTES: PRODUCTS INCLUDE COTTON LINT, COTTONSEED, COTTONSEED CAKE, AND COTTONSEED OIL.

RANK: U.S. #4; CALIFORNIA SHARE 8%.

Cotton in the field is a spectacle. The magnificent crop, with deep-green foliage and spectacular flowers of yellow and pink, is irrigated and relatively fast-growing in a summer-hot climate. Planted to

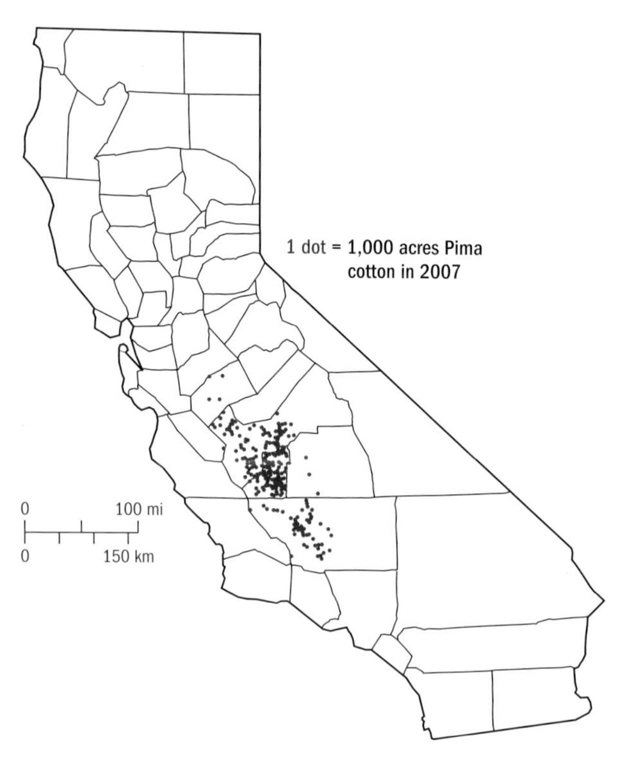

Map 17. Acreage in Pima cotton, 2007.

relatively heavy soils, cotton is overwhelmingly a crop of the southern San Joaquin Valley. The acreage and production of cotton in California is way down but is unlikely to go out entirely; that said, the peak days of cotton growing are likely in the past. For the historically minded or the nostalgic, there is a pity to this. Big cotton, like so much of California's mid-twentieth-century agricultural life, is documented in great photographs from the FSA, including the memorable work of Dorothea Lange. John Steinbeck's *Grapes of Wrath* is another touchpoint, and Walter Goldschmidt's *As You Sow* is a study of two nearby San Joaquin Valley towns that were very much tied up with cotton production and debates about social well-being and agribusiness. Cotton is a big part of California's agricultural history—just somewhat diminished in dollar value.

Historically, cotton was first grown in the 1910s, on land acquired by Standard Oil of California during the early days of the San Joaquin oil boom. The other acreage substantially in cotton was on the floor of what used to be Tulare Lake, a region of formidable cotton-based towns (Hanford, Tipton, Pixley, Delano, McFarland, Shafter) that was purchased and developed by the J. G. Boswell Company—headquartered in Corcoran, in Kings County—into California's largest single agricultural empire of the late twentieth century. Cotton in its California heyday was, for starters, about scale, size, and technological mastery. It was also about two crucial California realities: flexibility in the political control of water, and access to essentially unlimited acreage—certainly far more than the quarter-section (160 acres) of land that many Americans of the 1920s and '30s took to be their birthright. Acreage limitations were overcome by political force and the will of large landowners, including farmers but also corporate enterprises such as the Southern Pacific Railroad and Standard Oil of California, which controlled a colossal extent of land. Water came to damp down the dust through a similar political bypass, advanced by support for the interests of big growers from the federal Bureau of Reclamation, the U.S. Corps of Engineers, and state officials of a like mind.

For all the dominance of big cotton more than a half-century ago, cotton in 2007–2008 is a considerably reduced crop; 650,000 running bales were ginned (the units are less important than the comparison), down from 1,790,000 ginned in 2004–2005. The decline is not quite by two-thirds, but certainly production is far less than half of what was raised four years ago. The bright story is a partial rise in Pima cotton production, which California grows in greater quantity than any other state. Although not considered as fine as Egyptian extra-long staple cotton, Pima cotton has environmental and production advantages, and California gins nearly 94 percent of American-Pima cotton. Acreage in both varieties totaled 268,000 acres: hardly tiny, but down to almost exactly half the acreage grown in 2006.

The crucial feature of cotton is the boll, a soft staple fiber that grows around the seeds of the cotton plant. The boll forms after the cotton flower is fertilized—and it is curious that although anyone can recognize cotton when it is defoliated and showing its bolls in the field at the very end of its life, only a relative few could single out cotton in the earlier stages of its life. The crop is

Plate 98. The white, puffy appearance of the mature cotton boll is familiar, but the flowers of the juvenile plant are likely less so; seen here (with a visiting honey bee) in a field in Merced County.

a thirsty one, and irrigation is a must. Cotton is grown in a complicated process with efficiencies of scale that are crucial—which accounts for the production of cotton by relatively few growers with huge acreage in California. The cotton boll is harvested and gathered into truly huge bales. The size of bales has increased through the years; today's vast loaves of cotton are deposited in the field and a tarp with the producer's name thrown over the top, to prevent damage from the elements, but these may still be in the field months after harvest, waiting for prices to rise or rates at the gin to drop.

The ginning process breaks down the cotton and removes the seeds, which then go into separate storage, because cottonseed has other uses. Development of the cotton gin (credited to Eli Whitney, in 1793) was widely regarded as a crucial step that allowed for expansion of British control of world fiber production, and the upswing of economic power in the American South. The uses of cotton are too many to even suggest. It comes in varying forms and grades, with upland cotton generally producing short fibers, and the American-Pima yielding longer textile staple fibers, and Egyptian cotton the longest and softest of all. The close-by presence of a compatible cotton gin is crucial to growers, yet the number in California in 2006 is 61 active gins, down from 299 in 1963.

The name cotton is itself derived from the Arabic *(al) qutn*, which entered use around AD 1400. The United States now leads world exports. Cotton growing is, however, heavily subsidized with federal payments. Cotton growers in California are no exception to the rule of growers pocketing federal price and production supports, although the future is uncertain as doubt increases about the form of future five-year U.S. farm bills. Cotton has historically been among the very highest users of both water and pesticides in California, although new growing techniques, rising petroleum prices, and resistance to pesticide-heavy production systems make it more likely that growers will attempt to back down on the chemical and water footprint of California-grown cotton. Some companies, such as clothing manufacturer Patagonia, have made a point of purchasing only organically produced cotton for their high-end apparel.

Among the supporting elements in cotton growth is cottonseed. A by-product of the ginning process, cottonseed is overwhelmingly used as livestock feed (some 95 percent), but the remnant generally goes into the production of cottonseed oil, used for cooking, cosmetics, soap, and as a carrier for agricultural sprays; cottonseed was worth $84 million in 2008.

Home base for California cotton is the San Joaquin Valley. When Oklahoma and Dust Bowl migrants arrived in California in the 1930s, many found that the only employment available was grueling work as hand harvesters of cotton bolls. Along with oil and cattle, cotton built Bakersfield, and certainly the southern Valley. There is still a pronounced geographical concentration: Kings County ginned the most cotton in 2008, followed by Fresno, Kern, Tulare, and Merced. Even with 219,000 running bales produced, Kings County failed to break onto the list of the top five U.S. cotton producing counties: in 2008 it was seventh. Cotton lint was worth $326 million in 2008, down a heady

$275 million from the previous year. The running bale count is much decreased from California cotton's 1981 gravy days, when 3.5 million bales were produced—nearly five times the 2008 production. Rising and falling yields within the cotton industry are to a degree normal, but the line is trending down.

HAY *Medicago sativa*

Alfalfa, grass, oat, mixed, and other

Pls. 99, 119, 122, Part 3 closing photo

FAMILY: FABACEAE.

RANK: U.S. #1; CALIFORNIA SHARE 6%.

For the agricultural purist, there is nothing more beautiful than the smell of water mixed with the aroma of purple alfalfa blossoms—except perhaps the utterly palpable wealth and security implied by a stackyard fully loaded with substantial and tightly packed hay bales. Alfalfa should be immediately familiar to most passersby, with the plant exhibiting the characteristic botanical form of a member of the pea family. Alfalfa is indeed a respected perennial crop because of its ability to fix nitrogen into the soil and, over time, improve ground on which it grows. But alfalfa is not the sole hay crop, and hay is a collective category in California agriculture, produced in a value of $1.4 billion (2008), which includes alfalfa hay (950,000 acres), and oat hay, barley, grass hay, and other varieties that cumulatively came from another 570,000 acres. A great deal of hay, however, never leaves a farm or ranch, and is consumed locally by livestock in winter and spring, before warming starts pastures toward growth.

The value of hay is in preservation—or, rather, curing. Crops are carefully swathed, or cut down. They are allowed to dry on the field (rained-on hay is never a good thing), and adjacent rows may be gently rolled together to minimize tractor and baler travel and build better bales. Smaller rectangular bales are constructed by a tractor-drawn baler that gathers hay from a field and rams it with high compression to form a flake—an amount of hay typically given to a few sheep, or perhaps a cow or a horse in one feeding. Hay bales come in many sizes: small rectangular bales (80 to 120 pounds), larger ones weighing several hundred pounds, and the largest units—round bales or bale loaves that can weigh up to several thousand pounds—and which once

Plate 99. Alfalfa hay is worth more than $1.6 billion in revenue, with some harvest diverted into silage for dairy use. High water consumption goes hand in glove with multiple alfalfa cuttings.

built are not easily moved. Round bales, when assembled, actually have the hay wound around a central core. A look at a cylindrical bale on end will show an essentially continuous coil or spiral of hay going from the core outward. Both systems are efficient, and round bales or loaf bales can be set into a field, fenced off, and fed piecemeal to livestock later in the year.

The use of hay is straightforward: it is feed for livestock. Which livestock, and what hay goes to which sort of animal, are two important questions. The crucial producers are generally the counties that lie closest to the major destinations for hay. Bales of hay are awkward items to move long distances, since hay is bulky and heavy, doesn't benefit from travel, and is expensive to load up, move, take down from a truck, and restack. The very highest-quality alfalfa hay in California invariably goes to dairies, and the highest prices paid are often by dairies in the Chino

Valley, east of Los Angeles. Those feed yard–based sites have no pasture; the ground surface is well-drained concrete manure collectors, and under those circumstances, no one wants to waste hay. Dairy cattle generally get the alfalfa hay highest in total digestible nutrients, which is desirable for maximizing milk production.

Feedlot livestock receive what is available, which may be lower-grade alfalfa hay, or hay made, using the same techniques, from oats or barley or even wheat, although that is less common. Horses and pastured livestock can be fed less vaunted oat or grass hay, but often their owners give them the best: not a great idea, for generally sedentary animals, but a reality. And, finally, ranches necessarily feed their animals hay in winter, especially in parts of the state where the snow flies in winter months. On-farm or on-ranch use is common, and hay fed locally can include up to a quarter of total hay production. Hay can last through a year, but is best used before that, since it loses nutritive value with time.

The big producers of hay in California are San Joaquin Valley counties (Kern, Tulare, Merced, Fresno), plus Imperial County, where hay can be grown year-round, with cuttings going on almost continually. Grown on 1,748,000 acres (yes, 1.75 million), for 2008, hay brought in $2.1 billion. Of that, $1.66 million was from alfalfa hay, the rest from other hay crops. Silage, which is often made from feed corn but which can also be made from alfalfa, sorghum, oats, and other grain, generated another $338 million in 2008.

OATS *Avena sativa*

FAMILY: POACEAE.

RANK: U.S. #14; CALIFORNIA SHARE 2%.

California is chock full of oats—they beckon from every hillside, road edge, backyard, and fallowed field. Those, however, are *Avena fatua*, the wild oat, an introduction from Spain and the eastern Mediterranean that arrived in the eighteenth century and spread with vigor through hill and dale, especially into tilled but later-abandoned farmland. The oat under discussion in this entry is not the common wild oat *A. fatua*, but instead *Avena sativa*, the thoroughly domesticated common oat. Although superficially similar in appearance, the common oat has far larger groats (the seeds, which are eaten, often after being steamed and rolled), and the oat plant is a vigorous bearer whose products are widely used in feeding animals, either as grain or as hay.

Like corn, sorghum, and wheat, but unlike alfalfa, oats are a grass, and fit the classic definition of a grain. A minor California crop when compared to barley—and certainly to wheat—oats for grain were produced in a value of $5.1 million (2008), grown on 20,000 acres. The production was in smaller plots around California—Siskiyou, Fresno, Sacramento, Sonoma, and Modoc counties. Production in 2008 was one-fifth of the all-time high, in 1957.

The common oat is a cereal grown for seed. Although eaten as oatmeal or in a breakfast gruel, oats are most often fed to livestock. Because they are readily digested and less likely than other grains to cause founder or colic, oats—either as grain, or in the form of oat hay—are a preferred hay for horses. Oats are sown in spring or early summer, as early as possible, so they can grow before the summer heat puts them into dormancy. The grain is usually swathed in fall, and the grain removed. On dry years, oats will be left in the field, and livestock turned loose to glean what they can from the unharvested crop. The groats can be flaked, or rolled, to make feed, or they are milled to remove the oat bran from the flour; otherwise, milling would convert a crop to whole oat flour, which is not in common use in the United States.

OIL CROPS

Sunflower — ***Helianthus annuus***

Safflower — ***Carthamus tinctorius* L.**

Pls. 100, 101

FAMILY: ASTERACEAE (SUNFLOWER AND SAFFLOWER).

NOTES: COTTONSEED, ANOTHER OIL SOURCE, IS AN EXTRACT FROM REMNANTS OF COTTON PRODUCTION.

UNRANKED—BUT CALIFORNIA IS A MODEST PRODUCER.

Oil crops, in the USDA orthodoxy, include safflower, sunflower, and canola. Canola is not grown in California in commercial quantity, but the other two assuredly are. There could hardly be two more different crops, or two more distinct contributors to the California agricultural economy. And, curiously, the acreage

Plate 100. With its pivoting head familiar to backyard gardeners, the sunflower in full oil seed production mode is a startling crop: it's one thing to see a few sunflowers, but another to see 100,000.

of safflower and sunflower in 2006 was almost exactly the same, within a thousand acres of bearing acreage at 52,000 and 53,000 acres.

The sunflower is surely one of the most cheery of all crops; the Spanish term "girasol" is perhaps an even better description, since it literally means "pivots with the sun," something that younger sunflower heads indeed do. A stem of the sunflower can grow up to 10 feet in height, and the flower, more correctly described as a head, will grow up to a foot in diameter, with large edible seeds. The pattern of florets within the sunflower inspires a great deal of admiration from scientists who study natural patterns that exhibit mathematical perfection, with the left spirals and right spirals in the flower head examples of successive Fibonacci numbers—34 in one direction, and 55 the other—or, in a large head, perhaps 89 in one direction and 144 in the other. The explanation is that this is particularly efficient, a signature of organization in nature.

Sunflowers are native to the Americas, but traveled back to Europe with Pizarro, who encountered them in Peru. They were adopted quickly, and grew popular in Europe, especially in Russia, where consumption of sunflower oil was permitted in the Lenten season. When grown for oil, the heads are severed, and the heads with dried seeds shaken and stripped of the seeds, which are then crushed. The oil is relatively healthful, and has distinct advantages, not least of which is sustaining higher heat than olive oil.

In contrast to the toothsome sunflower, the safflower is a branched, thistle-like annual with sharp spines on its leaves. Plants grow up to five feet but are generally shorter—about three feet tall—in the field. As they grow, they are green and vigorous, but that vigor is not what is useful: it is the seeds, which are pressed to produce an oil used in cooking, for salad oil, and in the production of margarine. Two variant types of safflower actually produce quite different oil: one high in monounsaturated fatty acid, the other in polyunsaturated fatty acid. The market favors the former.

Plate 101. With its bright colors, immature safflower shares the cheeriness of sunflowers (also an oil seed crop). Safflower is allowed to dry in the field, however, and is later harvested for the seeds within the heads.

Seen in the field, safflower has a lovely yellow flower, but wandering into a planting produces hazards more dangerous than the poppy field in *The Wizard of Oz*: spines on the plant leaves snag and tear, especially as the plants mature and dry. And that they will—safflower is never harvested until the field is completely dried out, and a safflower field is easily recognized (aside from the yellow flowers) in fall, when field upon field of safflower is completely desiccated but not yet harvested. The view never ceases to startle—for California agriculture, which is so attuned to prompt care of crops, the dried out safflower field looks like an example of heartless neglect, yet is anything but that: combining will follow, and with seeds extracted, oil will flow from the presses. Or as an alternative, safflower, like sunflower, can be used for bird seed or can be ground and used as livestock and poultry feed. Oil is by far the most common use of safflower in California, however.

Together, sunflowers and safflowers were $66 million contributors to the California economy in 2008. Safflower brought in $33 million from Kings, Yolo, Colusa, San Joaquin, and seven other counties (2008), generally from in and around the Delta Region. Sunflower, for seeds and oil, was a $32.8 million 2008 producer, exclusively in the Sacramento Valley (Yolo, Solano, Glenn, Sutter, and Colusa counties).

PASTURE — Various species
Irrigated and range

Pls. 12, 13, 25, 44

FAMILY: POACEAE.

Pasture is discussed earlier, as part of the California Cornucopia, but deserves brief mention and some statistics. Pasture is rangeland where livestock graze—the livestock element is inescapable in the definition. Pasture can be in natural grass (rangeland), a natural meadow that is irrigated (irrigated pasture), or—far more common—irrigated pastureland with perennial grasses sown, at some point, in the pasture to provide quick-growing and nonseasonal feed so long as it, too, is irrigated.

Irrigated pastures (nearly one million California acres in 2008) can be watered any which way, from underground pipelines and gates to hand-moved sprinkler lines to aboveground rotary sprinkler heads to gently inclined flood-irrigated fields fed by siphon tubes that pull water from irrigation ditches. When irrigated in a relatively warm to hot environment, the limiting condition for the capacity of a pasture that is planted to suitable grass is simply the regular addition of water. In 2008, 51 of 58 counties in California reported a "harvested acreage" of irrigated pasture.

The alternative is rangeland, and that is simply privately owned land (which can also be leased to others who own livestock) where animals are grazed for at least part of the year. If managed properly, residual dry matter enough can be left on a pasture at the break from spring to summer to allow animals to graze what is left on the pasture during summer and the early fall months, until rain again starts to fall. But absent irrigation, there is no "alternative" rainfall in much of California—at least, not reliably so. All but three counties in California noted that they had "pasture, range" under paid use in their counties—the three missing are Del Norte, San Francisco, and Orange counties. The total area used for private rangeland was more than 17 million acres—nearly 18 percent of the area of the State of California. Gleaning a total of $181 million in income, that is hardly enough to compete with the giant crops—but rangeland pasturage of animals makes some other activities, such as dairying, raising cattle and calves, goats, or sheep, feasible.

Pasture is nearly a universal in California, and the animals grazing it can be cows, sheep, goats, horses, alpacas, llamas, emu, ostriches, and, of course, a wide variety of avifauna (geese, ducks, free-range chickens or turkeys) and wildlife, including migratory fowl. Finally, it bears noting that some of the value of rangeland and pasture is in the form of ecosystem services, the provision of habitat and buffer space around cities and suburbs. Plus, pastures are remarkably peaceable to look at and enjoy. That the owners of pasturelands make some money from their acreage is simply an added benefit in the pragmatic world of agricultural costs and returns.

POTATOES *Solanum tuberosum*

All potatoes, excluding sweet potatoes

FAMILY: SOLANACEAE (MEMBER OF THE NIGHTSHADE FAMILY).

RANK: U.S. #9; CALIFORNIA SHARE 3%.

Potatoes, sweet potatoes, and sugar beets are three California crops that are roots but that are also consummately high in sugar—which, of course, makes them of particular interest. A starchy crop, the name potato refers to both the plant and to its product, a node that forms underground as the potato plant is developing. The part of the potato that is eaten is not actually a bulb or a root, but rather a modified stolon—akin to runners from a strawberry—enlarged for use as a storage organ. Although capable of reproducing from seed (the potato will flower, given enough time), the plant reproduces readily when pieces of the potato, or eyes, are removed and replanted under appropriate conditions—the practice of vegetative reproduction makes it an easy crop to spread.

Although the potato is a native of Peru and Bolivia, in the Western Hemisphere (where 100 distinct potato cultivars might coexist in a single valley, each one valued), potatoes went eastward across the Atlantic with some of the early return voyages to Europe, and they were established in the Old World by 1540. Or rather, "the potato" was taken back, because worldwide, 99 percent of cultivated potatoes are descendant of a single subspecies of potato, *Solanum tuberosum* species *tuberosum*, that was indigenous to south-central Chile—and regarded there as a relatively uninteresting variant. If South America was the potato fancier's paradise, what went to Europe was depauperate. It was precisely an absence of variety in the stock of potatoes in Europe that led to blights in the nineteenth century, including the so-called Great Irish Famine that sent so many Irish to Ellis Island, and thence to New York and Boston.

Irish potatoes are the world market leaders. There are more than 5,000 varieties of potato in South America, and although added cultivars have spread northward in the last several decades, there is not a great deal of variety in potato demand outside the Andes: in California, most acreage is in White Rose, Russet Burbank, and Kennebec, all of them introduced before 1940. Much of the variation is associated with distinct harvest techniques, but some draw on newly introduced cultivars: new potatoes are thin-skinned examples pulled early from the ground and not given curing time after picking to harden their skin; fingerlings are an alternative cultivar that naturally grows thin and long; the Yukon Gold likely hails from Russia, or perhaps Scandinavia. Considerable effort is going into genetic modification of potatoes, among other things seeking the Holy Grail of the spud world: the perfect French fry potato, with high dry matter content.

California potatoes are as good as many, although spuds tend to do especially well in cooler climates and higher elevations on well-cultivated (and rock- and obstacle-free) soil. Kern, San Joaquin, Riverside, Siskiyou, and Modoc counties are all significant potato producers. Yields are down from previous years, and certainly declining from the peak of production in 1966, when 35 million hundredweight of potatoes was grown in California. The state holds only a small share of the market—less, even, than New York State—and is unlikely to pick up a great deal of added production; there is simply too much fervent competition, and there is a surfeit of other crops likely to bring in higher prices; these hold down valuable land where potatoes could otherwise be planted.

Growing on 41,900 acres in 2007, California's potato crop has stayed remarkably constant over the last decade, though well below the peak years of the mid-1960s. With $163 million in revenue for all types (potatoes, winter; potatoes, spring; potatoes, summer; potatoes, fall), it is no surprise that potatoes are heavy yielders, with nearly 15 million hundredweight of production. The main problem is price, which has dropped consistently since 1999. Nor is there much of any way to develop added value for potatoes, although since there is no significant vodka production so far in California, perhaps that might be a route. After all, 23 other countries in the world now produce local vodka varieties from potato crops.

RICE — *Oryza sativa*

Pl. 102, Map 18

FAMILY: GRAMINACEAE.

RANK: U.S. #2; CALIFORNIA SHARE 21%.

Two crops, alfalfa and rice, bring a furtive joy when they are encountered in the field during the heat of a scorching summer day. They share a common need, ready access to water, and, better still, are lush and green and smell good in a cross-breeze, with a combination of chlorophyll and musk peculiar to something growing with vigor. On occasion, an alfalfa field might also be in flower while irrigated, and that adds another mint-like scent;

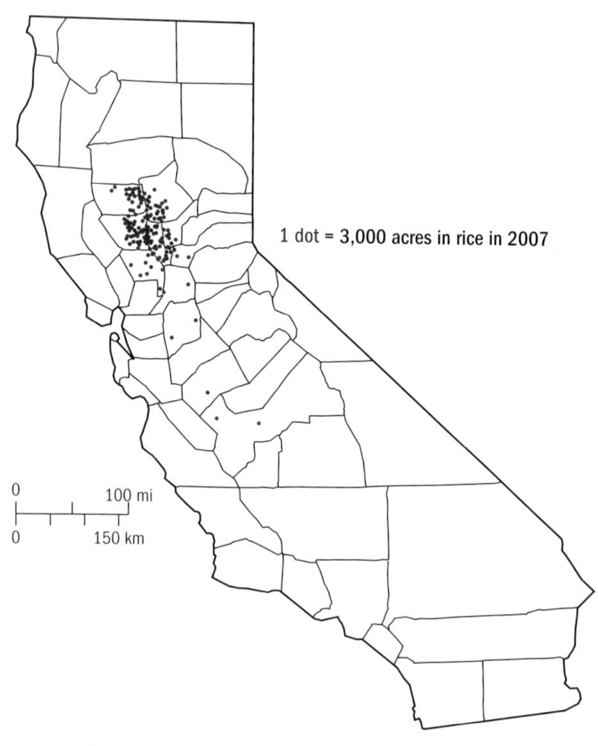

Map 18. Rice acreage, 2007.

what a rice field promises is a bounty of bugs that squish onto windshields with primitive force. Rice is instantly recognizable not as the crop itself, but for its growing environment: water, held in check by carefully formed levees that curve gracefully across a field, holding to a contour. The degree of engineering cannot be mistaken for anything else. A member of the grass family (like wheat, oats, and barley), rice is usually thought of as a paddy crop, but it is now routinely grown around the world in wide, open fields. California has an outsize bounty of rice, most of it in the middle Sacramento Valley, and it is a prolific and often-profitable crop.

In 2008, California produced $830 million worth of rice on 517,000 acres, and 2009 production is expected come in at more than 8,500 pounds per acre, reflecting a steady yearly rise in yield. Although total production saw a small drop from 2006 and 2007, the value of the crop was higher than either preceding year in 2008, thanks to rising prices. Rice could break into the top 10 commodities in California agriculture in another good year or two; only alfalfa hay had a higher value among field crops sold in 2008.

Rice is a consistent producer in its main Sacramento Valley home, where better than 90 percent of California rice is produced: Colusa, Butte, Sutter, Glenn, and Yolo counties (but other Sacramento and San Joaquin valley counties are contributors). Colusa County, in the north-center of the Sacramento Valley, in a recent survey reported 160,000 acres of rice. The distinguishing trait of good rice land is actually less than "perfect" agricultural soil; an impervious claypan or hardpan substrate is

Plate 102. Rice is another staple crop from the grass family and is an important food throughout the world. The rice here in Butte County overgrows the checks, or field divisions, that mark careful field leveling.

desirable for ricegrowing, because the crucial ingredient is relatively impermeable soil that will hold water; sandy or otherwise fast-draining soil would be quite undesirable.

To understand the advantages of the Sacramento Valley, a brief discussion of drainage is helpful. The central Sacramento Valley is bisected north-to-south by the Sacramento River, and from the eastern side of the valley come several major rivers that drain the gently sloping western edge of the Sierra Nevada.

Before human meddling reconfigured these streams with hydraulic mining, flumes, small dams for hydroelectric power generation, and big dams for reservoir storage and flood control, rivers flowed unchecked into the Sacramento River system. The famous Sacramento flood of 1862 put the capital city under a dozen feet of floodwater, and upstream saw far more inundation. Even in 1996, panic releases of water from the chain of Feather River and Shasta Lake reservoirs flooded the interior of the Valley. In any given spring or early summer during the pre-dam era, water carried an enormous load of fine sediment to the valley floor. Those alluvial deposits, built up through thousands of years, created soils with layer upon layer of clay and silt. As those layers compacted and hardened, clay and hardpan surfaces formed that water could travel through only ever so slowly, if at all. Modern-day growers can, should they wish, use deep ripping or subsoiling to break up "developed" layers, and those who wish to plant orchards or vineyards bring in a specialist to rip the ground. Even then, they subsequently water their crops with considerable care so that they don't drown roots in perched water. But ricegrowers not only do not mind impermeable soil, they benefit from it. They work the topsoil, level the ground carefully, and pour on water. A geological disadvantage for most agriculturalists is the ricegrower's firm friend.

Few California crops have benefited so much or as quickly as rice from advances in technology, production, and yield. Thanks to weather and improvements in everything from laser-leveling of the ground where levees are built to the development of semi-dwarf rice varieties that yield heavy crops, rice production has not yet hit an upper limit. That may change as concern about water supply for field irrigation grows. Almost all the water for rice comes from reservoirs in the Sierra Nevada, and the continuation of a series of drought years could drastically reduce water deliveries from the State Water Project or the federal Central Valley Project. That has happened before. Rice was planted on 500,000 acres in 1975, but two years later, in the second year of a hard drought, acreage dropped to 300,000 acres. With prices at a peak in 1981, the acreage rose to 605,000 acres. For 15 years, production has stayed between 450,000 and 550,000 acres. For now, however, the trend line on price and profit is clearly up, and in acreage, California is the second rice producer in the United States, after Arkansas.

The techniques associated with California rice production are constantly examined and revised by an aggressive research program driven by the California Rice Research Board (paid for by a marketing order that taxes each pound of rice produced to support research), the California Rice Experiment Station, a University of California facility near Richvale, in Butte County, where commercial ricegrowing began in 1912, and the Cooperative Extension and Ag Experiment Station scientists, also with UC, who disseminate research results. Being such a significant worldwide staple, rice was the first crop to yield a complete gene sequence. Many varieties of rice are produced in California: long grain, medium grain, short grain, aromatic (basmati, jasmine, della), Arborio (paella, Valencian), and sweet (mochi, waxy). But for growers, the favored current variety of California rice is Calrose, a "radiation-bred" strain of *japonica* rice developed in early 1970s California. Quite popular in the Pacific Islands, especially in Hawaii, Calrose is sought after, and is considered the suitable multi-purpose soft-grain rice for sushi use, risotto, Asian cuisine, and desserts. It is in plentiful domestic supply. Among the advantages of the newer rice varieties is a much shorter growing season. Early varieties mature in 130 days; the late varieties can go 175 days. Maximum yields of improved varieties can exceed 12,000 pounds per acre, though averaging more than 8,000 pounds per acre was once regarded as a watershed event. What is referred to as wild rice in the United States is the unrelated *Zizania palustris.*

Although each grower embraces small variations in technique, ricegrowing in California follows a fairly standard order. In February through early April, fields are dried out enough for equipment to level the field, something now done with assistance from laser-slaved tractors and scrapers, which read light emanating from the laser and automatically adjust blade height on the land-leveling equipment. This step used to be done by hand and eye, but with notably less accuracy. However, the efficiency of laser-leveling comes at considerable cost, and many growers contract with outside firms for this part of the process. Levees or check dams are inserted along contours, so when water is added to the fields, the water level can be held at a depth of two to four inches within the entire segment of the field. Any error will short some of the rice or drown other parts of the field.

Once the field is leveled, a shallow depth of water is added from a grower's irrigation water allotment, and pregerminated seed is flown onto fields and fertilized. Watching rice seed being sown from the air is like being in the audience of a particularly demanding ballet, with a lot of irreproducible and unimaginable pirouettes and airs above the ground. Water in fields is then deepened, and in September, the fields are drained, the grain dried, and harvested with combines. A final element of the rice crop is postprocessing. The rice is cleaned, hulled, dried, milled, polished (usually), and, if not claimed immediately, stored in large grain elevators that stand alongside railroad lines or sidings. Roughly paralleling I-5 at a few miles of distance through the Sacramento Valley are a series of these structures in towns such as Grimes, Arbuckle, Williams, Colusa, and the auspiciously named College City. These elevator and rice-processing complexes are not just functional architecture, they are monuments to a crop.

There is an aftermath. In the months following harvest, it used to be standard practice to set fire to the rice straw that is dropped by the harvester onto the field in the process of harvesting. In an attempt to provide waterfowl with habitat and feed, a number of growers now smash standing straw into the soil surface with large tractor-pulled cage rollers and then turn water back onto the field to provide feed for migrating waterfowl that used to be regarded as pests. With newer rice varieties that mature two months earlier than used to be the case, competition is far less an issue. Because not all grain is gathered in the combining process, that puts feed into the ground, and reflooding the field provides habitat for invertebrate animals that are an important protein source for waterfowl.

A concern among early rice producers was grain depredation by migratory wildfowl arriving as rice was setting its heads. The migratory birds could eat their way through a sizable part of the crop. Waterfowl traveling along the Pacific Flyway were accustomed, in nineteenth-century California, to having as much as four million acres of wetland at their disposal. As rice expert James Hill notes, the creation of wildlife refuges helped, as did the more self-interested establishment of several hundred thousand acres of duck clubs in the Sacramento Valley.

For birds, natural wetlands were an ideal habitat, rich in native grains and in the waterborne creatures that were prime feed. Growers who shift to postharvest flooding recreate a habitat somewhat like what waterfowl might have found in historic times. The birds are dependent on rice fields and the aftermath of other grain fields for food, and reflooding the fields has promise for solving the habitat dilemma. The only question is about sustainability: in some other environments, flooding rice fields around the year can render the soil anaerobic, and cause future production problems. Whether that will be an issue is not known yet—but certainly will be watched for.

SAFFLOWER

SEE: OIL CROPS (SAFFLOWER AND SUNFLOWER).

SILAGE

SEE: CORN, FEED; AND HAY (ALFALFA AND OTHER).

SORGHUM AND SUDAN GRASS

Sudan grass ***Sorghum vulgare*, var. *sudanense***

Pl. 103

FAMILY: POACEAE.

UNRANKED—CALIFORNIA IS A SIGNIFICANT PRODUCER.

Raised for grain, and, in Imperial County, for hay, sudan grass is a forage crop that generated $89 million in 2008 on a California area of 112,000 acres. Native to subtropical regions of Eastern Africa, sudan grass is a form of sorghum, itself a member of the grass (Poaceae) family. The sorghum crops are not differentiated in California, although there is variation in the seed stock planted and raised. Some sudan grass harvests go to silage, some are used for hay, some are used for grain, and some are used for food (including for the production of a molasses variant).

Grain sorghum in Texas is sometimes known as milo, and the crop, which is quite tolerant of heat, is planted where corn (or maize) might not survive. There are 105 distinct varieties of sorghum, which is both ancient and, clearly, varied in its world uses. Aside from its service as a cover crop, and alternate duty as livestock feed, sorghum may be best known as the grain fermented to make maotai, a Chinese distilled spirit of formidable punch and durability, said to have been President Richard Nixon's favorite drink.

The area planted to sorghum in California is about 16,000 acres, much of it in Tulare and Kings counties, and that acreage brought in about $11 million in 2008. The grain is combined, or alternatively it can be baled or made into silage. The great versatility of sorghum is its ability to serve as nearly whatever end product a farmer might wish for it to be, and that adaptability earns it friends in livestock country.

Plate 103. Sudan grass is almost always used to make grass hay, although a slowly developing industry may see it used, along with switchgrass, in the production of biofuels.

SUNFLOWER

SEE: OIL CROPS.

SUGAR BEETS *Beta vulgaris*

FAMILY: AMARANTHACEAE.

RANK: U.S. #5; CALIFORNIA SHARE 5%.

A sizable $65 million crop in California, sugar beets are produced for one thing and one thing alone: sugar content. The roots of sugar beets are chock-full of sucrose, even if there are few things quite as horrible-tasting as a slice of sugar beet popped in the mouth and chewed. As that first taste-test might suggest, sugar beets go through a thoroughgoing cycle of change before, suitably transformed, they emerge as white refined sugar.

A biennial crop, sugar beets produce a massive root their first year of growing, and if unharvested, will then generate flowers and seeds the second year, which will grow by feeding off the root. Few indeed are the roots that make it to a second year. Although a global crop, sugar beets are a marvel of sorts in modern food chemistry; they have largely replaced sugar cane in the United States as the sugar producer of first resort—in fact, cane sugar is beginning to gain style points for its relative rarity, a shift from the heyday of California and Hawaiian (C&H) Sugar, with its sizable cane processing plant in Crockett, along San Pablo Bay and just outside the Delta.

By the end of the first year, the sugar beet has a large underground storage root that can reach five pounds, which scores up to 20 percent sucrose by weight. Planted in spring and usually harvested in fall, sugar beets produce masses of leaves and a heavy crown. Once one of the most labor-intensive crops in California, several sugar beet varieties, especially those produced out of state, now are classed as genetically modified organisms (GMOs), and are especially bred to resist glyphosphate (Monsanto's Roundup), which is used on the field for weed control without significantly affecting yield or crop health. Whether California growers have adopted growing of the modified sugar beet, or whether GMO use is limited to the Pacific Northwest, is unresolved. Considering sugar's proliferation in processed foods, the origins of sweeteners is likely to remain controversial.

The most dramatic feature of sugar beet cultivation is the harvest phase. With a frost, which may come in late September or October in the San Joaquin Valley where much of the sugar beet acreage is, growing stops (the Imperial Valley rarely has killing frosts) for the season. Once sugar beet processing starts, it tends to continue with little interruption, and growers are expected to keep a steady flow of sugar beets moving into the processing plant until their supply is exhausted, since stopping and restarting the diffusers, pulping machines, and extractors is both costly and complicated. Carbonation removes impurities from the raw juice that has been extracted from the sugar beets, and purification presents a sizable challenge in organic chemistry. At the end, a syrup is generated that is called the thick juice; this is run through crystallizers, producing a mother liquor that is then boiled down, producing the first run of sugar, which is considered raw. A final refinement generates clean sugar, with molasses as a by-product that has a higher quotient of impurities than is considered acceptable in the sugar trade. The basic technique of sugar extraction from sugar beets was a development of French chemists during the Napoleonic Wars, when Britain blockaded France and prevented sugar shipments from the cane fields of the Caribbean. The practice spread inland, and a new way of forming an old, familiar food was born.

SWEET POTATOES *Ipomoea batata*

Pl. 31

FAMILY: CONVOLVULACEAE.

RANK: U.S. #2; CALIFORNIA SHARE 23%.

Yes, sweet potatoes and yams are two quite distinct crops. Sweet potatoes are native to the Americas and were discovered by Columbus, likely in the West Indies off the coast of Yucatan and Honduras. Yams are of African origin, a tuberous climbing vine from the genus *Dioscorea.* In value, the production of yams is a minimal contributor in California agriculture, but there are 40,000 acres of sweet potatoes (which are generally somewhat lighter-colored than yams), and that acreage produced a crop valued at $81 million in 2007. Merced, Fresno, and Stanislaus counties are the major producers, but Merced grows four-fifths of the acreage in the state.

As with the regular potato, the sweet potato is not a tuber, but an enlarged storage root. Sweet potatoes are established as transplants, by plant propagules known as slips that go into the ground from February into early April. Easily worked ground is best, since heavy soil slows plant growth. The plant has a low-growing form, spreading a sizable body of foliage. Harvest is usually complete by about Halloween or the first week of November. As with sugar beets, the production of sweet potato biomass that is exhumed from the ground is startling, and large parts of the sweet potato crop must be warehoused until already-stored backlogs start to clear.

The fresh market crop is actually substantial in the sweet potato market. As befits a crop loaded with useful nutrients, the sweet potato is easily microwaved, loaded with appropriate amendments, and eaten as a quick food. Demand in Asian markets is on the rise, a reminder of markets still to be developed. This is all just as well, since the sweet potato yield in California was at a record level in 2008, with 4.3 million tons. Recognition of the value of the crop is on the rise with the skying acreage.

WHEAT *Triticum* L.

Pl. 3

FAMILY: POACEAE.

RANK: U.S. #18; CALIFORNIA SHARE 1%.

Durum and winter wheat are each grown in California, and production totals are sizable. Produced in 26 counties in California in 2006, wheat is surely the most recognizable of any grain crop, and California has plenty on hand. Still, the total wheat production of California is barely 1 percent of the U.S. total. The specialty wheat that California does best is durum wheat, grown specifically for its semolina, which is used to make noodles. There is still considerable demand outside the United States for California wheat, but nothing on the scale of the past.

The story of wheat in California has to be a comparative account: then and now. The "then" is maybe easiest to understand by looking at the years when California had more than a million acres of wheat harvested (not planted): that would be every year from 1870–1905, again in 1919, again in 1975, and then in 1980–1982. The peak year, however, was 120 years ago in 1888, with a suspiciously round total of 3,000,000 acres harvested. To make the case clear, that would mean that 3 percent of the area of California was planted to wheat.

Of course, those peak years for the last three decades of the 1800s were the summit of the Bonanza Wheat era, when California wheat was the toast of Europe (perhaps literally). After 1905, Siberian and Russian wheat began to become more available, and the transportation costs to move California wheat to Europe took a toll. Those highs of twelve decades ago have never been equaled since. There can be a good bit of variability: the area harvested was 315,000 acres in 2007, but jumped to 626,000 acres in 2008, which brought in $479 million in combined durum and winter wheat income.

Wheat is produced mostly in the San Joaquin and the Sacramento valleys, which grow the large share of the fall-sown hard red wheat, the hard white wheat, and durum wheat. Irrigation is needed for most of the production, which is grown in rotation with other annual crops, including vegetables and other field crops. Where there is enough moisture, wheat is grown in the coastal regions, especially on the uplands of the Salinas Valley and the Central Coast hills. And since the Imperial Valley rarely fails to surprise, the desert agriculture region specializes in durum wheat for pasta production—the name "Desert Durum" is registered with the U.S. Patent Office. Wheat will ascend again, in California—but nevermore to the level of 1888.

REGIONS AND DISTRICTS

Physiographic diversity and cultural contrast are among California's most obvious identifying features; the famed susceptibility of its inhabitants to unorthodox religions, architectural, and political movements suggests as much. Yet there remains a conspicuous regional quality to the land and its people which stems from the political fact of the state, the isolation which the barriers of desert and distance have imposed, its distinctive combination of landforms, climate, and natural vegetation and, especially, the recency and rapidity of its settlement and the diversity and origins of its settlers.

JAMES J. PARSONS, "THE UNIQUENESS OF CALIFORNIA," 1955

Boundaries for dividing California agriculture into regions can be crafted in dozens of plausible ways; finding agreement on a single set of formal boundaries is far from a foregone conclusion. Assorted physical and cultural dimensions of California are suggested at the opening of this guide with the question "why is that raised there?" Unless committed already to memory, that text merits a second glance. Climate and precipitation regimes, a cultural schism 'twixt coast and interior, plus necessity, attitude, history, and distinct crops north to south, are realities that shape and shade the countenance of California agriculture. Crop distribution maps make the variation evident. Accepting the imperfection of any typology, this section dwells on commonalities instead of harping on the past or on minute—if also genuine—differences.

Dividing California into discrete sections requires recognizing two major urban areas, acknowledging multiple poles of heavy agricultural production, and respecting a pivotal role played by climate in the production of some crops, whether in the searing desert croplands, the wind-crossed Delta, or the fog-cooled edges of the Central Coast from Santa Cruz to Salinas to San Luis Obispo. Because agricultural statistics generally are delivered in county-sized blocks, those divisions are by and large retained, with one exception asked for: Mendocino County can be placed either in the North Bay (for wine grapes, tree crops, and sheep and goat dairies) or in the Emerald Triangle—since the crop that truly dominates northern Mendocino County is marijuana. Accordingly, Mendocino County is cracked apart, north half and south. With marijuana appearing in no official statistics, that act of geographical surgery has no effect on Mendocino County data—but makes understanding the two regions much easier. Counties adjoining the desert and the Sierra Nevada are also split east–west because their creation was geographically unnuanced, and parts of many of the transmontane counties (generally eastern parts with a minimum of crop acreage) fit best in the Sierran Realm region.

Aside from those nudges, Map 19, in two parts, is simplified from one developed by the Forest and Rangeland Assessment Program (FRAP) of the State of California, and the areas indicated in color show farmland in brown, grazing land and woodland in tan and light blue, urbanized acreage in gray, and open-water areas (for reference) in blue. The concentration

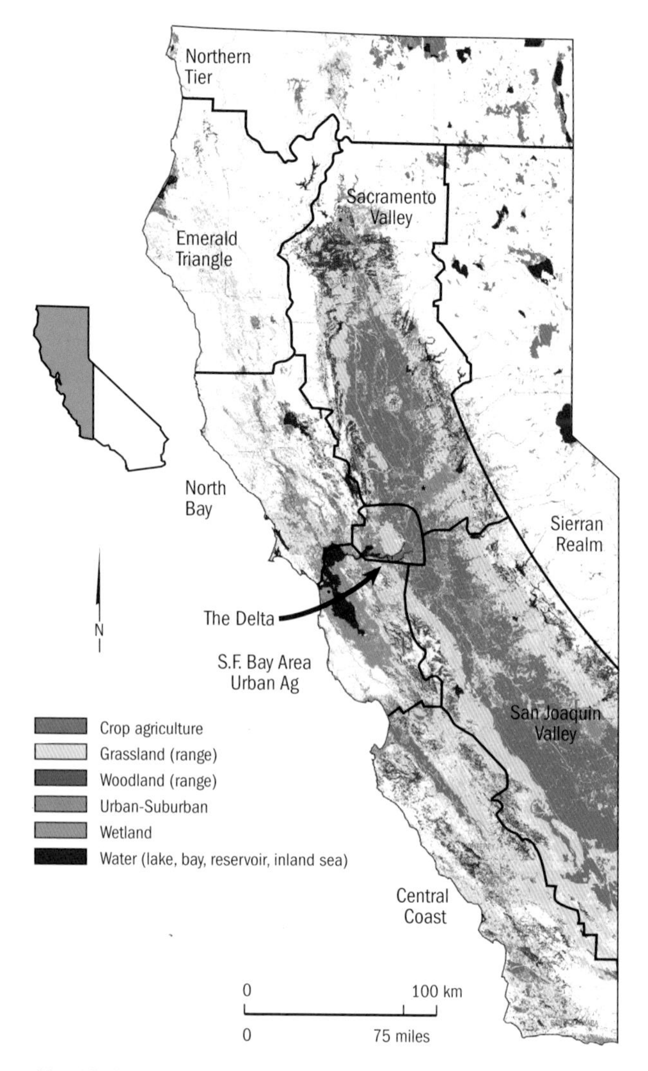

Map 19. Eleven regions of California agriculture.

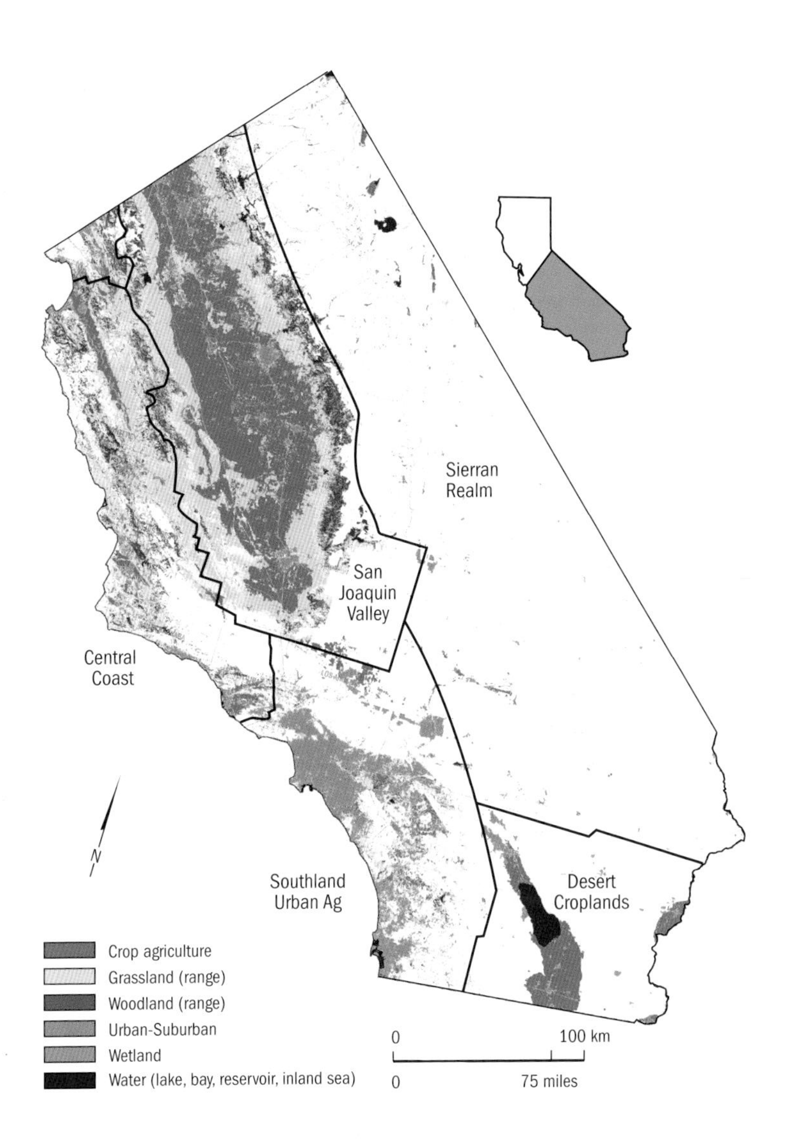
Sierran
Realm
San
Joaquin
Valley
Central
Coast
Southland
Urban Ag
Desert
Croplands
N
Crop agriculture
Grassland (range)
Woodland (range)
Urban-Suburban
Wetland
Water (lake, bay, reservoir, inland sea)
0
100 km
0
75 miles

of cropland agriculture in some areas—and its absence in others—is diagnostic. The finer details of ag region divisions may be argued; they are here for purposes of discussion.

Starting with Bay Area agriculture, our discussion expands outward, spiraling to areas lying within the gravity well of the San Francisco Bay Area in its role as a forceful prophet in agriculture, with a no less potent status as a market for California-derived agricultural products. In trying to sequence a discussion of regions and districts, we decided that although Southern California has easily three times the population of the Bay Area, it is the extended Bay Area that attends far more to agricultural themes. Why this is so raises interesting questions: Dominance in water supply? Is it the San Francisco Bay Area's role that is key in the outflow of exported California food products? Do physical barriers—mountains and aridity—that separate the Southland and San Diego from the rest of the state matter so much? Can Southern California's contemporary distance from agribusiness-scale production (actually only 60 miles away, across the Grapevine) be said to create disinterest? Are too many other attractive nuisances drawing avid attention in Southern California? For whatever reason, concern about crops and food is voiced more often in public debate and discourse in the northern half of the state. That is neither a slap-down of the south state nor a shout-out for the Bay Area: geographers understand that regions are discrete categories, none inherently better than another.

Bay Area Urban Agriculture

The hearth of much California agriculture, the San Francisco Bay Area is engaged with novel ways of producing food, profiting from agriculture, doing front-line developmental research, and seeking environmental protection. In ways both healthy and histrionic, Bay Area foodies shape what California grows and what the United States eats.

If other regions of the United States poke ungentle fun at Bay Area residents and chide them for excessive self-absorption, that does not change a fundamental truth: Bay Area urban agriculture is like the arrow of an old-style analog produce scale: It reads right—set something heavy in the pan and the indicator wildly bounces, then settles at an accurate weight, whether the issue is future forms of food delivery, ecological oversight, producer funding, or export economies—and not least because with the area's more than 7.4 million residents, Bay Area urban agriculture is a potent economic and cultural force.

Nine counties completely surround the two bays, San Francisco and San Pablo: Alameda, Contra Costa, Marin, Napa, San Francisco, Santa Clara, San Mateo, Solano, and Sonoma. Only six are officially mapped into this section—Napa, Sonoma, and Marin (plus Lake and southern Mendocino) are properly in the North Bay. But Marin County, in particular, wields a strong—even a pervasive—influence on Bay Area attitudes toward agriculture, especially in matters relating to environmental quality, philosophy, and the conservation of agricultural resources. So for this one category, consider allowing Marin County—an active agricultural producer and proselytizer—to rest lightly in both regional categories.

Plate 104. Long rows of specialty lettuces and leafy greens appear in organic farms in the Bay Area foodshed—here in Marin County, but equally in Capay Valley or other havens of careful production that supply the market dining industry.

Historically, the Bay Area called the shots for agricultural California in finance and banking, technology, transportation, entrepreneurial energy, ecological awareness, and education and intellect. In finance and banking, a defining force in California agriculture was branch banking, developed by A. P. Giannini, the founder of the Bank of Italy, which became the Bank of America in 1928. In loaning funds for agricultural projects, Giannini enabled farmers to obtain loans for seeds and equipment—something rarely possible before—and anticipated the later creation of production credit associations. Giannini's role as a supporter of innovation expanded with contributions to the University of California, Berkeley, which named a building, and later the A.P. Giannini Foundation of Agricultural Economics, after its banker–donor. Specialization within the Bay Area exists, too. Technology and transportation owe more to the East Bay, where Oakland was the terminus for the Transcontinental Railroad, completed in 1869, and the Port of Oakland became the shipping center and railhead for agricultural products bound for foreign shores that did not ship from the Port of Stockton or the Port of Sacramento.

Advances in agricultural technology in a dozen subfields were driven by the University of California, with innovations hailing from faculty researchers and Berkeley students, and from industrial manufacturers on the Berkeley flatlands and in Emeryville. Entrepreneurial energy could seem all but boundless. Financial, intellectual, and social capital concentrated in the Bay Area filtered north, south, and east through the American West and, some would say, powered through the Golden Gate and westward across the Pacific Ocean. In the nineteenth and early twentieth century, Berkeley was at least as much a pole for agricultural innovation as Stanford and SRI would later be for Silicon Valley and computer startups. Pump manufacturing and soil science, equipment design, developments in agronomics and organic farming, pest management research, plant breeding and forest genetics, labor management and financial ingenuity—all came from Berkeley, Oakland, and the near environs. For more than a hundred years, the Bay Area was a realm of smart, ambitious, hard, and sometimes scruple-challenged people who could and did make things happen.

With so much having occurred—hydraulic mining devastating the Delta, invidious acts passed to exclude the Chinese from owning agricultural land, rises and busts in given products, labor crises, land ownership battles, proposals posed to build dams of unlimited scope across the Bay and Delta—the Bay Area, by and large, was charged with figuring out how to fix things when they were broken, ecologically and philosophically. Although arguments aplenty abound, at root it is fact that the modern environmental movement, begun in the 1950s and '60s, was born in the Bay Area. With a profusion today of NGOs unequalled anywhere in the world, a great deal of the region's attention focuses on issues of environmental quality and recuperation. Repairing now-suspect agricultural, mining, and engineering decisions of the past is one path toward restoration and remediation—another is through continued innovation. Of late, much of that is focused on food, farming, and sustaining agriculture. With a region that in aggregate is the wealthiest enclave in the United

Plate 105. If children don't visit a dairy, there is always the cow train at Uesugi Farms in Morgan Hill, where agriculture does a stand-up business at the urban fringe.

States, there is a willingness to spend more to eat better, to invest in environmental quality, to support once-innovative, now customary, agricultural initiatives such as land trusts, conservation easements, Community Supported Agriculture, revived farmers markets, and a wealth of other activities, should those make life better and easier.

In 1993, the *San Francisco Chronicle* published a short observation that the last privately owned farm in the city and county of San Francisco had sold, and was soon to be cleared and built upon. There are plenty of farms and ranches across the Bay Area, but much of the region's history is a paean to massive change and urbanization. At the Bay Area's northern, eastern, and southern edges, agriculture is pushed away. But a voiding of agriculture does not go unmentioned or without protest. There is resistance, and a great deal of intellectual debate supports agricultural life, including a strain of cultural conservatism that tries to make continuation possible for area farming, ranching, and the sustaining of locally produced food—a movement encapsulated in the phrase "support for working landscapes." An appreciation for innovation and the world of well-cultivated foodstuffs moves radially outward from the Bay Area to touch the entire state. The prevailing hope afield is that agriculture will not simply become a cartoon-like activity, like something from a Margaret Atwood novel, where everything that future generations know about ranching and farming has come from a toy train in which children sit in hollowed-out fiberglass dairy cows, drawn by a miniature tractor that for all the world looks like it was lifted from a Shriner's Memorial Day parade.

North Bay

The North Bay is in no small measure Wine Country, a fine place to eat, buy, B&B, and drink wine. A hotbed of agricultural tourism—true, generally wine-related—is seen by those who know it well as a home to working landscapes where agriculture is taken seriously. Local growers with sufficient wherewithal (and deep enough pockets) work to develop stewardship practices, experiment with new ideas, and aspire to superlatives: best red or white wine, best restaurant, best apples, best land trust, best cheese . . . the idea is not elusive or illusory.

Four-and-a-half counties (the half is southern Mendocino County), each urbanized in its own way, hold down the agricultural fort in the North Bay. Culturally and economically allied with the San Francisco Bay Area, they are nonetheless producers, in an agricultural sense, and they are among the most dramatic and visible profit centers for California agriculture.

Marin, Sonoma, Napa, Lake, and Mendocino counties are longtime significant agricultural producers. The Gravenstein apple district in Sebastapol was a California legend, in part because famed Sonoma plant breeder Luther Burbank remarked, “It has often been said that if the Gravenstein could be had throughout the year, no other apple need be grown.” Today, with labor and market shortages, many a grove of apple trees goes unpicked, and an orchard floor is littered with windfall apples exuding the saucy, if nose-puckering, smell of fermentation.

Plate 106. A vista of grapes near and far at Armida Winery, west of Healdsburg, Sonoma County.

That said, North Bay counties are also recognized as boutique agricultural producers, delivering high-value foodstuffs to restaurants and providing staples for the supremely food-conscious Bay Area.

Not all the demand is for solid food. The North Bay is a destination for tourists interested in visiting the wine country, especially the dozens of viticultural appellations that make up Napa and Sonoma. And every year more visitors travel an added distance to Mendocino County, home to inventive vineyards, pear orchards, goat milk dairies, abundant cattle and calves, and California's first brewpub, the Hopland Brewery. The brewery opened in 1983—appropriately enough, since as late as the 1940s (though today no longer), Hopland was a center of hops production. Cheek-by-jowl across the Golden Gate Bridge from San Francisco, Marin County is a mainstay in the Bay Area foodshed, offering strong community support for local agriculture and conservation easement help through the Marin Agricultural Land Trust (MALT) and other entities for the preservation of working landscapes that include dairies, oyster beds, tree crops, and livestock ranches. There is pronounced enthusiasm for agriculture in these five counties—each day in summer, literally thousands of tourists journey north to visit the wine country, eat in tony restaurants, and buy local produce. U-pick farms advertise and find ready takers, whether the products are pumpkins, apples, plums, or strawberries. A laser-like attention turned on North Bay agricultural production is at a far remove from the way that ranching and farming are pursued in the coastal counties that lie south of the Bay Area.

Napa, Sonoma, Lake, and Mendocino counties offer a dramatically different story from that of Marin County and the rest of the Bay Area. Each of these counties has fought its own battles over development and has seen traditional crops disappear under the relentless arrival of aspiring landowners, exurban migrants, and wannabe vintners with truly enormous amounts of agriculture-subsidizing dollars to spend. Napa and Sonoma counties, and Lake County also, were traditional wine-growing regions in the nineteenth century; Sonoma, in particular, grew much besides (Napa was somewhat less diverse). These counties saw a huge spurt of growth in the 1970s with retirees arriving, many of whom were superlatively well-to-do refugees from Hollywood, or hailed from the vast financial services industry of the Northeast. Newly minted ag landowners, they were poised for personal redemption through agriculture. The draw of grape-growing caught on, and suddenly being a winery owner (and, presumably, a vintner) took over from, let us say, the cachet of owning an orange grove in Southern California, an equivalent activity in the Southland through about 1950. For the well-traveled and the even better capitalized, grape-growing in Napa and Sonoma resonated with echoes of Bordeaux or Burgundy, Tuscany or Spain's Rioja, except it could be done in close proximity to the Bay Area—even, if need be, within commute distance. To Napa and Sonoma they came—less so to Lake County, which has its share of wineries. Exquisite southern Mendocino County is today where some of the wineries that could not afford to stay in Napa or Sonoma (or that decided there was a better deal to be had north and west) went for surcease. Culture and agriculture met, mingled, and remastered the landscape.

Suffice it to say that as destinations, Napa and Sonoma have made it. Agricultural tourism (if wine tasting is considered that)

Plate 107. Pumpkins are brought from more distant growing sites and deposted in Nicasio, Marin County, where they are sold to visitors interested in selecting their own ornamental Halloween pumpkins.

is rampant, and there are together in Napa and Sonoma nearly 600 wineries. An interesting tension glimmers at the North Bay's northern and eastern edges, as Lake and southern Mendocino County decide just how they are going to develop—what they will, and will not, permit. Time will tell; there are potent forces against development in northern Mendocino, but sites such as Anderson Valley are seeing inquiries, and the quite interesting landscape there may find itself receiving offers that can't be refused.

If Marin is less productive in farm-gate value than the other four-and-a-half counties, it does have active dairies, rangeland, and a variety of small producers whose clients literally include some of the most famous restaurants in the United States. Marin County is also where concern grew in the 1960s and 1970s about incursions into acreage formerly devoted to agriculture. With details spelled out in a variety of excellent books, there is no need to go into depth, but suffice it to say that Marin agriculture risked disappearing as urbanization spread west of the Hwy. 101 corridor, and an extension of swarming exurban growth appeared imminent into northwest Marin, traditionally the realm of Italian-American farmers, ranchers, and dairy operators. The Marin Agricultural Land Trust was formed in the 1970s to fight development proposals. As opposition came to fruition, development slowed, then all but stopped (with attention instead redirected toward infilling), and a variety of less destructive activities were undertaken that would at least leave alone Tomales Bay, the Point Reyes Peninsula, and other areas in west Marin. An important precedent was set with that change.

The greatest pressures felt in North Bay counties are, paradoxically, wedded to the effects of overproduction—especially in vineyards—and to aspiring winery owners who attempt to push their vineyards up the hillsides, cutting out oaks and rangeland and replacing them with precarious plantings of pinot, cabernet, syrah, or zinfandel. If the terroir is potentially great, the economics are a fright. The grower goal is less to earn profit than to colonize and establish a foothold in the wine business; this, precisely, is what the worldly Stanford economic philosopher Thorstein Veblen once acutely dubbed "conspicuous display." County planners, however, face a reality of master plan revisions, untenable agricultural development, and potential property tax rebates when property and yearly farm income values fall for all.

The Delta

The Delta functions as a valve to control water entry and release for nearly 60 percent of California, including every water body or river that drains from the Sacramento–San Joaquin valleys. Sizably altered by human manipulation, the Delta has crucial cropland, with singular products grown on unique and prime soils. Its significance is by no means solely agricultural; in addition, the Delta is key to water supply for the urban San Francisco Bay Area and the main source for water pumped southward to Southern California. With infrastructure that includes aged levees at the edge of collapse, the status of the Delta as a conduit for California's water—and crops—is anything but secure, and should the Delta system collapse, two-thirds of California city residents risk going thirsty. Little wonder concern is deep.

The Delta envelops slices of Solano, Yolo, San Joaquin, Contra Costa, and an added chunk of Sacramento County. Technically, the Delta is one of the few examples worldwide of an inverted river delta—think of a 1,100 square mile fan, the narrow end pointing toward San Francisco Bay, and the spread spokes forming the Delta itself, creating a roughly triangular system of channels and islands locally called "tracts." Waterways draw from two source areas, the Sacramento and the San Joaquin valleys, although the rivers that bear the name of each valley are themselves fed by a dozen smaller tributary streams and rivers, predominantly draining west from the Sierra Nevada.

Many a Delta physical attribute is unique. Highly productive land, and a leader among equals in any roster of California's environments classified as unique farmland, deep Delta peat soils were formed by millennia of flooding in spring and early summer runoff from the Sierra Nevada. Layers of locally accumulated tules (bulrushes), sedges, and other plants, along with organic material washed to the Delta by Sierra runoff, were so often flooded that minimal exposure to air permitted the organic matter to decay only anaerobically—and an absence of oxygen is the crucial ingredient in the formation of the acid, yet water-retaining, peat.

The curious process of fresh peat soil formation was ceased with the establishment of dam systems that blocked runoff. Reservoirs built upstream in the Sierra foothills impounded water supplies for cities and agriculture. Ending sediment-rich flows was a problem for the Delta, which is further caught in an urban squeeze between the Bay Area and Southern California and the growing communities of adjacent urban Sacramento and Stockton, in San Joaquin County. Like the cumulative flows of the Colorado River in the Southwest, waters entering the Delta are grossly oversubscribed. With more water diverted, fish migrations grew more difficult. Annual upstream runs of steelhead and salmon were impaired by low flows and blocked channels; fish species native to the Delta by the 1990s were ever-rarer, and the chubs and smelt hanging on risked being sucked into pump systems established to divert water from the Delta to urban water districts both nearby and distant.

The masterful 1979 *California Water Atlas* includes a visionary map sheet by geographer Bill Bowen entitled "The Virgin Waterscape" that lays out what the conjoined Sacramento–San Joaquin valleys would have looked like in pre-European times. In evidence are great riparian gallery forests, seasonal

Plate 108. The role of Japanese and Chinese immigrants in building Delta levees and farming is reflected in artwork in Isleton, Sacramento County.

marshes and massive wetlands, and, in a wet late spring, a land surface tens of thousand of square miles in extent—entirely under water. Those water sources are thoroughly throttled, and even nature's own spigots seem to be depositing less snow pack in the Sierra and from Mt. Shasta to the north, a shortage that impairs the river flows through the Delta. Eventually, the Sacramento and San Joaquin join west of Antioch to flow, united, into San Pablo Bay, and eventually onward to San Francisco Bay. But east of that confluence is an intricate fan-shaped system with some 700 miles of waterways isolating islands that reach almost to Sacramento. Through the long span of California's prehistory, these channels flooded with either spring runoff or saltwater intrusions from the Bay. Since the 1850s, many a Delta island was topped by carefully formed levees up to 60 feet high, generally first built with Chinese labor and know-how. In the protected zone below the levees, the Delta's peat and alluvial soils are rich, and produce unusual crops including pears, asparagus, sweet corn, and a variety of tree and field crops.

Plate 109. A Delta car ferry crosses one of the meandering channels at Steamboat Slough that cut through ancient sediment at the junction of the Sacramento and San Joaquin Rivers, draining all of central California.

A 150-year history makes the Delta an interesting destination, if one where it is tricky to drive along the sinuous but narrow roads that run atop the levees. With farm machinery compacting soils and effectively no new organic material descending from the Sierra onto the land (thanks to upstream dams), much of the Delta is settling below sea level. The Delta tracts, as they are known, are kept dry only by their levees, some of which date nearly back to statehood. Wandering the Delta holding a GPS unit can be scary, with elevations in feet reading "–4," "–9," or "–14" below sea level. The sense of alarm is palpable during a lashing warm spring storm, as upstream flood gauges rise with snowmelt-choked water at a rate likely to give one pause.

A concern among geologists, hydrologists, groundwater experts, and water supply managers is that seismic damage or flooding could bring about widespread levee breaches and an ensuing destruction of the Delta's agriculture. That would trigger a crisis affecting fresh water supplies for Bay Area cities, a greater point of worry for urban planners than is the prospect of damage to prime or unique farmland. The water supply workarounds forced on cities by a Delta implosion would be contentious and complicated. Tens of millions of state and federal dollars are slated already for Delta levee repair, but late in 2009, work on new dams and the water distribution system, with a likely bonded indebtedness totaling over $11 billion dollars (with a "b"), was passed after prolonged discussions pursued with deathly seriousness by state legislators and the governor's office in Sacramento. The voters may, or may not, go along with the proposals, but there's no doubt that something is needed to balance the ecological and water requirements of the Delta, Bay Area, Southern California (Metropolitan Water District), and by no means least of all, (though they tend to feel slighted by the process) the water needs of San Joaquin Valley farmers.

Meanwhile, agriculture and watercraft-based recreation in the Delta live through ups and downs. Successful annual crops are not a sure thing; in years of heavy production but little labor, pears are harvested and allowed to ripen but, without a market, are dumped by the truck- and trailer-load down the side of levees. Growers worry about the long-term commitment to hand labor required to bring in the "grass" of an asparagus crop. When water flows from the mountains coming down the Sacramento and San Joaquin rivers are low, engineers, farmers, and city water districts fret about saltwater intrusion, with brackish or salty water literally infiltrating to flow upriver from San Pablo Bay. But then, when spring water flows are high, worry shifts to possible levee breaches and the permanent flooding of farmland that has been cultivated for 150 years. These are more than the frets, worries, and preoccupations normal to California growers. The Delta brings special concerns that show no sign of going away, except for a shift to the worse when another Delta tract is flooded and abandoned, or when an added plat of land succumbs to the pressures of suburban development or a farm family's resignation at having to make the best of a bad lot. Fans of the Delta are many, but they don't have an easy time of it—and hydrologists, geographers, and other experts look at those hundreds of miles of channels and worry, "What if . . . ?"

Sacramento Valley

Drivers crossing the Sacramento Valley on I-80, or motoring north along I-5, can quickly grasp the larger significance of the Valley with a small exercise: find a safe place to pull over, look intently at the landscape, close your eyes, and imagine everything you have just seen in late spring of a deep snowpack year, with a roiling cover of runoff as all the waters of northern California run through the Sacramento Valley. An 1862 account tells of everything the eyes could see, save only the Sutter Buttes, underwater. This is the drama of the Sacramento Valley—remarkable change, but potential for stunning productivity.

The Sacramento Valley is an agricultural behemoth—just not quite as huge as the San Joaquin Valley to the south. With much of 10 counties involved (Butte, Colusa, Glenn, Placer, Sacramento, Shasta, Sutter, Tehama, Yolo, Yuba), the Sacramento Valley is a realm shaped by government—the state capital is in the city of Sacramento—and, traditionally, by agriculture. No single ag product dominates, though a dozen distinct tree crops are especially important. Effects of urban expansion offer a significant force for change. But agriculture still best defines the Sacramento Valley, and it is agriculture that people think about when they mention the area.

Plate 110. Black plastic reduces weed production in strawberry fields on Brannan Island in Sacramento County. Fumigation of soil used to be a common occurrence but is now mostly banned.

In the production of rice, almonds, kiwifruit, olives, peaches and nectarines, pears, dried plums, kiwifruit, grapes, pistachios, and walnuts, the Sacramento Valley is a sumo-weight player.

The benefits of the Valley are considerable. More precipitation falls in the northern Sierra than south, and through eons of time, the Sacramento Valley was swathed in fertile soils washed from the Sierra Nevada and the Coast Ranges. Soils make for great annual productivity: Joan Didion, in *Notes of a Native Daughter,* only somewhat sardonically repeats the refrain of cant-and-response mimicking the unctuous voices from her childhood: "In what way does the Holy Land resemble the Sacramento Valley?" –"In the type and diversity of its agricultural products." At the southwestern edge of the Sacramento Valley is a zone of organic farming whose producers regularly appear in farmers markets in the Bay Area, affording them elevated prices for their output in the realm of specialist farming. The proximity of deep alluvial soils along Cache Creek and Putah Creek made Davis, in Yolo and Solano counties, the logical location for a University of California campus established in 1905 and originally devoted primarily to crop agriculture and animal science (although since the 1970s, agricultural engineering, veterinary medicine, and enology are at least equal contenders for prestige). The Cache Creek Indian Casino now lies upstream, so far without much effect on agriculture.

The Sacramento Valley in theory has a nearly unlimited capacity for agricultural development but is likely to see that chiseled down by claims from the south for water deliveries. Rights to water, the grower's manna, are better consolidated in the San Joaquin Valley, where farmers and irrigationists arrived in the mid-nineteenth century, before Europeans and Americans considered the Sacramento Valley fully habitable. The Valley was punctuated by intermittent floods and a pervasive fear of miasma—a deep fog—that was (wrongly) believed to spread malaria and tuberculosis. Varieties of eucalyptus trees imported from

Plate 111. Aluminum siphon tubes, each started flowing by a few skillful strokes of an irrigator's wrist, apply water to a bean field near the Sutter–Colusa county line.

New Zealand and Australia, planted for their fragrance and oils, were believed to be a prophylaxis for malaria—the trees actually drank up enough water to somewhat abate the risk, as Kenneth Thompson has described.

With time, drainage of wetlands, and improved knowledge of epidemiology and etiology, such superstitions faded, and the Sacramento Valley was seen instead as a healthful place, especially for tree crops. Glenn and Colusa counties were planted to wheat in the nineteenth century; acreage reverted to rangeland and duck habitat in the early twentieth century and later shifted to rice cultivation as specially developed rice species were adapted for local soil and moisture regimens. After the millennium, ingenious development continues apace, with efforts to expand into water-light plantings such as olives and, to some extent, grapes, which as crops of Mediterranean origin are less prone to gulp copious flows of irrigation water.

Ethnic variety flourishes among the Sacramento Valley grower communities, with specialization in crops a constant. The Sacramento Valley remains an interesting corner of California's agricultural world, especially with rice cultivation, plantings of super high density olives, stone fruits and their Sikh growers, and the extraordinary rangeland both on the western side of the Valley near Stonyford or the Dunnigan Hills—and in areas such as the rangelands of the Sutter Buttes, a system of volcanic necks and plugs, all privately owned but quintessentially beautiful.

Central Coast

For some visitors, the diversity, intensity, and uniqueness of Central Coast agriculture, along with a Spanish–Mexican history that dates back to the earliest European arrivals in Alta California, brings everything about California agriculture together. Monterey County's agricultural aorta, the Salinas Valley, never palled in the interest and accounts of John Steinbeck, who won a Nobel Prize in Literature for his writings about the area. Lettuce of every form, ranching, spinach, superlative wine, artichokes, strawberries, broccoli and brussels sprouts, wild boar, and free-range goat: in the Central Coast is something for everyone, including suitable back roads and mission after mission to explore.

Plate 112. Workers at a bean field in Monterey County take a lunch break.

Although traditionally the Central Coast is classified as a single region, splitting it north and south around the San Francisco Bay area, as done here, makes far more sense. From south of San Jose to San Luis Obispo and the Santa Inez Valley are zones of superlative, unique, and profit-ready agriculture. The vegetables, fruits, vineyards, and tree crops are a marvel and a visual feast, and have been that since the padres located their missions on propitious agricultural sites in the eighteenth and nineteenth centuries. The pattern of land claims and Spanish–Mexican land grants mark the land with imprints still on USGS topographic maps, but crops such as artichokes, garlic, strawberries, lettuce, spinach, kale, grapes, and cruciferous vegetables offer a crazy quilt of color as readily apparent from a vantage point high in the adjoining mountains as on Google Earth. The production values fully match the crop diversity, and it is no accident that Monterey County regularly ranks third in agricultural value produced in California; much of the lettuce, spinach, broccoli and cauliflower, strawberries, asparagus, and great wine hail from Monterey—this was John Steinbeck's terrain and inspiration.

Plate 113. To the left, broccoli; at the right, lettuce, growing below the grade of Cattleman's Road. Both suggest the scale of agricultural production in the Salinas Valley and Monterey County.

The Central Coast has an added charm: in the summer-dry hills of Paso Robles or the Sierra de Salinas is rangeland of no small significance to California's economy. Wines hailing from the Central Coast, and especially from the Adelaida Valley, between Hwy. 101 and the coast, are among California's most recognized, and are perpetual contenders for the world's best. Nor are all the superlatives food-related. The importance of UC Berkeley and UC Davis is matched by Cal Poly San Luis Obispo, the destination of choice for many a child from a California ranching or crop-growing family. Cal Poly has for years been the most difficult admissions ticket to pull in the California State University system. With specialties in agronomy and engineering, the school fields teams in both rodeo and surfing. Could anything be more "California"?

Just a bit farther south, past the wine-growing area of the Santa Ynez Valley, lies Santa Barbara, perhaps the most elegant town in California. Santa Barbara was the hearth of California's vaquero culture, which was famed, even after the territory became part of the United States, as home to the buckaroo: a ranch hand whose traits reflected a singular mix of California environments and Spanish–Mexican skill and temperament. Little wonder that aspiring cowboy Ronald Reagan bought a ranch near Santa Barbara and desperately wanted (and got) an invitation to ride in the annual circuit of the Rancheros Visitadores, a local group whose prestige and elitism eclipse even those of Northern California's Bohemian Club.

On the Oxnard Plain, and in the Santa Clarita Valley, barely north of the San Fernando Valley of Los Angeles, are vegetable growers who serve the Southern California urban markets that lie just a few miles away. Cattle and calves are a top contributor to California's agricultural economy, and the rugged coast from Monterey to the Oxnard Plain is righteous rangeland. The agricultural wealth of the Central Coast region, which starts just south of the San Francisco Bay Area and continues to the edges of the Los Angeles Basin, is both historical and of the here and now. The pivot is the Salinas Valley, a fault-shaped valley lying between two components of the Coast Ranges—the Sierra de Salinas, rising to nearly 4,000 feet west of the Salinas Valley; and, bounding the eastern side of the Salinas Valley, the Gabilan Range—1,000 feet lower, but also drier, and more formidable.

Wildness also defines the Central Coast. Wildfires regularly burn in the Coast Ranges, so untracked that fires must be fought by borate-dropping heavy cargo planes, by helicopters, and by elite fire corps that drop by air into the least hospitable circumstances imaginable. Here William Randolph Hearst established San Simeon—and less than 80 miles away, the Hacienda Milpitas Rancho, designed as a hunting lodge and a remote cow camp for Hearst by architect Julia Morgan on a site not 500 yards away from Mission San Antonio. "Wild" certainly is a byword; the mountains between Carmel and San Simeon were the last redoubt of the grizzly bear in California. Properties nonetheless being developed east of Paso Robles are using cluster development techniques, with houses bunched together and the remaining acreage of a property retained as actively grazed range—an interesting idea, but hardly as effective in providing wildlife habitat and untrammeled country as never permitting development in the first place.

San Joaquin Valley

Agriculture's most exciting venue in the world? Perhaps that is a matter of taste, but if agriculture whets your whistle, it's not difficult to voice enthusiastic notes singing praise of the San Joaquin Valley's agricultural bounty. Eight of the nine top-dollar agriculture counties in California are in the San Joaquin Valley. For scale and raw excitement, San Joaquin Valley agriculture reigns, from dairies and orange groves to cotton fields and sugar beet plantations. All that comes at a cost, though. The work is hard, and labor conditions are challenging; the sights and smells are something else.

There's no mistaking that the agricultural giant in California, without precedent in extent and power, is the north–south-trending system of the Sacramento and San Joaquin valleys that lie between the Sierra Nevada and the Coast Ranges on the west. The valleys, sutured together at the navel by the Delta, are on

Plate 114. The enormous milking carousels increasingly seen in large California dairies move in a slow circle, allowing passengers (such as cow 777) to take a gentle ride while being relieved of up to 40 pounds of milk—twice a day.

occasion referred to as "The Great Central Valley," a term with charm if tarnished accuracy, since there are two conspicuous watersheds, separated by the comparatively diminutive Delta region (which, though small in area is nonetheless the lynchpin of northern California's water supply). Sixty percent of California's surface area, from the Sierra Nevada crest to San Francisco Bay, is tied together by this intimate ecological interchange.

Although the Sacramento Valley has more than four million acres in agriculture, that total is utterly dwarfed by the 11 million ag acres in the San Joaquin Valley (not including rangeland in the grazed foothills). In total productivity in 2007, the San Joaquin included seven of the dozen most agriculturally productive counties in California: Fresno (1), Tulare (2), Kern (4), Merced (5), Stanislaus (6), San Joaquin (7), and Kings (9). Of the dozen, only Monterey (3), Ventura (8), San Diego (9), Imperial (10), and Riverside (12) are located outside the twin valleys. In the San Joaquin Valley, dairy, cattle and calves, citrus, almonds and pistachios, grapes and raisins, tomatoes, walnuts, poultry, cotton, and varied field crops manifest a scale and diversity seldom equaled elsewhere in the world.

Plate 115. A lone tree stands above a tomato field in San Joaquin County, east of Stockton.

The San Joaquin Valley, as geographer James J. Parsons once commented, is the most interesting place in the world—if you happen to be wired for the complexities of agriculture. If not—well, there is always hiking to enjoy close by in the Sierra Nevada foothills, rangeland with grazing cattle and sheep, cities of substance and tiny towns with a toehold on existence, and a dramatic petroleum landscape near Bakersfield and on the western side of the San Joaquin Valley, caught in spirit by the 2007 Oscar-winning film *There Will Be Blood.* The San Joaquin Valley, from Modesto and Fresno to Bakersfield and Buttonwillow, is home to unusual ethnic diversity, technological innovation and energy, landholdings of controversial size, and a great deal of wealth—and working-class poverty. It has serious fans: not only has the Oildale-born writer Gerald Haslam written well about life as a roughneck in and around Bakersfield, he is also a champion of the Bakersfield Sound, a distinctively non-Nashville strain of country music whose followers include Buck Owens and the Buckaroos, Lefty Frizzell, Merle Haggard (who was born there), with allegiance from a number of groups, including the Grateful Dead and the Rolling Stones. There is culture by the yard in the Valley—it's just not quite the same as what's au courant in Santa Monica or San Francisco.

Physically, the San Joaquin Valley is a complicated mixture of distinct smaller terrains, and these shape the morphology of agriculture. The Sierra Nevada, which ramps gently upward along the eastern side of the Valley, sends water downslope that is appropriated by irrigation districts dating to the nineteenth century. For years, water from the San Joaquin River, which gives the Valley its name, never went unclaimed long enough to reach the Delta. That drought ended in 2009 when a complex series of compacts set water inching northward. At press time, the water flow to the Delta was incomplete, but assurances of fulfillment are a matter of legal promise. With reservoir establishment, long riparian forest galleries that once grew next to rivers in the San Joaquin Valley dried out, were cut, and summarily removed, except at Caswell Memorial State Park, along the Stanislaus River, where several hundred acres of habitat reminiscent of the original forest were protected by a single landowner who liked the oak bottomlands and made sure they remained by donating it to become part of California's state park system in the 1970s.

The San Joaquin Valley includes two intermittently dry lakes, Tulare and Buena Vista–Kern. These terminal lakes, fed by streams from the southern Sierra Nevada, have pushed water north from the seasonal lakes to the San Joaquin River only in the wettest of years—otherwise, water settled and evaporated. On maps from the 1870s, these lakes stood at the southern end of the Valley; they now are drained and are routinely planted to crops. Tulare Lake is still home to some of the largest cotton acreage in California. Along the western edge of the San Joaquin, more or less where I-5 arrived in the 1960s and '70s, is a bench land where plantings of citrus, almonds, and other tree crops fringe the concrete Interstate for up to 40 miles without break. The same expanse of orchards and groves lies at the eastern edge of the San Joaquin Valley, especially in a fertile thermal belt from Ivanhoe to Porterville. There, tentative citrus plantings arrived in the 1890s, then poured in wholesale during the 1950s as citrus growers essentially vacated the Los Angeles Basin, abandoning acreage to asphalt to move north. Strathmore and Exeter are only two of the prominent citrus towns. In the region, well managed Tulare County dairies are interspersed at irregular distances among the citrus groves, which adds a hallucinatory aura as the signature redolence of orange blossoms meets manure in the night air. But such is San Joaquin reality: where citrus meets aerosolized manure, the smell is all of money. Porterville and Visalia were once livestock powerhouses, and to a degree remain so. Drought, though, has hit rangelands along the western edge of the Sierra Nevada, an upland still within the San Joaquin Valley, and has depleted rangelands far more than anyone would wish.

Although the San Joaquin Valley is an agricultural powerhouse, and host to many an agricultural event from Basque Festivals in Bakersfield to the International Agricultural Exposition in Tulare, not all is reliably rosy. Ag irrigation tailwaters can be toxic, as at Kesterson in the 1980s. Water supplies for irrigation are in short supply. Some San Joaquin Valley farmers in 2009 received only 10 percent of their usual water allocation—and have effectively closed shop, pruning almonds into dormancy, and ceasing irrigation of field crops. Air contamination associated with fine soil drift from cultivation, pesticide use, and pollutants blowing in from the San Francisco Bay area and the Southland, along with a deep and noxious valley fog that produces 70-car pileups in winter, are contributors to suspect air quality that had led to a diagnosis of chronic Valley Fever among some residents—a named and known ailment often debilitating and sometimes fatal. A place of notable human diversity and signature agricultural qualities, the San Joaquin Valley is not quite paradise on earth, although there is time aplenty for improvement.

Sierran Realm

Cattle, cull cows, calves, an apple here, lots of hay and alfalfa there, fruits and nuts, some spuds and the occasional onion patch. Beyond that, the Sierran Realm is tall mountains, rain shadow, isolated valleys (often at elevations nearing a mile above sea level), and suboptimum cropland—until lots of irrigation water is added. Thirteen of the 20 smallest ag producing counties are in the Sierran Realm.

With a great deal of tectonic vigor brought to bear, California is a mountainous state, dependent on winter rainfall and snowpack accumulations in the Sierra Nevada and portions of the

Plate 116. The drama of French plums (raised now as "dried plums," and formerly as prunes) is in their color, shape, and sweetness; seen here in Live Oak, in east-central Sacramento Valley.

volcanic Cascade Range that reach into the northeastern fifth of the state. Peaks rise above 14,000 feet, and California mountains are frequently in the public domain—a higher proportion of montane land is under the control of the U.S. Forest Service, the Bureau of Land Management, federal game preserves, or the National Park Service than any area, except for desert land in the southeastern corner of the state.

Mountains bring an added element, a pronounced rain shadow effect in the lee of the mountains that augurs lower precipitation totals there than on the wetter, western mountainsides; in our map, many of the Sierra Nevada counties, which run west–east from foothill to piedmont, are split down their middles, north-south, recognizing a divide that California's politicians at the time of statehood had trouble acknowledging. Agricultural bottomlands in the Sacramento and San Joaquin valleys are utterly dependent on waters stored as snowpack in the winter, and on dams lower down in drainages where the meltwater can be stored; without these, irrigation dries up and disappears. The Sierran Realm is bracing to visit, with startling and isolated local examples of agriculture but not much in the way of agricultural heavy hitting. As a keeper of the waters, however, the Sierran Realm reigns: snowpack is as much supreme in the winter as the wandering Sierran hiker or backpacker is in summer.

Although agricultural activities in the Sierran Realm are not common beyond western foothill vineyards and cold-tolerant fruits, a good bit of rangeland grazing takes place, drawing on the cycle of transhumance. Livestock grazing into the mountains during the summer months and returning to lowland areas as fall arrives is still a common practice. And in the lee (eastern) slope of the Sierra Nevada and Cascade Range in northern California are valleys that retain agricultural activities that include raising grass hay, alfalfa, and occasionally food crops such as potatoes or cold-hearty grains.

There are notable pockets of agricultural activity in the realm of the Sierra and the Cascade Range that tend to involve either large-scale ranching as practiced in the northeast of California, or the raising of hay, which is either fed to livestock or shipped to Southern California or the San Joaquin Valley for dairy use. The Sierra includes small pocket districts that take advantage of microclimates that favor local but valuable crops: firewood from the oak woodland, apples in the Apple Hill district of Placer County, and wineries in the Sierra Nevada foothills, where some grape varietals do well on distinctive local soils.

Emerald Triangle

When a region's main income source is an agricultural product that some residents loathe, from which others profit from in substantial ways, and that few people ever see growing because it is so valuable that the threat of theft is astronomical, is there a proper regional description? Marijuana growing shares certain traits with other kinds of Emerald Triangle agriculture: it is haphazard, and it is rarely obvious to travelers.

Although it may seem unfair to pigeonhole a significant part of California with one crop, at least the candidate is a big one. Marijuana produced in Humboldt, Trinity, and northern Mendocino counties completely dominates economic produc-

Plate 117. Although a few agricultural products are produced in the Emerald Triangle region, marijuana plants eclipse more than the distant horizon, with a grower (not street) value of $19–40+ billion dollars, depending on who is doing the estimating.

tion in a region that has been known for decades as the Emerald Triangle. Marijuana is a recognized economic force that regularly surfaces in news accounts, legislative discussions, and considerations of the disparities between the northern and southern parts of California. Tales told locally of hospital emergency rooms, big-box retailers, feed stores, and veterinarian offices where payment is rendered in hundred-dollar bills peeled from rubber-banded rolls are too many to all be fictitious. Other agriculture in the Emerald Triangle is not prolific: a few cattle, small herds of sheep and goats, berries and tree crops, and plenty of pastures and fields grazed by dairy animals or cut for hay in a good year.

Humboldt, Trinity, and northern Mendocino counties, which make up the Emerald Triangle, constitute a land of growing marvels. Most are trees, especially forests of species that were heavily cut over once in the late 1800s, then hit even harder after the 1906 earthquake in San Francisco, when in particular the redwood forests of the Emerald Triangle were cruised again, cut heavily, and much-milled to rebuild San Francisco after its savaging by tremors on the San Andreas fault. With the rise of the environmental movement in the 1970s and '80s, most resident timber companies cut hard, recognizing rising resistance, then downsized, often parting with mills and yards, closing entire company towns. Finally, they sold off timberlands, many by that point cut to the nubbin. The remaining pockets of old-growth forest were all but inaccessible, and in small groves there the tallest trees in California remain, according to experts such as Stephen Sillet of Humboldt State, whose work in canopy science has discovered previously unknown ecosystems of epiphytes and other arboreal species that coexist above 300 feet in the individual tall trees that constitute the remaining forest giants.

As the resource scientist Lynn Huntsinger has noted, Native Americans once made use of the complex ecosystems produced by tall trees, wildfires, and clearings among the forests to maintain a sophisticated economy that Yurok and Hoopa tribal members today might describe as their traditional variant on farming. Intermixed with fishing, indigenous practices once produced a rich economy, now endangered with fisheries, forests, and farming equally under threat. Small hay crops, fruit orchards (especially apples), and commercial crops of berries and field crops dot the region.

In Humboldt Bay is California's second large aquaculture area with working oyster beds, which offers a companion to Marin County's Tomales and Drake's Bay. Humboldt Bay oysters are pursued at a larger and more commercial scale than those of Marin County, but with mixed success, and certainly without the agritourism cachet of the oyster farms of Marin.

In the Emerald Triangle, however, the largest crop by far goes largely unseen: marijuana. To refer to marijuana cultivation as "growing weed" is a gross misnomer nowadays. As Michael Pollan, among many, has noted, the cultivation of marijuana has likely seen more change, and attracted more brilliant agronomic practitioners, than any agricultural activity in the last 50 years. Advances in growing marijuana are at an unimaginable scale, especially for someone who first ran into the results in the 1960s or '70s. The cannabis grown now is five or six times as powerful, in THC content, as anything that could have been experienced during the 1967 Summer of Love in San Francisco. Upgraded marijuana takes its toll (*see* Marijuana), but also supercharges the Emerald Triangle economy during the fall harvest season. Whatever happens to the legal, or illegal, status of marijuana will have a significant effect on the economy of the three-county geographical area.

Northern California

How can the wettest part of California be short of water? Forests find enough to grow and shade a deep understory. For crops, the agrimorphology of the state is insidious: water can sometimes be diverted for hay, for horseradish, and for a few other crops in this winter-wet climate. But complex geology and substrate make for well-drained and thirsty soils, and massive mountains impose rain shadow effects and poor crop-growing conditions. Fish like water, too, and this is country that includes big rivers dependent on sizable annual precipitation and downstream flows. Arguments over water rights, dams, and fisheries will be an ongoing struggle.

Northern California is necessarily restrained in its agricultural ambitions. The region caps the state, extends from the Pacific Ocean, stretches eastward along the California–Oregon boundary, and ends where California meets the Nevada border. Once described by Ernest Callenbach as "Ecotopia," this

Plate 118. Recognized for its great chemical powers, and a favorite internationally, horseradish is an important (if small) crop grown mostly near Tulelake, Siskiyou County, and in nearby Alturas.

country—which he had extending into Oregon—diverges culturally from other parts of California. In fact, where I-5 leaves California, Oregon governor Tom McCall is alleged once to have permitted the posting of a sign—"Welcome to Oregon, Now Please Go Home"—to discourage cross-border migrants (a story that if untrue has nevertheless been retold a million times). A large part of coastal California north of the Emerald Triangle is sparsely inhabited and highly forested, with a Native American presence and landownership more pronounced than in anywhere else in the state. Broken up topographically, with canyons and mountains and more wilderness than other parts of California, it is sublime country in which to be a tree, but not thriving in any traditional economic sense; too many industries have pulled away, shut down, and sold out. There is a ruggedness that continuing residents cherish—but not much conventional agriculture. The communities tend to be petite, resource extraction–related, and faring far less well now than a generation ago.

Pasture and rangeland are locally important, but not a blip in terms of statewide production, and as with the Emerald Triangle, tree crops exist, mostly given over to apples and cider. Some of the same marijuana crop that flourishes in the

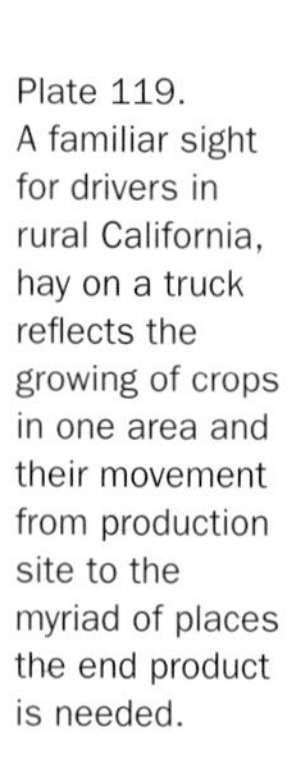

Plate 119.
A familiar sight for drivers in rural California, hay on a truck reflects the growing of crops in one area and their movement from production site to the myriad of places the end product is needed.

Emerald Triangle exists, similarly reserved out of sight, if not entirely out of mind. Production is not huge, however, of any crops aside from trees, and the human populations are limited, especially as logging and mill towns lose their industries and have to turn elsewhere in search of an economy to replace forest products.

Atop the state, drawing on Klamath River waters (though perhaps not for long), are the crops of Tulelake, including horseradish, potatoes, and dry-farmed grain. Some forest grazing takes place in the forests, but Northern California is similar, in some respects, to the great spine of the Sierra Nevada batholith, which is not big agricultural country either. Far more important is stock-raising in the mountains a hundred miles to the west, which wrest the first water from storms arriving on the California coast. The main agricultural profit center in Siskiyou and Shasta counties, curiously, is raising strawberry nursery plants, which once well started are shipped southward for transplant into fields (*see* Strawberries). From Mount Shasta south, this is a region that since Spanish–Mexican times has provided a water supply for the coastal and interior valleys, and Shasta Dam is just a part of the water infrastructure of the federally-controlled Central Valley Project.

Southland Urban Agriculture

Once the world's most perfect urban–suburban agricultural landscape, the extended Southland in 2010 is home to over 20 million residents. With an agricultural landscape so beautiful it was conveyed worldwide on a billion orange crate labels, Los Angeles, San Diego, and the Inland Empire have faced every development imperative a region can brave, and have by and large caved. "Why?" is the perfect question to ask. Ag gems remain, if hidden, and the profits mount.

Plate 120. Once an agricultural giant producing dozens of crops, San Diego County remains important in nursery and greenhouse crops, bedding plants, tomatoes, and avocados, seen here in a young planting.

The landscape south of Oxnard in Ventura County, where the Central Coast region ends, is increasingly given over to suburbia, and lots of it. Leaving the Coast Region's Ventura County, or heading southward from the San Joaquin Valley, brings on Thousand Oaks, Agoura Hills, and the San Fernando Valley: a world of suburban cities effectively balkanized by barrier freeways. You don't grow this stuff—developers must build it. Although movement can cease during rush hour (which can stretch for hours), Greater Los Angeles proper is never far away, and a driver passes places whose names are inscribed in the plaques and ledgers of California's (if not the world) Agricultural Hall of Fame: Anaheim, Azusa, Cucamonga, Pasadena, Ripon, Riverside. Crops grown in the Southland in 1930 were many, and dependent only on water supply; by 2007, the story was considerably changed.

In the 1910s, William Mulholland brought water to Los Angeles from the Owens Valley, 200 miles away, with a reach that was altogether legal but that today seems frightfully unseemly to skeptics who cannot believe that a metropolitan region's infinite checkbook will trump any local desire to keep on farming almost every time. That is a lesson Los Angeles taught every other metropolitan region in the American West. The Los Angeles Department of Water and Power needed water, bought it, and delivered the water through the sophisticated Owens Valley aqueduct—on time, and under budget—accelerating a process of L.A. Basin change from suburbanizing ag land to a new and defining form of sprawling city in the time span of less than a single generation.

Settled by Euroamericans as a proto-farming paradise, the Southland's arable acreage (a substantial area) was farmed with human energy in the nineteenth century. A second wave of migrants, once the land ownership situation was settled with the reallocation of usurped Hispanic lands to recent Anglo arrivals, came in the twentieth century. Organized groups established agricultural colonies in the Southland, based on cautious control of irrigation water and careful urban layout of communities: the Mormon San Bernardino, the German-designed Anaheim, and the "Model Colony" of George Chaffey in Ontario are just a few of the examples. The successors who came to Southern California in the 1900s and '10s were gentry—former farmers or climatic refugees—ready to evacuate from the Dakotas,

Iowa and Nebraska, Missouri and Ohio. They prospered, and so did a mixed town-and-agriculture community, largely based in nearby orchards, vegetables, and dairies—and taking advantage of lots of readily available cheap public transportation using trolleys or interurban railways. Agriculture in Los Angeles Basin attracted tens of thousands of newcomers who believed they were participating in a novel West Coast blend of urban agriculture. All this was at a great economic remove from the end-of-their-rope Oklahoma migrants who 20 years later ended up in Kern County during the Dust Bowl.

Southland groves, orchards, and cropland from the 1930s are now long gone, except for the occasional dozen orange trees or loquats or other fruits. Some are pockets of suburban trees once advertised in advice to readers columns as appropriate for a backyard orchard. Those give inklings of the past, but not a great deal of hope for the future; eventually, even fruit trees disappear in the constant sequence of merges and linked concrete that defines the reality of the daily freeway commuter. Agriculture is pretty much an afterthought, with the occasional joyous exception—as in foothills of the San Gabriel and Mount San Jacinto ranges, where isolated apple orchards have eked out an existence for a century or more. Urbanization was active before agriculture came; it's a few hearty orchardists who jumped at an opportunity and inserted agriculture amid the millions.

Southern California is a huge region, graced with many smaller production areas within a massive urbanized population. As a whole, the Southland has almost 18 million residents, and San Diego County another 2.8 million. Together, the Bay Area and extended Southern California approach a population of 29 million residents, three-quarters of the state's population. Los Angeles and Orange counties were once a scenic agricultural paradise—the orange crate vistas were no chimera—but those days are largely memory, although a few orange groves, dairies, and exotic fruit production areas remain in relict sites. Dairies in Chino are part of the milkshed, with cows that rarely set foot on grass or pasture, living instead on slotted concrete pads, eating high-protein alfalfa hay trucked south from the Owens Valley or Nevada—an act not lacking in irony, since that is where L.A. reached for its water a hundred years ago.

Los Angeles, Orange, San Bernardino, Riverside, and San Diego are by no means the smallest producers in California agriculture—they actually rank twenty-eighth, twenty-sixth, eighteenth, twelfth, and ninth in value of sales. What are striking are the dominant crops: trees and shrubs, woody ornamentals, propagative materials, potted plants, bedding plants, nursery stock, and foliage plants. Those are tools of the greenhouse, nursery, and landscaping industry—not a problem, but hardly emblems of production agriculture. Then again, in some of the county commissioners' reports appear orchard fruit, root vegetables, avocados, strawberries, milk, eggs, and tomatoes. And even Los Angeles County (#28) is a larger ag producer than any county in the San Francisco Bay Area, except for Santa Clara (#27), which is one spare spot above L.A.

North County San Diego is stunningly productive, ranking ninth as an agricultural commodity producer in the state, with three times the combined value of agricultural production of Los Angeles and Orange counties. San Diego's agriculture is remarkably concentrated in the north part of the county, where nurseries provide Southland landscapers and florists with bedding plants, ornamental trees and shrubs, and flowers for nurseries. Near Temecula and Fallbrook the scale of agriculture grows larger with the appearance of avocado orchards, citrus groves, and well-kept beds of tomatoes and strawberries. High prices advanced for agricultural land that can be developed are always on offer, and there is not much more land that can be put to the plow, so pressure on farmland is sure to increase.

Desert Regions

The "casual visitor" to Indio or Mecca or the Imperial Valley is a rarity indeed. Those who arrive have a purpose, and it had best be agricultural: the gadfly sojourner and the desert rat traveler are memories of long ago. The glories of desert agriculture include crops grown in heat using water imported by aqueduct. The results include dates and melons, alfalfa and carrots, broccoli and lettuce, cattle on feed and their calves. There is no apology tendered for Imperial Valley or eastern Riverside ag; it's there to pay, and pay big—which it had better do, or else.

Eastward of San Diego lies the true desert. Crops and livestock products raised there depend on aridity and the heat of the

Mojave and Sonora deserts, which bakes the desert Coachella and Imperial valleys during summer and provides the warmth for growing winter vegetables that keep meat and vegetables on dinner tables through the United States in months when the rest of the country finds its loins collectively wrapped in the grip of arctic cold. There are three areas of critical agriculture in the desert regions: Imperial County (all of it), Blythe (along the Colorado River), and the Coachella valley, usually reported as "eastern Riverside County" in the official accounts. Each is a center for hot-weather crops—or crops profitable only in areas that stay warm in California's winter growing season.

Imperial County, east of San Diego, continues to the Arizona border, south to Mexico's Baja California, and north through the Salton Sea. Just north is Riverside County with the resorts of Palm Desert, La Quinta, and Indian Wells. The Palm Springs vicinity is not solely golf courses; it includes Oasis, Thermal, Mecca, Indio, and the Coachella Valley. In terms of agrimorphology, these are contenders in any contest for California's most eccentric and exciting spots. Tenth in the roster of Riverside

Plate 121. Not all is tongue-in-cheek—Shields Date Gardens in Indio advertises the "Romance and Sex Life of the Date" and delivers on its promise at a visitor center combined with a well-stocked store.

County's crops is dates—eighth is grapefruit—and both owe their existence to a slender agricultural belt that offers the closest climatic analogy in California to the productive oases of North Africa and the Near East.

In far eastern Riverside—northeast of the Coachella Valley, alongside the border with Arizona—is Blythe and the Palo Verde Valley, with 100,000 acres in crops. At hand are the Southland's loneliest remaining commercial orange groves, but given over mostly to an enormous hay raising empire. A region of surface salt deposits, the terrain is difficult, and the summer heat hits like a body blow below the belt. But desert agriculture gives America its produce in winters when the northern states must make do with reduced day length, cold from Alaska and Canada, and a cover of snow and ice. Agriculture in the California desert comes then into its own and, along with imports from the Southern Hemisphere, provides the dietary variety Americans expect and demand. Of course, shipping costs—in dollars and carbon footprint—are not insignificant, but they go in lockstep with a consumer insistence on having strawberries in January, lettuce in March, and carrots year-round.

The Imperial Valley is pivotal in California agriculture. Crisscrossed by the Colorado River Aqueduct, which in a good year hauls a million acre-feet of water from the Colorado River to slake the thirst of urban Southern California, the Imperial Valley is Hades-hot during the summer months, and anything but frigid in winter. The All-American Canal, a second Colorado River tap, delivers water to the Imperial Valley, especially to provide water for agriculture that keeps Imperial County the tenth-largest producer in state agriculture. But water is only one part of the equation. Heat is the other. A pervasive tepid-to-hot regime allows production of vegetables crops even through the winter. An average high of 105 degrees F in July is matched to an average high of 59 degrees F in January; killing freezes are more rare than willing Southland pedestrians. In that climate thrive lettuce and broccoli, alfalfa and onions, and miscellaneous vegetables that cannot be winter-grown in Monterey, Ventura, Yolo, Sacramento, and Stanislaus. With the simple and definitive advantages of heat and water, crops are the godsend grown in the desert region.

Annual rainfall is slight, although the Imperial Valley picks up rainfall from the late-summer Arizona Monsoon, bringing

Plate 122. Imperial County is not the largest producer in farm-gate value, but what it produces is significant: cattle, alfalfa (seen here), lettuce, broccoli, and "vegetables"—quite the powerhouse assortment.

subtropical moisture north from Mexico. In the Imperial Valley, evapotranspiration routinely tops 84 inches a year, twice the rate of the Sacramento Valley. Farmers use pan evaporation studies and—in some cases—subscriptions to satellite imagery that can sense water, fertilizer, and pesticide effectiveness to merge science and agriculture in a search for profit and water economy. The Imperial Valley is simply not for the lily-livered; it is a supremely technical place for agriculture, which makes a gritty character all the more intriguing, and it has a huge Hispanic population, along with many an Anglo grower or ranch owner who will speak perfect Mexican Spanish—something not always prevalent elsewhere in California. To live in the Imperial Valley is a commitment, not a lark. Farming the California desert region is high-stakes agriculture, and the successful growers aren't dabblers or dilettantes: they go all in.

A curious sidelight, but hardly irrelevant, is the Salton Sea, substantially below sea level, which exists because a mistake in the 1910s allowed a water diversion to escape and fill the Salton Sink, a topographic depression—literally, a graben, or down-dropped block in a seismically active zone. A tourist attraction in the 1920s, the Salton Sea fell on harder times after World War II, and for years, photographers and essayists have recorded jaw-dropping images while roaming Imperial County and the Coachella Valley. Because of the nature of agricultural chemicals and fertilizers used with utter abandon in generations past and the relinquishment of adjacent properties as the level of the "sea" is intermittently raised and lowered with agricultural tail water and the occasional wet year, the photographic portraits that emerge often are not heartening.

Through much drama, agriculture in the desert region remains a crucial link in California food production. High-value crops and agribusiness-scale growing dominate, along with a formidable presence of the Border Patrol (a division of the U.S. Immigration and Customs Enforcement, or ICE) and cities and towns that signal distinctive borderland qualities with stoicism and some resignation. Collectively, these qualities give agriculture in the desert regions a distinctive feel: square, hulking, and formidable.

EPILOGUE

Looking toward the Nature of Agriculture in California

> *A venerable tradition in cultural geography is the scholarly attention devoted to ordinary landscapes or regions. No theory, however, explains the magnificent heterogeneity of elements that compose a place . . . . For me few places are more exciting than California's San Joaquin Valley, especially on a blistering hot afternoon in late summer. It has been called "the world's richest agricultural valley," a technological miracle of productivity where dog-eat-dog competition is at its keenest. Yet even Californians tend to take this big, flat world for granted.*
>
> — JAMES J. PARSONS, "A GEOGRAPHER LOOKS AT THE SAN JOAQUIN VALLEY," 1986

A field guide conveys the formative details about what is done where, why, how, and when. We've tried to deliver images, ideas, a body of facts, and with that there isn't always space for leisurely discussion and analysis. A great deal more could (and should) be said—or better yet, discovered—with further work instigated by readers as they explore agricultural landscapes.

By and large, this book draws on the oldest definition of "culture": it is about cultivation, about how to grow things, about rearing animals and raising crops under controlled conditions, refining and improving knowledge, tracking the politics of production and labor, and understanding how markets are made and maintained. The backstory and data for this is inevitably mustered from unwieldy parts. But acknowledging that something is "complicated" is not necessarily a limiting factor in

Plate 123. When in blossom, cherry trees (here in a commercial orchard near Morgan Hill) offer a reminder of why the tree is valued for its ideal mix of aesthetics and agriculture.

telling its tale. If through time this book is hard worn by being drawn repeatedly from car side pockets and seatbacks, unholstered like a gunfighter's six-shooter, there will be no denying its success. Our hope is to have delivered a recipe that is used and adapted, and enjoyed for any pleasure that the mixed ingredients might convey.

There is another meaning of culture we have touched on only a little—and never head-on—in this guide. The other "culture" is not about soils or sun exposure or biodynamic farming or the delivery of summer water through an irrigation ditch. The second meaning of culture is the life and society, the ways and behaviors, and the separate existences led, and choices made, by people whose existence is intimately agricultural, a trait the genuine geographer Peter Kropotkin noted in a fine 1892 book,

Plate 124. Habitually curious, dairy cows kept outdoors in the often-dry San Joaquin Valley climate come to examine visitors at the Johann Dairy, near Fresno.

The Conquest of Bread. How is the infrastructure that maintains California's water system paid for and supported? How does slow food contrast, in form, function, and delivery, with fast food? Is the vaunted market dining that is beloved of locavores in California and elsewhere a laudable return to roots, or is it agricultural elitism? What are the ties between flavor and health—or are they sometimes inversely related? How is ethnicity—let us say, among Dutch-derived dairy families, among Sikh clingstone peach growers, or among the 32,000 Hmong residents of Fresno County, so many of them involved in agriculture—expressed on California agricultural landscapes (much as Italian surnames dominated nineteenth century agriculture in Marin and Sonoma counties)? These are not questions that must be answered now, but each offers the plot for a perfect detective story. We have, in this effort, been as much collectors as we have been researchers preparing photographs and words about agriculture. Our parting tip is an easy one: to figure out how to find information is just as much fun as knowing the answers. Join us in the search.

Plate 125. A sign of a crop on the upswing, these bottles of unfiltered extra virgin olive oil from Sciabica's California Olive Oil are on display at the Ferry Plaza Farmers Market in San Francisco, linking producer with end consumer.

There are benefits. The quite wonderful historian Donald Worster, born in Needles, California, has spent much of his scholarly life so far examining agriculture in various manifestations, and the intensely complicated choices and contradictions in policy, diet, and (especially) land uses that are forced upon us by a disconnection from the natural world. But, like Jim Parsons—quoted at the beginning of this section—Worster offered a declaration of faith, in a 1982 essay for *Agricultural History*, that merits brief quotation:

> Put a historian down anywhere in the state, and he or she will find something profound to say about modernity. My own choice is to be deposited by the side of a concrete-lined irrigation canal in Kern County, by a stream that is not a stream, where no willows are allowed to grow or herons or blackbirds nest. That intensely managed piece of nature tells us a great deal about contemporary rural life and land use, some of it profound, some of it disturbing, all of it indicative of a worldwide momentum.

For showing and telling you about these and more matters, there is a second, a companion, volume to the *Field Guide to California Agriculture*. We hope you'll be interested in that more essay-driven look at California agriculture, advancing theme by theme, rather than crop by crop. A quick tip of the hat toward the larger vision of that book might go like this: agriculture is everywhere around us, and learning how it supports us and costs us is crucial. David Grene, a famed professor at the University of Chicago and a life-long farmer, wrote in *Of Farming & Classics* that "the lack of knowledge . . . of the role of the domestic animals as providers of food produces in city men and women a special blankness of sensibility." In your use of this book, we hope you will have felt engaged with your surroundings. Our goal is that readers catch the excitement, even succumb to exuberance, as they venture into the croplands, the rangelands, and the ever so highly humanized landscapes of California agriculture. We'll be there with you.

REFERENCES

Seven Essential References **(ordered by date)**

Wickson, Edward J. 1889. *The California Fruits and How to Grow Them: A Manual of Methods Which Have Yielded Greatest Success; With Lists of Varieties Best Adapted to the Different Districts of the State.* San Francisco: Dewey & Co. and the Pacific Rural Press.

Hutchinson, Claude B. (ed.). 1946. *California Agriculture: By Members of the Faculty of the College of Agriculture University of California.* Berkeley and Los Angeles: University of California Press.

Kahrl, William L. (ed.), with William A. Bowen and Marlyn L. Shelton. 1979. *The California Water Atlas.* Sacramento: The Governor's Office of Planning and Research, State of California.

Jelinek, Lawrence J. 1982. *Harvest Empire: A History of California Agriculture,* 2nd ed. San Francisco: Boyd & Fraser Publishing Company.

Scheuring, Ann Foley (ed.). 1983. *A Guidebook to California Agriculture by Faculty and Staff of the University of California.* Berkeley and Los Angeles: University of California Press.

Parsons, James J. 1986. A Geographer Looks at the San Joaquin Valley. *Geographical Review* 76(3):371–389.

Johnston, Warren E., and Alex F. McCalla. 2004. Whither California Agriculture: Up, Down, or Out? Some Thoughts about the Future. Paper SR041, 79 pages; Giannini Foundation of Agricultural Economics, University of California, Office of the President. Berkeley: Giannini Foundation, University of California. http://repositories.cdlib.org/giannini/srs/SR041.

Online Sources: Data

Bureau of the Census (Census of Agriculture), especially maps, historical data, and current crops
www.agcensus.usda.gov/Publications/2007/Online_Highlights/County_Profiles/California/
www.agcensus.usda.gov/Publications/2007/Full_Report/Volume_1,_Chapter_2_County_Level/California/index.asp
www.agcensus.usda.gov/Publications/2007/Full_Report/Volume_1,_Chapter_2_US_State_Level/index.asp
www.agcensus.usda.gov/Publications/2007/Online_Highlights/Ag_Atlas_Maps/Crops_and_Plants/index.asp

California Department of Food and Agriculture. 2009. *California Agricultural Resource Directory, 2008–2009.* (Use the edition available online—a superlative resource.) Sacramento: State of California.
www.cdfa.ca.gov/statistics/

California Farmland Mapping Program (CA)
www.conservation.ca.gov/dlrp/FMMP/Pages/Index.aspx

University of California, Davis (agricultural costs and returns)
www.coststudies.ucdavis.edu

USDA Main Site
www.usda.gov/wps/portal/usdahome/

USDA National Agricultural Statistics Service. 2009. *California County Agricultural Commissioners' Data, 2008*
www.nass.usda.gov/statistics_by_State/California/Publications/AgComm/Detail/index.asp

USDA National Agricultural Statistics Service (NASS)—California, especially crop and livestock reports
www.nass.usda.gov/Statistics_by_State/California/Publications/recent_reports.asp
www.nass.usda.gov/Statistics_by_State/California/Historical_Data/index.asp
www.nass.usda.gov/Statistics_by_Subject/index.asp

Online Sources: Thematic

Agricultural History (academic journal)
www.aghistorysociety.org

Agricultural Marketing Resource Center
www.agmrc.org/commodities_products

American Livestock Breeds Conservancy
www.albc-usa.org
American Farmland Trust
www.farmland.org
California Agriculture (University of California publication)
http://californiaagriculture.ucanr.org
California Certified Organic Farmers (CCOF)
www.ccof.org
California Crop Calendar
www.wrpmc.ucdavis.edu/Ca/CaCrops/calendar.html
California Department of Fish and Agriculture
www.cdfa.ca.gov
California Farmer (private)
http://farmprogress.com
http://magissues.farmprogress.com/CLF/clfindex.html
California Rare Fruit Growers
www.crfg.org/xref.html
Center for New Crops and Plant Products, Purdue University
www.hort.purdue.edu/newcrop/default.html
Culture Sheet.Org
www.culturesheet.org/home/
Environmental Working Group Farm Subsidy Database
http://farm.ewg.org/farm/dp_text.php
The Great Valley Center
www.greatvalley.org
Local Newspapers, also the *New York Times*, especially Verlyn Klinkenborg; the *Los Angeles Times*, the *Sacramento Bee*, and the *San Francisco Chronicle*; and, in particular, articles by George Raine and Olivia Wu (on food).
Slow Food USA
www.slowfoodusa.org
www.slowfood.com (international)
USDA Agricultural Research Service
www.ars.usda.gov/main/main.htm
USDA Germplasm Resources Information Network
www.ars-grin.gov/cgi-bin/npgs/html/index.pl?language=en
USDA Natural Resources Conservation Service
www.nrcs.usda.gov

Topical Books and Articles

Arak, Mark, and Rick Wartzman. 2003. *The King of California: J. G. Boswell and the Making of a Secret American Empire.* New York: Public Affairs.

Berry, Wendell. 1977. *The Unsettling of America: Culture and Agriculture.* San Francisco: Sierra Club Books.

Berry, Wendell. 1990. The Pleasures of Eating. In *What Are People For? Essays by Wendell Berry* (145–152). New York: North Point Press, Farrar Straus and Giroux.

Blank, S. C. 2000. Is This California Agriculture's Last Century? *California Agriculture* 54(4):23–25.

Brechin, Gray, and Robert Dawson. 1999. *Farewell, Promised Land: Waking from the California Dream.* Berkeley and Los Angeles: University of California Press.

Broek, Jan O. M. 1932. *The Santa Clara Valley: A Study in Landscape Changes.* Utrecht: N. V. A. Oosthoek's Uitgevers.

Broome, Janet C., and Keith Douglass Warner. 2008. Agro-Environmental Partnerships Facilitate Sustainable Wine-Grape Production and Assessment. *California Agriculture* 61(3):133–141.

Brunke, Henrich, Rolf A. E. Mueller, and Daniel A. Sumner. 2008. California Wine and the EU Wine Policy Reform. AIC Issues Brief, Number 34. Agricultural Issues Center. Berkeley: University of California, Giannini Foundation of Agricultural Economics.

California Magazine. 1986. *The Best of California: Some People, Places, and Institutions of the Most Exciting State in the Nation, as Featured in* California *Magazine, 1976–86.* Santa Barbara, CA: Capra Press.

Carle, David. 2004. *Introduction to Water in California.* California Natural History Guides, No. 76. Berkeley and Los Angeles: University of California Press.

Cash, Sean B., and David Zilberman. 2003. Environmental Issues in California Agriculture. In *California Agriculture: Dimensions and Issues.* Paper ISO31; Giannini Foundation of Agricultural Economics, University of California, Office of the President. Berkeley: Giannini Foundation, University of California. http://repositories.cdlib.org/giannini/is/ISO031.

Cleland, Robert Glass. 1941. *Cattle on a Thousand Hills: Southern California, 1850–1880.* San Marino, CA: The Huntington Library.

Cleland, Robert Glass. 1946. *California Pageant: The Story of Four Centuries.* New York: Alfred A. Knopf.

Collector, Stephen. 1991. *Law of the Range: Portraits of Old-Time Brand Inspectors.* Livingston, MT: Clark City Press.

Columella, Lucius J. M. 1941–1955. *On Agriculture* in 3 volumes: Books I–IV edited and translated by Harrison Boyd Ash; Books V–IX translated by E. S. Forster and Edward H. Heffner; Books X–XII (*On Trees*) edited and translated by E. S. Forster and Edward H. Heffner. Cambridge, MA: Harvard University Press, Loeb Classical Library.

Cooper, Erwin. 1968. *Aqueduct Empire: A Guide to Water in California, Its Turbulent History and Its Management Today.* Los Angeles: Arthur H. Clark.

Daniel, Cletus E. 1981. *Bitter Harvest: A History of California Farmworkers, 1870–1941.* Ithaca, NY: Cornell University Press.

Dasmann, Raymond F. 1966. *The Destruction of California.* New York: Collier Books, Macmillan Limited.

DiLeo, Michael, and Eleanor Smith. 1983. *Two Californias: The Truth about the Split-State Movement.* Covelo, CA: Island Press.

Donley, Michael W., Stuart Allan, Patricia Caro, and Clyde P. Patton. 1979. *Atlas of California.* Portland, OR: Academic Book Center.

Downey, Sheridan (Senator). 1947. *They Would Rule the Valley.* San Francisco: Irrigation Districts Association of California.

Economist. 2009. Of Ossis and Wessis: California Is Now Divided More East-West Than North-South. *The Economist* (London), Dateline: Berkeley; 23 April, pp. 38–39.

Eisen, Jonathan, and David Fine (eds.), with Kim Eisen. *Unknown California: Classic and Contemporary Writing on California Culture, Society, History, and Politics.* New York: Collier Books, Macmillan Publishing Company.

Fradkin, Philip L. 1995. *The Seven States of California: A Natural and Human History.* Berkeley and Los Angeles: University of California Press.

FRAP [California Department of Forest and Fire Protection, Fire and Resource Assessment Program]. 2003. *The Changing California: Forest and Range 2003 Assessment.* Sacramento: The Resources Agency, State of California.

Galarza, Ernesto. 1964. *Merchants of Labor: The Mexican Bracero Story.* Charlotte, NC: McNally & Loftin.

Gettman, Jon. 2006. Marijuana Production in the United States, 2006. *The Bulletin of Cannabis Reform* 2:1–29.

Goin, Peter (ed.), and Ellen Manchester. 1992. *Arid Waters: Photographs from the Water in the West Project.* Reno: University of Nevada Press.

Goldschmidt, Walter R. 1947. *As You Sow: Three Studies in the Social Consequences of Agribusiness.* Montclair, NJ: Allanheld, Osmun & Co. Publishers.

Goodhue, Rachael E., Richard D. Green, Dale M. Heien, and Philip L. Martin. 2008. Current Economic Trends in the California Wine Industry. *Agricultural and Resource Economics Update,* Agricultural Issues Center, 11(4). University of California, Giannini Foundation of Agricultural Economics.

Gregor, Howard F. 1957. The Local–Supply Agriculture of California. *Annals of the Association of American Geographers* 47(3):267–276.

Gregor, Howard F. 1963a. Industrialized Drylot Dairying: An Overview. *Economic Geography* 39(4):299–318.

Gregor, Howard F. 1963b. Regional Hierarchies in California Agricultural Production, 1939–1954. *Annals of the Association of American Geographers* 53(1):27–37.

Gregor, Howard F. 1970. *Geography of Agriculture: Themes in Research.* Englewood Cliffs, NJ: Prentice–Hall, Inc.

Gregory, James N. 1991. *American Exodus: The Dust Bowl Migration and Okie Culture in California.* New York: Oxford University Press.

Grene, David. 2007. *Of Farming & Classics: A Memoir.* Chicago: The University of Chicago Press.

Grigg, David. 1995. *An Introduction to Agricultural Geography,* 2nd rev. ed. London: Routledge.

Gudde, Erwin G. 1998. *California Place Names: The Origin and Etymology of Current Geographical Names,* 4th ed., rev. and enl. by William Bright. Berkeley and Los Angeles: University of California Press.

Guthman, Julie. 2004. *Agrarian Dreams: The Paradox of Organic Farming in California.* Berkeley and Los Angeles: University of California Press.

Gutiérrez, Ramón A., and Richard J. Orsi (eds.). 1998. *Contested Eden: California before the Gold Rush.* Berkeley and Los Angeles: University of California Press, in association with the California Historical Society.

Hart, John Fraser. 1991. The Perimetropolitan Bow Wave, *Geographical Review* 81(1):35–51.

Hart, John Fraser. 2003. Specialty Cropland in California. *Geographical Review* 93(2):153–170.

Haslam, Gerald W. 1990. *Coming of Age in California: Personal Essays*, rev. ed. Walnut Creek, CA: Devil Mountain Books.

Haslam, Gerald W. 1994. *The Other California: The Great Central Valley in Life and Letters*, rev. and enl. ed. Reno and Las Vegas: University of Nevada Press.

Haslam, Gerald W. 1999. *Many Californias: Literature from the Golden State*, 2nd ed. Reno and Las Vegas: University of Nevada Press.

Helphand, Kenneth. 2006. *Defiant Gardens: Making Gardens in Wartime*. San Antonio, TX: Trinity University Press.

Hill, Mary. 1984. *California Landscape: Origin and Evolution*. Berkeley and Los Angeles: University of California Press.

Hine, Robert V. 1953. *California's Utopian Colonies*, rev. ed. Berkeley and Los Angeles: University of California Press.

Hornbeck, David, with Phillip Kane and David L. Fuller. 1983. *California Patterns: A Geographical and Historical Atlas*. Palo Alto, CA: Mayfield Publishing.

Hundley, Norris, Jr. 2001. *The Great Thirst: Californians and Water—A History*, rev. ed. Berkeley and Los Angeles: University of California Press.

Igler, David. 2001. *Industrial Cowboys: Miller and Lux and the Transformation of the Far West, 1850–1929*. Berkeley and Los Angeles: University of California Press.

Isenberg, Andrew C. 2005. *Mining California: An Ecological History*. New York: Hill and Wang, a division of Farrar, Straus and Giroux.

Jacobsen, Rowan. 2007. *A Geography of Oysters: The Connoisseur's Guide to Oyster Eating in North America*. New York: Bloomsbury.

Johnson, Stephen, Gerald Haslam, and Robert Dawson. 1993. *The Great Central Valley: California's Heartland*. Berkeley and Los Angeles: University of California Press, in association with the California Academy of Sciences.

Johnston, W. E., and Carter, H. O. 2000. Structural Adjustment, Resources, Global Economy to Challenge California Agriculture. *California Agriculture* 54(4):16–22.

Kahn, E. J., Jr. 1985. *The Staffs of Life*. Boston: Little, Brown and Company.

Kahrl, William L. 1982. *Water and Power: The Conflict over Los Angeles' Water Supply in the Owens Valley*. Berkeley and Los Angeles: University of California Press.

Kazan, Elia (director). 1955. *East of Eden*. Screenplay by Paul Osborn, starring James Dean, Raymond Massey, Julie Harris,

Burl Ives, Richard Davalos, and Jo Van Fleet; 115 minutes, Warner Brothers.

Kelley, Robert L. 1959. *Gold vs. Grain: The Hydraulic Mining Controversy in California's Sacramento Valley; A Chapter in the Decline of the Concept of Laissez Faire.* Glendale, CA: The Arthur H. Clark Company.

Kelley, Robert L. 1989. *Battling the Inland Sea: American Political Culture, Public Policy, and the Sacramento Valley, 1850–1986.* Berkeley and Los Angeles: University of California Press.

Kropotkin, Peter, with an introduction by Charles Weigl. 2008 [1892]. *The Conquest of Bread.* Oakland, CA: AK Press.

Leopold, A. Starker, and Tupper Ansel Blake, with contributions by Raymond F. Dasmann. 1985. *Wild California: Vanishing Lands, Vanishing Wildlife.* Berkeley and Los Angeles: University of California Press.

Lillard, Richard G. 1966. *Eden in Jeopardy: Man's Prodigal Meddling with His Environment.* New York: Alfred A. Knopf.

MacCurdy, Rahno. 1925. *The History of the California Fruit Growers Exchange.* Los Angeles: G. Rice and Sons.

Masakazu, Iwata. 1962. The Japanese in California Agriculture. *Agricultural History* 36(1):25–37.

Masumoto, David Mas. 1996. *Epitaph for a Peach: Four Seasons on My Family Farm.* New York: HarperOne.

Masumoto, David Mas. 2009. *Wisdom of the Last Farmer: Harvesting Legacies from the Land.* New York: Free Press / Simon & Schuster, Inc.

McNamee, Thomas. 2007. *Alice Waters and Chez Panisse: The Romantic, Impractical, Often Eccentric, Ultimately Brilliant Making of a Food Revolution.* New York: Penguin.

McWilliams, Carey. 1942. *Factories in the Field: The Story of Migratory Farm Labor in California.* Boston: Little, Brown and Co.

McWilliams, Carey. 1973 [1946]. *Southern California: An Island on the Land.* Santa Barbara, CA, and Salt Lake City, UT: Peregrine Smith, Inc.

McWilliams, Carey. 1976 [1949]. *California: The Great Exception.* Santa Barbara, CA, and Salt Lake City, UT: Peregrine Smith, Inc.

Mora, Jo. 1949. *Californios: The Saga of the Hard-Riding Vaqueros, America's First Cowboys.* Garden City, NY: Doubleday & Company, Inc.

Morgan, Dan. 1979. *The Merchants of Grain: The Power and Profits of the Five Giant Companies at the Center of the World's Food Supply.* New York: Viking Press.

Nabhan, Gary Paul. 2002. *Coming Home to Eat: The Pleasures and Politics of Local Foods.* New York: W. W. Norton & Company.

Olmstead, A. L., and Paul W. Rhode. 2003. The Evolution of California Agriculture, 1850–2000. In *California Agriculture: Dimensions and Issues.* Paper ISO31; Giannini Foundation of Agricultural Economics, University of California, Office of the President. Berkeley: Giannini Foundation, University of California. http://repositories.cdlib.org/giannini/is/ISO031.

Parsons, James J. 1955. The Uniqueness of California. *American Quarterly* 7(1):45–55.

Parsons, James J. 1972. Slicing Up the Open Space: Subdivisions without Homes in Northern California. *Erdkunde* 26(1):1–8.

Parsons, James J. 1977. Corporate Farming in California [Geographical Record Note]. *Geographical Review* 67(3):354–357.

Parsons, James J., and Paul F. Starrs. 1988. The San Joaquin Valley: Agricultural Cornucopia. *Focus* (American Geographical Society) 38(1):cover, 7–11.

Perry, Claire. 1999. *Pacific Arcadia: Images of California, 1600–1915.* New York and Oxford: Oxford University Press.

Pollan, Michael. 2001. *The Botany of Desire: A Plant's-Eye View of the World.* New York: Random House.

Pollan, Michael. 2006. *The Omnivore's Dilemma: A Natural History of Four Meals.* New York: Penguin Press.

Preston, William L. 1981. *Vanishing Landscapes: Land and Life in the Tulare Lake Basin.* Berkeley and Los Angeles: University of California Press.

Rawls, James J., and Richard J. Orsi (eds.). 1999. *A Golden State: Mining and Economic Development in Gold Rush California.* Berkeley and Los Angeles: University of California Press, in association with the California Historical Society.

Reisner, Marc. 1986. *Cadillac Desert: The American West and Its Disappearing Water.* New York: Penguin Books.

Reuter, Peter, and Franz Trautmann (eds.) 2009. *A Report on Global Illicit Drug Markets, 1998–2007.* Netherlands: Trimbos Institute and RAND Europe.

Richardson, Paul. 2007. *A Late Dinner: Discovering the Food of Spain.* New York and London: Scribner.

Sackman, Douglas Cazaux. 2005. *Orange Empire: California and the Fruits of Eden.* Berkeley: University of California Press.

Saloutos, Theodore. 1975. The Immigrant in Pacific Coast Agriculture. *Agricultural History* 49:82–201.

Schlosser, Eric. 2001. *Fast Food Nation: The Dark Side of the All-American Meal.* New York: Houghton Mifflin Co.

Shillinglaw, Susan. 2006. *A Journey into Steinbeck's California.* Berkeley, CA: Roaring Forties Press.

Siebert, Jerome (ed.) 2003. *California Agriculture: Dimensions and Issues.* Paper ISO31, 304 pages; Giannini Foundation of Agricultural Economics, University of California, Office of the President. Berkeley: Giannini Foundation, University of California. http://repositories.cdlib.org/giannini/is/ISO031.

Smith, J. Russell 1987. *Tree Crops: A Permanent Agriculture.* Washington, DC: Island Press.

Snyder, Gary. 1983. *Axe Handles: Poems.* San Francisco: North Point Press.

Spirn, Anne Whiston. 2008. *Daring to Look: Dorothea Lange's Photographs and Reports from the Field.* Chicago: University of Chicago Press.

Starrs, Paul F. 1988. The Navel of California and Other Oranges: Images of California and the Orange Crate. *California Geographer* 27:1–42.

Stein, Walter J. 1973. *California and the Dust Bowl Migration.* Westport, CT: Greenwood Press, Inc.

Steinbeck, John. 1939. *The Grapes of Wrath.* New York: Viking Press.

Steinbeck, John. 1952. *East of Eden.* New York: Viking Press.

Steinbeck, John, with an introduction and notes by John H. Timmerman. 1995 [1938]. *The Long Valley.* New York: Penguin Books.

Stoll, Steven. 1998. *The Fruits of Natural Advantage: Making the Industrial Countryside in California.* Berkeley: University of California Press.

Swinchatt, Jonathan, and David G. Howell. 2004. *The Winemaker's Dance: Exploring Terroir in the Napa Valley.* Berkeley: University of California Press.

Taber, George. M. 2005. *Judgment of Paris: California vs. France and the Historic 1976 Paris Tasting That Revolutionized Wine.* New York: Scribner.

Theobald, David M. 2001. Land-Use Dynamics beyond the Urban Fringe. *Geographical Review* 91(3):544–564.

Thompson, Kenneth. 1961a. Location and Relocation of a Tree Crop: English Walnuts in California. *Economic Geography* 37(2):133–149.

Thompson, Kenneth. 1961b. Riparian Forests of the Sacramento Valley, California. *Annals of the Association of American Geographers* 51(3):294–315.
Thompson, Kenneth. 1969. Insalubrious California: Perception and Reality. *Annals of the Association of American Geographers* 59(1):50–64.
Thompson, Kenneth. 1970. The Australian Fever Tree in California: Eucalypts and Malaria Prophylaxis. *Annals of the Association of American Geographers* 60(2):230–244.
Thompson, Kenneth. 1975. The Idea of Longevity in Early California. *Bulletin of the New York Academy of Medicine* 51(7):805–816.
Treadwell, Edward F. 1981 [1931]. *The Cattle King: A Dramatized Biography*. Santa Cruz, CA: Western Tanager Press.
Vaughan, John G., and C. A. Geissler. 1997. *The New Oxford Book of Food Plants*. Oxford, New York: Oxford University Press.
Vaught, David. 1999. *Cultivating California: Growers, Specialty Crops, and Labor, 1875–1920*. Baltimore: The Johns Hopkins University Press.
Vaught, David. 2007. *After the Gold Rush: Tarnished Dreams in the Sacramento Valley*. Baltimore: The Johns Hopkins University Press.
Vink, Erik. 1998. Land Trusts Conserve California Farmland. *California Agriculture* 52(3):27–31.
Wallach, Bret. 1980. The West Side Oil Fields of California. *Geographical Review* 70(1):50–59.
Wiley, Peter, and Robert Gottlieb. 1982. *Empires in the Sun: The Rise of the New American West*. Tucson: University of Arizona Press.
Wilson, Bee. 2008. *Swindled: The Dark History of Food Fraud, from Poisoned Candy to Counterfeit Coffee*. Princeton, NJ: Princeton University Press.
Wollenberg, Charles. 1985. *Golden Gate Metropolis: Perspectives on Bay Area History*. Berkeley: Institute of Governmental Studies, University of California.
Worster, Donald. 1982. Hydraulic Society in California: An Ecological Interpretation. *Agricultural History* 56(3):501–515.
Worster, Donald. 1985. *Rivers of Empire: Water, Aridity, and the Growth of the American West*. New York: Oxford University Press.
Yogi, Stan (ed.). 1996. *Highway 99: A Literary Journey through California's Great Central Valley*. Berkeley, CA: Heyday Books, in conjunction with the California Council for the Humanities.

ADDITIONAL CAPTIONS

COVER. Although not a commercial giant, kale embodies so much that is significant, nationwide and internationally, about California agriculture. An Old World cultivar, kale adapts well to the state's Mediterranean climate. As a cruciferous vegetable, kale is considered particularly healthy and is a frequent visitor in restaurant menus; it is a signature crop found in farmers markets and community-supported agriculture.

PAGE i. The radish is not a significant commercial crop, in part because it is so easily raised in a home garden. But radishes stand out as one of the classic garnishes that will forever appear in farmers markets.

PAGE ii–iii. Through a multiple-month harvest, delectable thistles are sliced from a plant that has the cheerful good looks of a sprawling alien invader. Workers sporting hooded sweatshirts, and carrying red frames into which 'chokes are tossed, complete the unearthly image. The artichoke is a signature California crop, and a profitable one.

PAGE vi. The loose skin characteristic of the mandarin is a feature often passed on in hybrids, as here in an Indio mandarinquat, seen in the research orchard of the University of California's Lindcove Citrus Research Station.

PAGE ix. In a bird's-eye view that varies dramatically with the season, there is a particular glory to oak woodland rangeland: variously a supple green carpet dotted with oaks or a summer-dry golden expanse, a result of the state's predominantly Mediterranean-type climate. There is no more characteristic California landscape than this one, devoted to range livestock, and wildlife. Little wonder that suburban housing is perennially poised to intrude on rangeland in an epic contest of wills: rancher versus suburban developer.

PAGE xiv. Sitting in a driver's seat with cushions worn bare may be monotony for insiders native to the world of farming, but there is an elemental attraction that infects even urban ingenues who see an empty seat beckoning in a windrowed hayfield. The farm tractor, though nowhere near as muscle-bound as a locomotive or a jet engine, exudes an innate magnetism of prepotent low-gear power.

PAGE xviii. Slicing through a Cara Cara orange at the University of California's Lindcove Citrus Research and Extension Center, local nurseryman

Angelberto Sanchez shows the fruit's consistency and advantages. If agricultural improvements in California come in part from grower-supported research, many of the remaining advances hail from university-affiliated researchers, experts in the ecology and agronomy of a given crop, who investigate within their specialty areas and share the results with commercial growers.

PAGE xxiii. No hyperbole is involved in this sign's claim that nearby Salinas is the "Salad Bowl of the World"—although the adjoining crop is all strawberries. Monterey County is reliably among the top five California counties in farm-gate production, and lettuce leads the produce pack. Nearby are plantings of artichokes, broccoli, brussels sprouts, peppers, wine grapes, cabbage, rangeland, and mesclun mix.

PAGE xxiv–1. Mild in flavor, and without any of the perceptible "heat" of chili peppers, the bell pepper comes in many colors and forms, a variety that diversifies salads and other dishes.

PAGE 72–73. Corn, known as "maize" in British English, is the Western Hemisphere's singular contribution to world food stocks. Other crops of the Americas—chili peppers, tomatoes, and potatoes—are significant among foodstuffs, but corn once fed indigenous populations in the Americas. As sweet corn, today it enlivens summer diets; as feed corn, it provides carbohydrates and protein essential to livestock feed and production.

PAGE 92–93. Two lettuce varieties, divided by an emblematic four-inch aluminum irrigation pipe, signal profit, change, and diversity along River Road, west of the Salinas River in Monterey County.

PAGE 94. The concept of a strict regimen governing transit through a farm field may appear as incongruous as the *Butch Cassidy and the Sundance Kid* protest: "Rules? In a knife fight?" But rules there are, posted here as a reminder that urbanites who descend on a U-pick field don't have free run. This sign lays down the law in Brentwood, in eastern Contra Costa County.

PAGE 365. Alfalfa, a sophisticated, protein-dense livestock feed that has traveled the world from the Near East to North Island, New Zealand, is also a spectacular hay crop. Alfalfa is cut with a mechanical swather, allowed to dry in place, and then either converted to silage or baled into stout units in which the leafy legume holds its nutritive value for months on end until a bale is broken up and fed.

PAGE 366–367. Perennially popular with fans of Mexican food in California, *nopales* (cactus pads) and *tuna* (fruit) appear on prickly pear cacti grown on benchland above the Salinas River plain, near Soledad, California.

PAGE 416. Few activities are so grueling as harvesting fruit, especially when the fruit is one as delicate and as easily bruised as cherries are. Ascending and descending the rungs of a ladder a few hundred times a day is inevitable—and part of the story of agricultural labor, and grower profit, in California.

INDEX

Page references in **boldface type** refer to main discussions of crops, products, families, genera, species, or regions.

ABOUT THE AUTHORS

Photo by Jean Dixon

Peter Goin (right) is Regents & Foundation Professor of Art at the University of Nevada. Peter spent his formative years abroad, in Indonesia, Turkey, and Brazil, before returning to the United States to complete his education. After stretches living in Fairfax, Virginia; Washington, D.C.; and Oxford, Iowa, Peter moved to San Francisco, and ultimately to Nevada, where he lives today.

Peter's photographs are exhibited in more than 50 museums, nationally and internationally, and he is author or coauthor of better than a dozen books that include *Tracing the Line: A*

Photographic Survey of the Mexican-American Border, Nuclear Landscapes, and *Stopping Time: A Rephotographic Survey of Lake Tahoe.* His *Humanature* explores and documents, through photography and text, the premise that nature is a cultural construct, a theme echoed in the edited volume *Arid Waters.* The coauthored *A Doubtful River* examines checkered human and watershed transformations that came with Nevada's Newlands Reclamation Project, the first federal irrigation effort. *Changing Mines in America* reinterprets the legacy and importance of mining landscapes in the United States. With Paul Starrs, Peter produced *Black Rock,* a seminal work that reaches out to embrace a lyrical and poetic vision of a spectacular desert region in northern Nevada.

Peter's family includes longstanding Californians who began their permanent residence in California in 1919, although earlier ancestors arrived in Oregon by wagon train in 1853. Peter's father, during his time at UC Berkeley, worked as a seasonal farm worker in the lemon groves.

Paul F. Starrs (left) is a member of the faculty in Geography at the University of Nevada, where he is Regents & Foundation Professor. A resident of both Nevada and California, Paul's journey to work across the Sierra has taught him the turns and tarmac of every intriguing Sierra Nevada back road. Born in Bordeaux, France, the child of U.S. diplomats, Paul spent much of the 14 years that followed in France, Spain, and Guatemala. He is author of *Let the Cowboy Ride: Cattle Ranching in the American West* and, with Peter Goin, of *Black Rock,* an intimate look at a remote northwestern edge of Nevada. In the last three years, Paul has written about film noir, Las Vegas, the artist Ed Borein (of Santa Barbara), Mexican miners, rock art, the oak woodlands of Spain and Portugal, cowboy poetry and literature, Mormon church architecture, Ry Cooder's CD *Chávez Ravine,* and ranching—the last in both worldwide and North American contexts.

A student during the mid-1970s at the famed Deep Springs College (California), Paul continued working as a ranch hand in Nevada and eastern California when he concluded his time at Deep Springs, watching over several hundred cattle and calves grazing rangelands up to 13,500 feet elevation. After earning an honors undergraduate degree at UC San Diego researching intentional communities and utopian thought, Paul continued his education with a PhD in geography at UC Berkeley. Through his graduate school years and afterward, Paul quartered the agricultural landscapes of California with his mentor, Berkeley geographer Jim Parsons—without question the foremost master of California geography of the last century. This field guide is a book Jim would have loved to write himself; instead, he inspired us to persevere in the work.

Authorship note: Peter and Paul work together on their book projects, and in a work such as this one, their contributions to authorship should be considered equivalent. Authorship order is reversed for each book that Paul and Peter write and reflects no priority of one author or the other.

Composition: Publication Services, Inc.
Text: 9.5/12 Minion
Display: Franklin Gothic typefaces
Printer and binder: Golden Cup Printing Company Limited

Field Guides

Sharks, Rays, and Chimaeras of California, by David A. Ebert, illustrated by Mathew D. Squillante

Field Guide to Beetles of California, by Arthur V. Evans and James N. Hogue

Geology of the Sierra Nevada, Revised Edition, by Mary Hill

Mammals of California, Revised Edition, by E.W. Jameson, Jr., and Hans J. Peeters

Field Guide to Amphibians and Reptiles of the San Diego Region, by Jeffrey M. Lemm

Dragonflies and Damselflies of California, by Tim Manolis

Field Guide to Freshwater Fishes of California, Revised Edition, by Samuel M. McGinnis, illustrated by Doris Alcorn

Field Guide to Owls of California and the West, by Hans J. Peeters

Raptors of California, by Hans J. Peeters and Pam Peeters

Field Guide to Plant Galls of California and Other Western States, by Ron Russo

Field Guide to Butterflies of the San Francisco Bay and Sacramento Valley Regions, by Arthur M. Shapiro, illustrated by Tim Manolis

Geology of the San Francisco Bay Region, by Doris Sloan

Trees and Shrubs of California, by John D. Stuart and John O. Sawyer

Pests of the Native California Conifers, by David L. Wood, Thomas W. Koerber, Robert F. Scharpf, and Andrew J. Storer

Introductory Guides

Introduction to Energy in California, by Peter Asmus

Introduction to Air in California, by David Carle

Introduction to Fire in California, by David Carle

Introduction to Water in California, by David Carle

Introduction to California Beetles, by Arthur V. Evans and James N. Hogue

Introduction to California Birdlife, by Jules Evens and Ian C. Tait

Weather of the San Francisco Bay Region, Second Edition, by Harold Gilliam

Introduction to Trees of the San Francisco Bay Region, by Glenn Keator

Introduction to California Soils and Plants: Serpentine, Vernal Pools, and Other Geobotanical Wonders, by Arthur R. Kruckeberg

Introduction to Birds of the Southern California Coast, by Joan Easton Lentz

Californian Indians and Their Environment: An Introduction, by Kent G. Lightfoot and Otis Parrish

Introduction to California Mountain Wildflowers, Revised Edition, by Philip A. Munz, edited by Dianne Lake and Phyllis M. Faber

Introduction to California Spring Wildflowers of the Foothills, Valleys, and Coast, Revised Edition, by Philip A. Munz, edited by Dianne Lake and Phyllis M. Faber

Introduction to Shore Wildflowers of California, Oregon, and Washington, Revised Edition, by Philip A. Munz, edited by Dianne Lake and Phyllis Faber

Introduction to California Desert Wildflowers, Revised Edition, by Philip A. Munz, edited by Diane L. Renshaw and Phyllis M. Faber

Introduction to California Plant Life, Revised Edition, by Robert Ornduff, Phyllis M. Faber, and Todd Keeler-Wolf

Introduction to California Chaparral, by Ronald D. Quinn and Sterling C. Keeley, with line drawings by Marianne Wallace

Introduction to the Plant Life of Southern California: Coast to Foothills, by Philip W. Rundel and Robert Gustafson

Introduction to Horned Lizards of North America, by Wade C. Sherbrooke

Introduction to the California Condor, by Noel F. R. Snyder and Helen A. Snyder

Regional Guides

Natural History of the Point Reyes Peninsula, by Jules Evens

Sierra Nevada Natural History, Revised Edition, by Tracy I. Storer, Robert L. Usinger, and David Lukas